CONTEMPORARY YUGOSLAV PHILOSOPHY: THE ANALYTIC APPROACH

Nijhoff International Philosophy Series

VOLUME 36

Aleksandar Pavković
editor

Contemporary Yugoslav Philosophy: The Analytic Approach

KLUWER ACADEMIC PUBLISHERS
DORDRECHT / BOSTON / LONDON

Library of Congress Cataloging in Publication Data

Contemporary Yugoslav philosophy : the analytic approach / edited by
Aleksandar Pavković.
p. cm. -- (Nijhoff international philosophy series)
Bibliography: p.

1. Philosophy, Yugoslav--History--20th century. 2. Analysis
(Philosophy) I. Pavković, Aleksandar. II. Series.
B4844.A52C66 1988
199'.497--dc19 88-24072
CIP

ISBN-13: 978-94-010-7770-5 e-ISBN-13: 978-94-009-2821-3
DOI: 10.1007/978-94-009-2821-3

Published by Kluwer Academic Publishers,
P.O. Box 17, 3300 AA Dordrecht, The Netherlands.

Kluwer Academic Publishers incorporates
the publishing programmes of
D. Reidel, Martinus Nijhoff, Dr W. Junk and MTP Press.

Sold and distributed in the U.S.A. and Canada
by Kluwer Academic Publishers,
101 Philip Drive, Norwell, MA 02061, U.S.A.

In all other countries, sold and distributed
by Kluwer Academic Publishers Group,
P.O. Box 322, 3300 AH Dordrecht, The Netherlands.

TABLE OF CONTENTS

NOTES ON CONTRIBUTORS

Miloš Arsenijević (1950) is a Docent in the Philosophy Department of the University of Beograd. He graduated and obtained his doctorate in philosophy from this university. His main interests are in metaphysics, philosophy of mathematics and of physics and in philosophy of psychology.

Jovan Babić (1951) teaches ethics in the Philosophy Department of the University of Beograd. He holds a master's degree in philosophy from the same university. His main interests are in applied ethics and meta-ethics, in particular utilitarianism and Kant's ethics.

Vojislav Božičković (1950) is a doctoral candidate at the University of Melbourne. He graduated and obtained his master's degree in philosophy from the University of Beograd. His main interest is in philosophical logic.

Nikola Grahek (1949) is a research fellow at the Institute for Philosophy of the University of Beograd. He holds a master's degree from this university. His main interests are in epistemology, philosophy of mind and political philosophy.

Mane Hajdin (1959) is a Lecturer in the Department of Psychology and Philosophy, University of Papua New Guinea. A graduate of the Philosophy Department of the University of Beograd, he holds a Ph.D in philosophy from McGill University. His interests are in ethics, philosophy of law, social and political philosophy, and in philosophy of language.

Aleksandar Jokić (1960) is a doctoral candidate at the University of California at Santa Barbara. He received an M.A. in philosophy from this university and is a graduate of the Philosophy Department of the University of Beograd. His main interests are in philosophy of science, epistemology and moral philosophy.

Svetlana Knjazev-Adamović (1931) is Professor of Logic at the University of Beograd. She graduated and obtained her doctorate in philosophy from this university. Her main interests are in epistemology, especially in theories of truth, and in meta-ethics.

Aleksandar Kron (1937) is Professor of Logic at the University of Beograd. He graduated from the University of Beograd and holds a doctorate in philosophy from the same university. His main interest is in formal logic, in particular, relevance logics and quantum theory logic.

Živan Lazović (1956) teaches epistemology at the Philosophy Department of the University of Beograd. He holds a master's degree in philosophy from this university. His main interests are in the philosophy of action, philosophical logic and epistemology.

Nenad Miščević (1950) is Professor of Philosophy at the University of Zadar in Yugoslavia. A graduate of the Philosophy Department of the University of Zagreb, he obtained his doctorate in philosophy from the University of Ljubljana. His main interests are in the philosophy of language and philosophy of action.

Staniša Novaković (1930) is Professor of the Methodology of Science at the University of Beograd. He graduated and obtained his doctorate in philosophy from this university. His main interests are in epistemology and philosophy of science.

Aleksandar Pavković (1950) is Lecturer and Head of Slavonic Studies at Macquarie University in Sydney. He taught epistemology and philosophical logic at the University of Beograd from 1976 to 1985. A graduate of Yale University, he obtained his doctorate in philosophy from the University of Beograd. His main interests are in epistemology and, more recently, in Yugoslav political thought.

Matjaž Potrč (1948) is Professor of History of Philosophy at the University of Ljubljana. He graduated and obtained his doctorate in philosophy from the same university. His main interests are in the philosophy of mind and the philosophy of language.

Aleksandar Pražić (1951) teaches philosophy and methodology of science at the Philosophy Department of the University of Beograd. He holds a master's degree in philosophy from the same university and his main interests lie in epistemology, philosophy of science and in history of ideas.

Svetozar Sindjelić (1946) is a research fellow at the Institute for Philosophy of the University of Beograd. He holds a master's degree from this university and his interests lie in the philosophy of science and philosophy of language.

Heda Šegvić (1956) is a doctoral candidate at Princeton University. She graduated from the University of Beograd and her main interests are in Ancient Greek philosophy and in political philosophy.

PREFACE

The aim of this collection is to present the work of Yugoslav philosophers who approach philosophy in a analytic manner. I have sought contributions from all philosophers of whom I knew to be working in this tradition. Not all sent their contributions; but I should say that the majority did. As a consequence of so wide an appeal, the papers published here exhibit not only a variety of topics but also differences in quality. This, I think, is to be expected given that the aim is to present the work of a group of philosophers who share only a poorly defined and general approach to philosophy.

Of many people who have helped me bring out this collection, I can mention only a few. Jovan Babić gave me, as usual, sound advice and helped me to establish and maintain contact with the contributors in Yugoslavia. Miloš Arsenijević rounded up the recalcitrant contributors and constantly recruited new ones. And Svetlana Knjazev encouraged me – perhaps unwittingly – to persevere by her yarns of the old times in which there was no analytic philosophy in Yugoslavia.

And Down Under, Steve Glaister checked the English, Nancy Simmons patiently typed the manuscripts into a rather temperamental word-processor, Mirjana Djukić proof-read them and Dean Davidson of the Macquarie University Computing Centre helped us to get them printed. To all of them many thanks.

I dedicate the book to my students from the University of Beograd, who, whether at home or abroad, I trust will carry on.

Sydney, 22 January 1988 *The Editor*

INTRODUCTION

ALEKSANDAR PAVKOVIĆ

The establishment of a Communist party rule in a country, so far as our experience can tell, is followed by the establishment of Marxism in the role of the official or dominant philosophy in the country. Marxism as an official philosophy leaves little if any room for a non-Marxist approach to philosophy. In this respect the Yugoslav Communist party and the official philosophy it established is no exception. In spite of this, in the last two decades various non-Marxists approaches to philosophy have been tolerated – some would say, have flourished – in Yugoslavia. This collection is a testimony of the work done by Yugoslav philosophers in one non-Marxist tradition: the analytic tradition. How this approach to philosophy came to flourish in Yugoslavia is the topic of this introduction.

This topic and my approach to it will naturally restrict the scope of the introduction. I am one of the philosophers who share this approach to philosophy, and not a historian of Yugoslav philosophy; so I shall make no attempt to approach the subject in a comprehensive and historical manner. For instance, I shall not attempt to relate in any systematic manner this approach to philosophy to any other nor shall I try to evaluate its relative significance in the development of philosophy in Yugoslavia. My account will certainly be incomplete in other ways too. For example, I shall not tackle the question I consider crucial for the rise of analytic philosophy in Yugoslavia: Why in the early seventies did the Marxists both in the academia and outside it, condone the appointment of analytic philosophers to a series of academic and research posts in Beograd, Zagreb, Zadar and Ljubljana? Had these appointments not been made, only a few philosophers in Yugoslavia would now consider themselves analytic philosophers. The question is probably outside the scope of history of philosophy proper; but it just shows how difficult it is in discussing Yugoslav philosophy to keep within the confines of this discipline.[1]

For the purposes of this collection, it proved unnecessary to define what is meant by the 'analytic approach' or 'analytic tradition': all of us contributing to it seem to agree that we share an approach to philosophy and all of us are happy to call it 'analytic'. Moreover, it seems that other philosophers (and even politicians) in Yugoslavia who discuss the work of this kind, tend to lump it under the heading of 'analytic philosophy'. This, of course, does not show that there is an approach which we all share. But at

A. Pavković (ed.), Contemporary Yugoslav Philosophy: The Analytic Approach, 5–26.

least one can say that we sincerely believe to be working in the tradition which starts with Frege, Russell, Moore and early Wittgenstein and which is nowadays continued in the works of Quine, Davidson and Dummett. This will not enlighten the uninitiated; but this is how the tradition is often described both in Yugoslavia and elsewhere.

THE PREDECESSORS

Only a few philosophers in Yugoslavia before the Second World War were acquainted with the works of Frege, Russell and Wittgenstein (and none seem to have read Moore). In fact, philosophers in Yugoslavia became acquainted with this tradition through the works of logical positivists of the Vienna Circle. Even before, the nineteenth century German positivists such as Wundt and Laas have been of some influence on Serbian philosophers. Their influence is most obvious in the works of Dragiša M. DJURIĆ (1871–1941), the Professor of Ethics and History of Philosophy at the University of Beograd who was a pupil of Wundt.

In his first book Problem subjekta i objekta (The Problem of Subject and Object) (Beograd, 1912) Djurić expounds a curious amalgam of dialectical materialism and phenomenalism. Every form and method of knowledge is, according to him, given within the realm of phenomena: we get to know the nature only in its phenomenal form (ibid.,p.22). Empirical facts offer the only criterion of truth. In consequence, any form of metaphysics, whether backed by a religion or not, is of no help in our cognitive endeavour. This antimetaphysical view is coupled with a causal materialist view of the mind: all mental phenomena are caused by physiological processes and every mental process is causally correlated with a physiological process in the subject's nervous system. The psycho-physiological correlation is a result of the evolution of the nervous system. Acquisition of knowledge also results from the causal processes in the nervous system and so cognition is a subject of scientific inquiry as any other process is. This thoroughoing naturalism is coupled further with Marxist dialectical materialism: in his opinion, we can discover by experience the underlying dialectical laws of the development of nature.

Djurić's reductive phenomenalism is even more apparent in his later work Filozofija fenomena sad (The Philosophy of the Phenomenon Now) (Beograd, 1925). This is a highly complex reduction of all temporal relations to their allegedly phenomenal basis – the phenomenon of the moment now. Here Djurić makes no attempt to incorporate his reduction into a dialectical materialist framework; and so the relation between dialectical materialism and his reductive phenomenalism is left unexplained.

Logical positivism was also introduced into Yugoslav philosophical literature within a dialectical materialist framework. This was the work of two philosophers of science from Zagreb: Zvonko RICHTMAN (1901–1941) and Rikard PODHORSKY. In the book Nauka, život i tehnika (Science, Life and

Technology) (Zagreb, 1932) written with a fellow chemist Ivan Brichta, Podhorsky introduces the logical positivist view of scientific theories: a scientific theory is a logical system based on hypotheses. These hypotheses are never fully confirmed as true because there is always a possibility of finding evidence disconfirming them; and the experiment is the only certain means of confirming or disconfirming a hypothesis.[2] In his review of this book, ("Jedna knjiga o nauci", Književnost, 1933, no. 2) Richtman finds the underlying dialectical approach to the philosophy of science especially appealing.

But attempts to incorporate logical positivist views within a materialist dialectical framework came under attack from Ognjen Prica who was acting as the spokesman of the Yugoslav Communist Party. In his article "Suvremena filozofija i prirodne znanosti" (Contemporary Philosophy and the Natural Sciences) (Znanost i život, no. 1, 1936) he claimed that logical positivism is a form of idealism which Lenin has already rejected as "Machism". In consequence, any attempt to incorporate such views into a materialist framework is futile and is only a concession to the bourgeois way of thinking. In their reply Richtman and Podhorsky (Znanost i život, no. 2-3, 1937) point out that even Engels thought that eventually nothing would be left of philosophy but logic and dialectics. They think that philosophy should take over from the sciences the critical, scientific approach and so banish any metaphysical speculation. The dialectical materialism is an elastic framework within which such an approach could easily be incorporated; accordingly, scientific approach to various philosophical problems – e.g. the existence of matter or of spirit – could be taken over by dialectical materailists. Prica's further reply to this made it clear that the Party would not tolerate any such revisionism and Richtman was proclaimed a renegade.

In his last two essays[3] Richtman argues that modern physics and not philosophy has succeeded in offering a unified theory of the nature and that therefore the task of philosophy should be reduced to an inquiry into the logical syntax of language and a historical inquiry into the development of sciences. Physics has freed itself of all metaphysics and other sciences should strive to do the same; the critical scientific approach should be extended to the problems of our everyday life. Any scientific hypothesis or theory may be rejected if evidence is found disconfirming it; the only principle not open to rejection is the very principle of verification. In accordance with this principle, modern scientific philosophy examines the status of and the evidence for the general belief in the existence of the external world; unlike Ernest Mach, the logical positivist neither doubt nor deny the existence of such a world. In consequence, logical positivists are not idealists. This, of course, does not make them into materialists. But in his last essays Richtman does not concern himself with the relation of logical positivism to dialectical materialism; the latter is not even mentioned.[4]

In spite of some familiarity with the work of logical positivists, most philosophers in Yugoslavia before the Second World War seem to have been unacquainted with philosophical analysis as a specific method of tackling philosophical problems. Only Kajica MILANOV (1905–1986) employed this method in his work on theory of knowledge. In his book Osnovni problemi teorije saznanja (The Main Problems of the Theory of Knowledge) (Beograd, 1937) Milanov distinguishes analysis of the meaning from interpretation of meaning or sense. By analysis we get to the main components or features of e.g. knowledge or reality. An interpretation of knowledge or reality should offers us both a clarification and a justification of the concept in question. In this sense, analysis is only a preliminary to a philosophical interpretation; but in a wider sense, philosophical analysis would comprise both the analysis and the interpretation in Milanov's sense. When interpreting a proposition or a thesis, Milanov holds, we substitute for it another proposition or thesis which should make the initial one better understood and even better justified. In this way by interpreting a proposition, we may try to find its grounds (ibid.,pp.80–82).

As an example of analysis, Milanov offers Frege's analysis of meaning into sense and reference; the famous example of the Morning and Evening Star thus enters Yugoslav philosophical literature as an example of philosophical analysis. As for philosophical interpretation, Milanov thinks that Wittgenstein's Tractatus and Reichenbach's works supply numerous examples of this procedure. In his opinion, gnoseological interpretation should show us how and under what assumptions it is possible to know something and under what conditions is knowledge valid (ibid.). As he understands it, such an interpretation is not exhausted by an analysis of the meaning of knowledge claims but it should offer an account of what the objects of knowledge are and of what knowledge is constituted.

From his point of view, the type of account which Neurath and Carnap offer of the protocol sentences is not only inadequate as an account of the basis for knowledge but is positively misleading. According to Milanov, the analysis of the meaning of protocol sentences cannot yield an account of the objects of knowledge; at most such an analysis can help us grasp what the starting point of knowledge is. Logical positivists in their analysis of basic sentences greatly contributed to the clarification of the problem of the starting point of knowledge but not of the status and validity of knowledge in general. In order to tackle this problem, Milanov seems to be ready to employ the methods of phenomenology; here his concern is how judgment - viewed as an act of judging - constitutes the objects of knowledge.

Milanov's book caused no controversy; perhaps this was because he had not attempted to place his work within any current ideological framework. In any case, until early sixties we find no philosophical work published in Yugoslavia in which philosophical analysis is employed as a method of clarifying and solving philosophical problems.

THE WHOLESALE REJECTION

The victory of the Yugoslav Communist Party in 1945 brought Communist philosophers into control of philosophy in the whole country. As a result, the newly appointed Communist professors of philosophy - Boris Ziherl in Ljubljana and Dušan Nedeljković in Beograd - set about to cleanse philosophy of all bourgeois influence and to make it a thoroughly Marxist Leninist discipline. A victim of this purge was Kajica Milanov: in 1947 he was summarily dismissed from his university lectureship and forced to leave the country.[5]

According the Leninist doctrine, the bourgeois philosophy reached its peak with Hegel and then went into an inexorable decline. A Marxist should closely follow the decline in order to be able to criticize various bourgeois errors and thus keep his philosophy correct and uncontaminated. In keeping with this practice is the critical study Savremena filozofija Zapada (Contemporary Philosophy of the West) (Beograd, 1952) read by its author Dragan M. Jeremić at the first (and last) Congress of Philosophers of Serbia held in 1951 in Beograd. In this lengthy review of living and dead philosophers in the West, Jeremić finds a school and a label for every one of them; Russell, Wittgenstein, Carnap, Ayer and Stuart Chase belong to the school of "semanticism" which originated from logical positivism. These philosophers treat all philosophical problems on the semantic level alone and argue that it is senseless to try to describe objective reality. Any philosopher ignoring the objective reality, as Lenin has conclusively shown, is a subjective idealist; and the "semanticists", these contemporary subjective idealists, thwart the scientific inquiry of the reality, defend the old capitalist social system and stand on the way of the revolutionary struggle against it (ibid., p.107). Mathematical logic, as developed by Russell and others, became only an organ of sophistry because it lacked the objective and dialectical aspects necessary for any true logic.

In short, there is nothing to learn from the work of the philosophers of the analytic tradition; moreover, their philosophy is reactionary and it will ultimately be destroyed with the destruction of the capitalist social system. This is the view to which the Dušan Nedeljković and Boris Ziherl clung to throughout their lives; fortunately, their influence on philosophy in Yugoslavia began to wane in early fifties and was ended in 1960.[6] In 1960 at a philosophical conference at Bled, Lenin's copy theory of knowledge - to which these two philosophers subscribed - came under a sustained attack by a group of neomarxists - the group which later attained fame as the Yugoslav Praxis group.

According to Lenin's copy theory of knowledge, knowledge is a subjective copy - image - of the objective reality. This, the neomarxists argued, is not only too narrow and simplistic a view, but is not a truly Marxist view at all. The central philosophical concept for Marx - of the Economic and Philosophical Manuscripts - is the concept of praxis; so, a true Marxist views knowledge as an aspect of human praxis and not as a copy of anything. Any doctrine based on so wrong and unMarxist a view of knowledge and truth, must itself be wrong and unMarxist. As this neomarxist view prevailed at the Bled conference, so ended the dominance of the Marxist–Leninist orthodoxy in Yugoslavia; this, in turn meant the end of the wholesale rejection of post–Hegelian Western philosophy. The neomarxist Praxis philosophers were ready to learn from non–Marxists including analytic philosophers. According to them, the analytic approach to philosophy may help to clarify – if not solve – a number of traditional philosophical problems. And so in early sixties the analytic approach to philosophical problems was revived within a Marxist framework.

Although it was revived within a Marxist framework, not all neomarxists in Yugoslavia are ready to tolerate it. As their orthodox colleagues, these neomarxists – such as Veljko Korać, the staunchest opponent of analytic philosophy today – believe that Hegel's philosophy is unsurpassed in the bourgeois world. But unlike their orthodox colleagues, these modern Marxists find the analytic approach dangerous because, according to them, it strives to eliminate metaphysics and with it philosophy. According to Korać, analytic philosophy denies to philosophy the status of science and of theory. "The whole enterprise [of analytic philosophy]", writes Korac, "could be explained only as a snobbish academic play, as a peculiar academic wisecracking. Apart from this, it is a specific form of technocratic ideology which hides behind the mask of the neutrality of philosophy or science."[7] As a useless play analytic philosophy is not worth any study; as a dangerous ideology, it should be simply rejected without much a do. As we shall see below, adherents of this view of analytic philosophy, have recently called for the containment of its spread and its eventual eradication from the Yugoslav universities. Clearly, to reject analytic philosophy <u>in toto</u> it is not necessary to be an orthodox Marxist–Leninist.

ANALYTIC APPROACH WITHIN A MARXIST FRAMEWORK.

More works in analytic philosophy were translated and published in Serbo–Croat in the early sixties than in any other period. Wittgenstein's <u>Tractatus</u> (1960) and <u>Investigations</u> (1966), Russell's <u>Human Knowledge</u> (1961) and <u>History of Western Philosophy</u> (1962), Ayer's <u>The Problem of Knowledge</u> (1961) and Moore's <u>Principia Ethica</u> (1963) were among the major works published in the period. The philosophers who initiated this translation programme were the three neomarxists who were mainly responsible for the revival of interest in analytic philosophy: Mihailo MARKOVIĆ and Svetozar STOJANOVIĆ from the University of Beograd and Gajo PETROVIĆ from the University of Zagreb.[8] In fact, it was Gajo Petrović who introduced the term 'analytic philosophy' into Yugoslav philosophical terminology; according to him, the term should cover the philosophy in the analytic tradition ranging from the works of Frege, Russell and early Wittgenstein to the present day; 'logical positivism' – often used in Yugoslav literature to refer to this approach to philosophy –

should be applied only to those analytic philosophers who used this label for themselves.[9] Petrović's introductions to Wittgenstein's Tractatus and Ayer's Problem of Knowledge are systematic accounts of the development and background of the views of each author: their aim is to acquaint the reader and not to forewarn him. Naturally, Petrović makes it clear that he does not espouse his author's views on the nature of philosophy and how it should be approached. In connection with Wittgenstein's comparison of his philosophy to the scales which one throws away after having climbed to the top, Petrović asks "How could we climb to or descend to the deepest sense of the world on the rotten scales of senselessness"?[10] Not only did Wittgenstein fail to give a coherent account of what he had been doing, but his view of philosophy as an activity of showing something unsayable is, according to Petrović, too narrow a view of philosophy. And Ayer's attempt to solve the problem of knowledge was bound to fail because he ignored the role of praxis. In Petrović's opinion, praxis is a way of human existence, a "proof" of the existence of the external world, and the final "criterion" of objective truth. In spite of his failure, the analyses Ayer offers are, according to Petrović, "subtle, piercing and spirited". In short, the work of Wittgenstein and of Ayer contains a variety of valuable insights; they fail to solve the problems they address partly because they ignore the conceptual resources of the dialectical materialist theory, the resources which Petrović finds in the very concept of praxis.

In his Dijalektička teorija značenja (The Dialectical Theory of Meaning), (Beograd, 1961) Marković also discusses some theories of meaning advanced by analytic philosophers; unlike Petrović, he does not use the term 'analytic philosophy' and does not distinguish this approach to philosophy from any other. But on his account, the founders of the contemporary functionalist theory of meaning are Wittgenstein and Ryle; and Frege, Russell and Carnap significantly contributed to the realist theory of meaning. From Marković's point of view these theories – as well as any other non-dialectical ones – are inadequate; contrary to the functionalists' view, not all meaning is found in the use of the terms; and the realist theory, according to Marković, cannot explain the meaning of the terms such as 'a centaur' and 'the realm of freedom'. An adequate theory of meaning should satisfy the demands of objectivity and elasticity; and so far only a dialectical materialist theory satisfies both. Starting from the crucial concept of social praxis, such a theory strives to encompass all aspects of meaning – the aspects which are highlighted by the functionalist, realist and other theories of meaning; this is how the elasticity requirement is fulfilled. As for the objectivity requirement, this is satisfied by regarding as material any referent which we consider objective; thus various mental phenomena about which we talk are, on Marković's theory, regarded as material (ibid., p.207). The meaning, on this theory, is defined as a six term relation encompassing the speaker, the sign system he uses, the realm of his mind, the external world about which he talks and the community of the users of his language. The insights offered by analytic and other non-dialectical-philosophers are valuable in so far as they can be incorporated into such a dialectical materialist theory. The work of some of them – e.g. Wittgenstein and Ryle – is of value because of its anti-dogmatic stand too: they have successfully refuted various wrong and dogmatic theories and thus cleared the way for a dialectical theory of meaning.

In contrast to Petrović and Marković, Stojanović's approach is not – at least not explicitly – dialectical: in his textbook on meta-ethics Savremena meta-etika (Contemporary Meta-Ethics) (Beograd, 1964) he offers a systematic account of a variety of meta-ethical theories without attempting to incorporate them into a dialectical meta-ethical theory. The two main contemporary approaches to meta-ethics are, according to him, the cognitivist and the anti-cognitivist one; his own views are, he says, moderately anti-cognitivist. A moral statement, on his account, has cognitive, performative and expressive meaning but only the last is necessary for any moral statement; this view is closest to that of P.H. Nowell-Smith which Stojanović calls the 'praxicist anti-cognitivism'. In spite of this term, Stojanović does not claim that his – or any other meta-ethical theory – is a Marxist theory. As he puts it in his later work Izmedju ideala i stvarnosti (Between Ideals and Reality) (Beograd, 1968), his examination of meta-ethical theory is only a methodological and epistemological preliminary to his work on normative ethics which, indeed, is a Marxist ethics (ibid, p.5).

Although interested in various theories propounded by analytic philosophers, the three Yugoslav philosophers show no particular interest in the analytic approach to philosophy. A first brief account of the methods employed by analytic philosophers appears in the book Filozofija Lavovsko-Varšavske skole (The Philosophy of the Lvov-Warsaw School) (Beograd, 1964) by Svetlana KNJAZEV. In the concluding chapter of the book dealing with the Polish philosophy, Miss Knjazev points out how different the logical positivist approach is from that of the "British analysts" – Moore, early Russell, and Ryle, later Wittgenstein, Austin and Strawson. The former insist on demonstrating the senselessness of metaphysical and even epistemological statements and work on a systematic reduction of statements about the external world to the statements of immediate experience. The latter are not primarily interested in the elimination of metaphysics and the reduction of one level of discourse to another; their interests are wider and their methods more varied. Moreover, 'the British analysts' interested in the ordinary language analysis differ in their approach from the older generation of Moore and Russell: they do not attempt to define the meaning of words but rather describe a variety of meanings that a word can have in various contexts. The Polish philosophers of the Lvov-Warsaw school, Miss Knjazev maintains, share an approach similar to that of the older generation of the British analysts.

A number of Miss Knjazev's essays published in the sixties deal with Marxist topics. For example, in her essay "Predmet i problemi teorije saznanja" (The Subject-Matter and the Problems of the Theory of Knowledge)[11] she analyses the concept of praxis and concludes that only a narrow concept of material praxis may be used as a direct criterion of truth of only some statements; if so, praxis (in this sense of the word) cannot be used as a criterion of the truth of other statements such as those of mathematics and logic (ibid,.p.232). This was clearly a criticism of her Marxist colleagues' view of the concept of praxis as the basis for the solution of all kinds of philosophical problems.

Staniša NOVAKOVIĆ did not seem to share this view; and yet in his book Problem metafizike u savremenoj analitičkoj filozofiji (The Problem of Metaphysics in the Contemporary Analytic Philosophy) (Beograd, 1967) - the first book in Serbo-Croat to deal exclusively with analytic philosophy - analytic philosophy is approached from a Marxist standpoint. His is a thorough and systematic review of various views on metaphysics ranging from the early logical positivists to the present day 'ordinary language philosophy' and philosophy of science. In particular, Novaković is interested in the criteria of demarcation between science and metaphysics and in the role of metaphysical statements in science. Having shown the deficiency of the logical positivists' attempt to deal with these problems, he approaches it in a dialectical way. According to him, there is no sharp division between purely metaphysical and purely non-metaphysical statements; instead there is a continuum at one end of which are statements on immediate experience and on the other completely unverifiable metaphysical statements akin to those we find in poetry. Neither logical positivists nor Stalin and his followers are on the right track in their attempt to eliminate metaphysics from science and philosophy; for metaphysical theories - such as the Ancient Greek atomic theory - often precede and are later transformed into scientific theories. In stimulating scientific discovery, metaphysics helps us to participate in "the struggle for the realization of the humanist ideals which are laid in the foundation of the original Marx's philosophy."(ibid., p.156). It seems that Novaković, like Stojanović before him, believed that the insights provided by analytic philosophers may help us realize the true Marxist ideals.

By the end of the sixties, the philosophers who wrote on analytic philosophy changed their approach and interests: Petrović, Marković and Stojanović lost interest in this approach and concentrated on the more obviously Marxist themes while Miss Knjazev and Novaković abandoned the Marxist standpoint. But a number of recently published books and articles still approach analytic philosophy from an explicit Marxist standpoint. Such is the case with the two books by Jelena BERBEROVIĆ from the University of Sarajevo : Znanje i istina : osnovni smerovi analitičke filozofije (Knowledge and the Truth : the Main Directions in Analytic Philosophy) (Beograd, 1973) and Filozofija Ludwiga Wittgensteina (The Philosophy of Ludwig Wittgenstein) (Sarajevo, 1978). In the first book Mrs. Berberović classifies the types of views found in analytic philosophy into phenomenalist, physicalist, pragmatist and the ordinary language analyses; her assignment of John Dewey to the pragmatist analytic group shows how broad her classification is. Having thus classified analytic philosophers she proceeds to give a systematic account of the views of one philosopher from each group viz., Ayer, Carnap, Dewey and Wittgenstein. In her view, analytic philosophy in general does not reach the high standards of Hegel's philosophy; but it still offers subtle albeit partial insights into various philosophical problems. And although in her comprehensive account of Wittgenstein's philosophy she shows an admiration for Wittgenstein's philosophical flair and his critical attitude to dogma, the absence of a humanist stance and the failure to employ the concept of social praxis are, in her opinion, serious flaws in his approach to philosophy.

In his brief discussion of Quine, Davidson, Dummett, Austin and Grice, Nenad MIŠČEVIĆ finds no flaws of this kind. Yet his introductory text on philosophy of language - Filozofija jezika (The Philosophy of Language) (Zagreb, 1981) - seems to have been written from a Marxist standpoint. Not only is the long final chapter given over to the Marxist philosophy of language but in the epilogue Marx's early work is invoked in order to justify the interest in contemporary non-Marxist philosophy of language: "The interpretation of the world [in this philosophy] is starting to participate in its change by opening, in the tissue of the metaphysical text of the fossilized dominant discourse, the pores in which there is a promise of something new" (p.228).

Analytic philosophy in Yugoslavia certainly did not start to change the world; but it seems to have had some influence on the neomarxist approaches to a variety of philosophical topics. Some of its results were incorporated into wider dialectical theories; some of the theories offered by analytic philosophers were used by neomarxists to clarify their favourite concepts and views. More importantly, perhaps, through the works of these neomarxists Yugoslav students of philosophy got a glimpse of another approach to philosophy and of another set of standards for the evaluation of philosophical work. And those who thus became interested in doing philosophy in the way it is done by analytic philosophers, were able to pursue their interests outside any Marxist framework.

ANALYTIC APPROACH OUTSIDE THE MARXIST FRAMEWORK.

In Yugoslavia analytic philosophy was initially not discovered only by Marxists. Among the first - if not the first - essays on analytic philosophy outside a Marxist framework is the essay "Vitgenštajnova logička analiza jezika" (Wittgenstein's Logical Analysis of Language) by Mihailo DJURIĆ published in 1958 in the journal Delo. After a brief account of the difference between the Tractatus and the Investigations, Djurić points out that in both Wittgenstein holds that philosophy is not concerned with the construction of theories but is an activity of clarification of thought. To this view Djurić has nothing to object. His main objection is that Wittgenstein ignores the possibility of treating philosophical discourse as a meta-discourse - on the level of a meta-language; had he considered this possibility, Djuricć thinks that he would not have been tempted to say that when one does philosophy, the language goes on holiday. In short, in contrast to Wittgenstein, Djurić does not see why in philosophy we could not offer, in a meta-language, theories concerning the working of an object-language.[1 2]

The same journal published in 1960 an analysis of the language of poetry entitled "Kratki spis o jeziku" (A Short Note on Language) by Jovan HRISTIĆ. Hristić compares the language used in poetry and in describing everyday experiences and notes that the reference of words in poetry is much more diffuse and that the items to which these words refer are not on the same ontological level as everyday experiences. But as in the everyday life, language is an instrument in poetry too; it is an instrument for the exploration of the as yet unexplored regions of our possible experiences.

Delo in which these essays were published is primarily a literary journal; so is Letopis Matice Srpske which in the early sixties published the essays of one of the most talented Yugoslav philosophers Jovan ĆULUM.[13] In the essay published in 1957, "Dva primera rdjavih pojmova" (Two Examples of Bad Concepts) Ćulum notes that philosophers use ordinary language which is completely unsuited for their use; and in consequence philosophy is full of contradictory statements and linguistic dogmatism. Moreover, philosophers are particularly prone to professional delusions because they usually write a book about another book which in turn was directed against yet another book and so on. His first example of a bad concept is the one used in the statement "Everything is identical with itself". This statement implies or presupposes that every entity in the world is a unique exemplar. But, Ćulum now argues, if we conceive of everything as a partially unique exemplar we shall be led to an infinite regress; and if we conceive of everything as an absolutely unique exemplar we shall be led to admit that it is impossible to conceive such a thing because it is ex hypothesi unrelated to anything. The second example is that of the pair of full and empty. If we conceive of something absolutely full, the way we are forced to do it presupposes the emptiness. Although his arguments here are far from conclusive, the clarity with which they are presented, the critical attitude towards philosophical professionalism and the lack of any appeal to authority makes this essay unique in the Yugoslav philosophical tradition; moreover, at that time Marxism-Leninism with its ritual appeal to authority, its cumbersome jargon and disregard for argument was still the order of the day.

In another essay "Beleška o pojmu determinizma" (A Note on the Concept of Determinism), he claims that the general thesis of determinism that everything is caused by something due to its generality cannot be verified by experience and so we should try to verify it in the realm of thought. He then proposes to relate this thesis to our activity of explanation; if so, the thesis should be understood as the claim that everything is explained by something else. This paraphrase of the original thesis also suggests a reason why the thesis is held to be universally true; it is universally true because it is a tautology – it only asserts that everything is explained by an explanation. Namely, an explanation is by definition an explanation by something else; on this definition there is simply no explanation if we explain something by itself. So it is logically impossible to conceive of an undetermined something simply because it is logically impossible to conceive an explanation which does not explain by something else. But if this is correct, the thesis of determinism is irrefutable simply because it is empty. This, Ćulum suggests, is often the case with philosophical theses.

This philosophical note displays Ćulum's preferred type of analysis: to transform "the propositions expressed in ontological terms into propositions formulated in the terms which expressly refer to certain human activities" (Introduction to Filozofske beleške, p. 21). Ćulum assumes that the propositions which refer to the realm of our experience and thought are clearer and more understandable than those which refer to the world outside this realm. The same type of analysis is employed in the essay "Velika shema filozofije" (The Great Scheme of Philosophy); how is a realist to distinguish his view of the external world from a Berkeleyean solipsist who believes that the world consists of his own sense-impressions? No attempt to characterize the external world in terms of the causes of the sense-impressions or in terms of the explanation of the regularities of sense-impressions, succeeds in elucidating – at least in Ćulum's opinion – the realist concept of the external world. The only way to do so is to paraphrase the realist's thesis by a proposition referring to a human activity, viz. the activity of predicting future events. The claim that the external world is independent of the subject's sense-perception, according to Ćulum, is for the purposes of this analysis equivalent to the claim that one is successful in predicting the future; realism is thus a claim about our successful predictions and not a claim about the world. Since the solipsist would not want to deny this claim, such an analysis shows that there is in effect little difference between a Berekleyean solipsist and a realist. But a solipsist differs from an illusionist who believes that everything is a dream: the latter has to presuppose that there is a distinction between dreams and wakeful states while the former presupposes no difference between sense-impressions and the external objective world. A Berekeleyean solipsist does not claim that everything is an illusion – because the latter concept presupposes the concept of veridical perception – but only that all statements about the external world are explicable in terms of sense-impressions and our ability to predict the future.

In the essay "Primer lažne suprotnosti" (An Example of a False Opposition), Ćulum in a similar way tries to show that there is no significant opposition between Kant's view of space as an a priori intuition and the objective realist's view of space as an entity existing independently of human mind. This series of essays on the alleged differences between realists' and idealists' views seem to have been part of the programme of showing these differences to be imaginary and the contrast between idealism and realism – so dear to the orthodox Marxists – unfounded. This unique programme in Yugoslav philosophy was cut short by the suspension and dismissal of Ćulum from his post at the University of Beograd; in the early sixties the neomarxists at the University of Beograd were not as yet ready to tolerate blatant non-marxism in their midst especially if propagated in so lucid and attractive way. For Ćulum's essays are unique not only in the clarity and simplicity of their arguments : they are unsurpassed masterpieces of Serbian philosophical prose.

The role and status of metaphysics was a topic of considerable interest to Yugoslav philosophers. The first attempt to discuss it outside a Marxist framework appears in the essay of Vojislav STOJANOVIĆ "Jedna analiza metafizičkih iskaza" (An Analysis of Metaphysical Propositions) (Ideje, no.6, 1971). In the essay, Stojanović argues that general characterizations of metaphysical propositions as either a priori or universal empirical propositions are both wrong. More attention should be paid to more specific characterizations such as the claim that metaphysical propositions express syntactical conventions or that they are general analogies from experience. He considers in some detail a number of such specific proposals, only to reject them all. At the end he argues that metaphysical propositions have a specific normative meaning and that they often express values; their meaning is thus mainly prescriptive and expressive. They also contain descriptive suggestions e.g. in the form of descriptive projections to another possible world. This view of metaphysical propositions is, Stojanović holds, only a hypothesis which he constructed by elimination of its rivals. It stands in need of substantiation by further research into the role which metaphysical propositions play in various philosophical systems.

The same painstaking care in the statement of the argument and the same tentative approach we find in Stojanović's essay "Istina i dobro" (Truth and the Good).[14] In the essay, Stojanović examines the similarities between the anti-attributivist arguments concerning truth and the anti-naturalist arguments concerning good. To clarify the basic meaning of "...is true" and of "...is good" one can refer to the type of speech act in which these two terms are used; the first is used in confirming or accepting something while the second is used in choosing or recommending something. Perhaps the metaphysical unity of truth and goodness is just in that these speech acts could be employed in an attempt to clarify the meaning of the two terms. The careful examination of arguments, construction of relevant counterexamples and the appeal to semantic or pragmatic theories of language pioneered by Vojislav Stojanović characterizes much of the current work of Yugoslav analytic philosophers.[15]

Meta-ethics, as we have seen, was regarded as an area of concern primarily of analytic philosophers. And it was one of them, Igor PRIMORAC, who was the first Yugoslav philosopher consistently to defend a deontological view of ethics. In his first essay on this topic[16] Primorac argues against Urmson (in the "Saints and Heroes") that the nature of morals and of moral speech is such that saintly and heroic actions are necessarily outside the domain of ethics; consequently such actions cannot be cited in an argument against deontological ethics. He admits that one can rightly say that such acts are worthier and even better than other acts but these terms are not used in their moral sense. This essay led to the first polemics on meta-ethics in Yugoslav philosophy. In a rejoinder to Primorac's essay[17] Svetlana Knjazev claimed that it is not sufficient to say that the use of a term such as "good" is not moral but one should show in what non-moral sense the term is used; moreover, it seems that Primorac's claims concerning the scope of ordinary moral speech have to be substantiated by an empirical inquiry using statistical methods; this allows philosophical arguments to degenerate into statistical inquiries. In his reply[18] Primorac refuses to accept the demand for an explanation in what sense one uses the term 'good' when speaking of saintly or heroic actions and reiterates his view that the moral evaluation of actions is an evaluation in terms of the performance of duties. However, he admits that his claim concerning the scope of moral speech requires empirical substantiation and in so far as this is lacking his argument is inconclusive. This essay is the first systematic examination in Serbo-Croat of the arguments based on the appeal to the ordinary language. For her part Miss Knjazev returned to this theme in the study "Etičke teorije i pojam 'žrtvovanja' (Ethical Theories and the Concept of Sacrifice) (Filozofske studije, Beograd, 1974) in which she first shows how varied the use of the word 'sacrifice' is and then argues at length that neither utilitarian nor deontological theories can explain our moral intuitions about sacrifice.

The next step in Primorac's defense of deontological ethics is also based on an appeal to the ordinary language - an appeal to the ordinary use of the word 'duty'.[19] In short, his main argument seems to be that the statement "The action which leads to the greatest amount of good is a duty" is not analytically true, and that, therefore, statements about one's duties cannot be analyzed in the way utilitarians would prefer. In consequence, utilitarianism as a general meta-ethical theory is not true. And in opposition to Svetozar Stojanović's total neutrality thesis, Primorac maintains that the refutation of a meta-ethical theory such as utilitarianism has clear consequences for the normative ethical theory (while the converse does not hold).

Some consequences of this kind Primorac draws in his book Kazna i prestup (Punishment and Offense) (Beograd, 1978); the arguments deployed in the seven essays collected in the book are mainly advanced against the utilitarian and in support of the retributivist view of punishment. In one of the essays Primorac argues that it is not analytically true that to punish someone is to hurt him because he has committed a crime. In consequence, if precisely formulated, this thesis can be maintained on purely retributivist grounds. For a retributivist, breaking a relevant law or rule is both necessary and sufficient ground for punishment; for him such an infringement by itself deserves punishment. In another essay, dealing with the punishment of innocents, Primorac points out that a deductive analysis of an ethical theory discovers its implications for a variety of moral situations; we can then test these implications against our moral intuitions or reactions and thus decide for or against the ethical theory in question. A clear case is an ethical theory such as utilitarianism which allows for the punishment of the innocent; the Gulag archipelago is an institution for the punishment of the innocent which is justified by considerations congenial to utilitarianism. Such a test by our moral intuitions would lead us to the choice between the good and the just; in case of so uncompromising a choice Primorac would clearly choose the justice. As the works of Jovan Babić, Aleksandar Pražić, Aleksandar Jokić and Aleksandra Marković testify, the use of analytic techniques in the discussion of normative ethics which Primorac pioneered, is of growing interest among analytic philosophers in Yugoslavia.[20]

Apart from a series of articles on the problems of normative ethics, Jovan BABIĆ has written a study on J.S. Mill: "Utilitarizam i sloboda" (Utilitarianism and liberty) (Dometi, no. 10, Rijeka, 1985). In the study he reviews a variety of recent interpretations of Mill's arguments for liberty and shows that there are widespread doubts in the utilitarian basis of some of his arguments. Although some of Mill's arguments are utilitarian, Babic points out that Mill's argument that freedom is required to allow for the unhampered pursuit and discovery of truth, cannot easily be interpreted within a purely utilitarian framework. Moreover, in his arguments for liberty Mill makes a number of judgments which are not based on utilitarian considerations e.g., that every use of force, as the use of force, is unjustified; and Mill often presupposes that liberty is an intrinsic value which stands in no need of further justification. This study which highlights the non-utilitarian aspects of Mill's views, is the first systematic examination in Serbo-Croat of Mill's arguments for liberty.

Aleksandar PRAŽIĆ's little book Reč o slobodi (A Discourse on Freedom) (Beograd, 1978) is not easy to classify within a single philosophical discipline; ostensibly it is a book on political philosophy but a part of it seems to be concerned with philosophical archaeology - the discovery of the hidden meanings in the works of Hegel. In his preface Pražić insists that his is a purely conceptual inquiry unrelated to any humanist value system or to any theory of philosophical anthropology. The reason why so much space is taken up by the examination of Hegel is found in his desire "to test the analytic tools on the hardest material which the history of philosophy offers" (ibid.,p.5). In its literal sense, a statement of the form "A is free in respect of x" should be analyzed as "the situation in which A finds himself/herself contains nothing which would hinder A in respect of x and forces him/her to do or suffer non-x"; and A can here be only a practical being. When in using the predicate "...is free" we deviate from this use (e.g. when we speak of the freedom of the press), we are using the term in a non-literal sense through which we often try to project our attitudes and interests. Philosophers, maintains Pražić, in this way often project their own particular interests. And so he examines the uses to which Plato, St.Paul, Kant and finally Hegel put this word. In Hegel's claim that only within a state a man will find the realization of his freedom, Pražić finds the tendency to project an extreme interest of Hegel - the interest to make himself (or the paradigmatic man) feel and be at home everywhere. In other words, in human freedom Hegel sees the elimination of anything that is foreign to humanity and so to realize this freedom the whole of human environment has to be made rational. Once we grasp the special interests and attitudes philosophers project into their concepts of freedom, we should be able to understand what the continuous controversy about human freedom is all about. This somewhat skeptical view of the philosophers' motivation recurs throughout Pražić's works and puts him apart from most of Yugoslav philosophers.

The interests of Neven SESARDIĆ from the University of Zagreb have centered not so much on the motivation as on the arguments philosophers put forward. In his first essay "O izvjesnosti" (On Certainty)[21] he defends Moore's views of certainty from Wittgenstein's criticism in On Certainty; in his opinion, Wittgenstein is wrong in saying that we cannot understand the basic propositions (which Moore favours) when uttered in a non-philosophical context - to assert any such proposition is simply not absurd. And in contrast to Wittgenstein he holds that one can determine whether a proposition is true or not even outside of a language game within which it is usually employed. Sesardić's main anti-skeptical arguments are deployed in his essay "Problem vanjskog sveta" (The Problem of the External World);[22] his main aim is not so much to show how skeptical doubts in the existence of the external world are slight, but to expose the invalidity of skeptical arguments such as the Argument from Illusion and Descartes' Evil Daemon argument. In a rejoinder to this essay, Aleksandar PAVKOVIĆ tried to show that Sesaradić's objections to the introduction of sense-data are unfounded and that he misreads the Evil Daemon argument. In this prolonged polemics, for the first time in Yugoslav philosophy the classical skeptical arguments were given a close and fairly technical scrutiny.[23]

Sesardić's anti-skepticism seems to be supported by his belief that scientific inquiry will eventually resolve many of the philosophers' puzzles. How it may resolve the puzzle of the relation of mind and body Sesardić sets out to show in his book Fizikalizam (Physicalism) (Beograd, 1984). The neurophysiology will eventually show that mental events are in fact events in the brain of the subject; but until it does so, Sesardić tries to refute various objections to this mind-brain identity hypothesis. Thus he argues that subject's reports on his own experiences may be corrected by later scientific discoveries and that, in any case, the greater authority that the subject has over the reports on his own experiences does not stand in the way of a reduction of mental to brain events. He also rejects Kripke's claim that to be in the same epistemological situation as when feeling pain, means to experience pain and Kripke's conclusion that physicalism is, therefore, untenable. In discussing various functionalist objections to physicalism, Sesardić also rejects Davidson's thesis of the nomological anomaly of the mental as well as the claim that psychology requires an intentional language. His sympathies clearly lie with the Quinean programme of the elimination of intentional terminology and its replacement with descriptions of dispositions to behave in a certain way; such descriptions, he hopes, will be systematically related to the structural states in the brain which cause them.

The mind-body problem and Sesardić's solution to it has now been a topic of philosophical controversy among analytic philosophers in Yugoslavia for some time. Sesardić's views on the corrigibility of the subject's reports on his own experiences were indirectly challenged by Nikola GRAHEK in his study "Nepogrešivost i čulno opažanje" (Incorrigibility and Sense-perception);[24] while rejecting the thesis of the absolute incorrigibility of subject's reports on his sense experiences, Grahek argues for the sovereignty of the subject in the field of his own experiences. Contrary to the Sesardić, Grahek claims that in some cases a neurophysiologists would not be obliged to correct a subject's report on his own experiences even when it deviates from the reports of other subjects and contradicts a scientific hypothesis. In short, Grahek suggests that neither our ordinary intuitions nor the requirements of neurophysiological inquiry support the demand that a subject should yield to the scientist on the question of the former's experiences. As the series of articles the mind-body problem published in 1986 and 1987 in the Zagreb journal Filozofska istraživanja and the Ljubljana journal Anthropos testify, the debate on this and related topics is continuing.[25]

Modern theories of meaning - from Austin to Davidson and Dummett - were briefly but accurately presented in Miščević's introductory book on the philosophy of language. Some of the classical essays on the theory of meaning by Frege, Russell, Austin and Grice were translated and published in various journals in Beograd as well as in Rijeka (where Miščević edits the journal Dometi). And it was Nenad Miščević who wrote the first critical study in the philosophy of language in Serbo-Croat: Od misli do jezika (From thought to language) (Rijeka, 1987). The central question discussed in the book is that of attribution of thought: What kind of theory could best explain the fact of thought and of our ability to express it in a language. Such a theory, argues Miščević, has to be be given within a naturalist framework. But this does not mean that our folk-psychological views have to be completely abandoned in favour of a representationalist theory such as the one favoured by Jerry Fodor. A critique of this kind of theory leads Miščević to list a number of desiderata for any theory of thought and talk. And such desiderate are satisfied by the theories which would explain how our languages establishes the referential link between thought and the outside world, how it "fills" thought with content and how it gives the thought the intentional features it possesses. In the final chapter Miščević argues for the need of philosophy to rely on the natural and social sciences and not on a priori intuitions. Miščević thus seems to share Sesardić's view that analytic philosophy continues and complements the work of the sciences.[26]

Yet Miščević, like many other analytic philosophers, does not share Sesardić's view of the need to expose the confusions and inconsistencies of Marxism and Marxist philosophers. Since his brief note on Marxist orthodoxy in 1983 Sesardić has written a series of essays and a short book on Marxism and its foibles. In his latest essay "Razmišljanja o praxis filozofiji" (Reflexions on the Philosophy of Praxis) in Theoria (1987) Sesardić was the first to point out the intolerance of the Praxis neomarxists to other approaches to philosophy. Although Sesardić's work on Marxism could hardly be regarded as part of the analytic tradition, it certainly concerns analytic philosophy in Yugoslavia. From a Marxist point of view, the question – What is to be done with a non-marxist philosophy, such as the analytic one, which refuses to be incorporated into Marxism or to disappear? – is always pertinent. Perhaps the epilogue is the right place to canvass some of the recent answers to this question.

THE EPILOGUE: WHAT IS TO BE DONE WITH ANALYTIC PHILOSOPHY?

A group of academics from Beograd formed in 1981 the Section for analytic investigations in philosophy within the Philosophy Society of Serbia; as their first chairman they elected Professor Aleksandar Kron, a logician who has for years quietly encouraged analytic philosophers all over Yugoslavia. A similar section was formed a few years later in Ljubljana within the Philosophy Society of Slovenia.[27] Although some Beograd philosophers objected to the foundation of the Section (on the ground that it artificially splits the unified enterprise of philosophy), the analytic section organized a series of talks on a variety of topics such as the Hilbert programme or the problem of capital punishment. This quiet – some would say boring – academic life was interrupted in the spring of 1985 by a damning Party document: "The Ideological and Political Aspects of the Development of Social Sciences" issued by the Commission for Ideological and Theoretical Work of the Central Committee of the League of Communists of Serbia. It contained a series of charges against analytic philosophy together with warnings and suggestions as to what should be done about it. The authors of the document charged: first, that as the dominant philosophy of the Anglo-Saxon world, analytic philosophy has "no roots in the tradition of our [i.e. Serbian] culture"; second, that although "it cannot have a permanent future because it is so anachronistic", it can do considerable spiritual harm and create a peculiar anti-philosophical atmosphere in Beograd; and, thirdly and most importantly, as analytic philosophy is far from being ideologically neutral, it "destroys the very presuppositions for the existence of Marxism itself". The authors of the document then warn that such developments cannot be left uncontrolled and go on to demand of every philosophically-bent Marxist to work on the synthetic Marxist philosophy and to contain the spread of analytic philosophy.[28]

After the distribution of the document to Party cells, a group of analytic philosophers joined by the neomarxists Marković and Stojanović asked, in an open letter, the President and the Steering Committee of the Serbian Philosophy Society to publicly protest against this attack. As no action was taken on this request, Svetlana Knjazev published an article in the literary weekly <u>Književne novine</u> suggesting the names of the authors of the document and pointing out how similar this document is to the orthodox Marxist-Leninist attacks on bourgeois philosophy and literature. In reply to her the same weekly published articles by Veljko Korać (who was named as a possible author of the document), Miladin Životić, Branko Pavlović and Slobodan Žunjić – all of whom were former or present members of the Philosophy Department at the University of Beograd. Their views as to what analytic philosophy is greatly varied; but all of them shared a strong disapproval of and even revulsion from analytic philosophy. For Žunjić, for example, analytic philosophy is a contemporary form of positivism and positivism is, of course, "a reined in and finished off reflexion". In Životić's opinion "analytic method throws out of philosophy all the essential questions of our existence as the questions of the possibility of another mode of being." For Pavlović, the analytic approach degrades philosophy to the status of a technique; and for Korać it is simply a foreign ideology. Korać and Životić explicitly endorsed the Party document while Pavlović and Žunjić refrained from discussing it, politely ignoring its existence.[29] Apart from Miss Knjazev, who, in her replies to the critics, did not pull her punches, Aleksandar Pražić, Miloš Arsenijević and Rastko Jovanović replied to the critics of analytic philosophy.[30]

The Party officials stayed out of this melée and the Party, so far as I know, issued no further documents or instructions on analytic philosophy. Perhaps the Party can trust Beograd philosophers to keep this approach to philosophy in check; and perhaps the Party is right to do so.[31] Meanwhile, judging from the papers published in the Zagreb journal Filozofska istraživanja, the Ljubljana journals Anthropos and Acta Analytica as well as the Beograd journal Filozofske studije, analytic philosophy is thriving both in Yugoslavia and among Yugoslav philosophers abroad. This collection, I hope, is yet another sign of this.[32]

NOTES

1. I should mention another deficiency of my account: my poor understanding of Slovenian prevents me from offering an account of analytic philosophy among the Slovenians.

2. The main source of my account of the works of Richtman and Podhorsky is Božo Kovačević's essay "Marksizam, filozofija i znanost u radovima Richtmana i Podhorskog" (Marxism, Philosophy and Science in the Works of Richtman and Podhorsky) Gordogan, no. 7 and 8 in 1981 and no. 10, 1982. Zagreb.

3. "Prevrat u naučnoj slici sveta" (The Coup in the Scientific Image of the World), Pečat no. 4, 1939, and "Suvremena naučna misao nije idealistična" (Contemporary Scientific Thought is not Idealist), Pečat, no. 13–15, 1940.

The main editor and contributor to Pečat was Miroslav Krleža a well known writer who was, at about the same time, proclaimed a renegade by the Communist Party. Thus, this journal probably had the most talented group of renegade Communist intellectuals at the time.

4. Although rejected by the Communists, Richtman was shot by the Croatian separatists Ustaše at the outbreak of war in Yugoslavia together with prominent Communist intellectuals including his main opponent Ognjen Prica; Podhorsky survived the war but stopped doing philosophy.

5. He first fled to Austria and then to Australia where in 1954 he became a lecturer in philosophy at the University of Tasmania in Hobart; there he was involved in the unfair dismissal of Professor Orr.

6. Dušan Nedeljković was summarily dismissed from his chair in 1954 but retained his post in the Serbian Academy of Sciences. Boris Ziherl, who was retired later, had little influence outside Slovenia. Dragan M. Jermić, later a Professor of Aesthetics at the University of Beograd, abandoned the rigid Marxist orthodoxy of his postgraduate years.

7. Filozofija i njena istorija (Philosophy and its History), Zagreb, 1978, p. 41 (the chapter is entitled "The Campaign of Analytic Spirit Against the Historical Heritage of Philosophy"). Veljko Korać is Professor Emeritus of History of Philosophy at the University of Beograd.

8. One should mention that they were the first philosophers after the Second World war to go to the West – in this case Great Britain – to do research; Marković also obtained a doctorate in philosophy from London University.

9. "Logički pozitivizam i princip verifikacije", first published in 1959; reprinted in Petrović's collection of essays Od Loka do Ejera (From Locke to Ayer), Beograd, 1964, p. 155.

10. Ibid., p. 153.

11. Published in the collection Problemi filozofije marksizma (The Problems of the Philosophy of Marxism), Beograd, 1967.

12. In a short period of time after the war, Djurić was a private pupil of Kajica Milanov. Apart from analytic philosophy, Djurić wrote on Ancient Greek philosophy and on the sociology of Max Weber. He has since lost interest in analytic philosophy and became one of the most speculative Yugoslav philosophers strongly opposed to the analytic approach.

13. The neomarxists had the editorial control over the two philosophical journals in Belgrade Filozofija and Filozofske teme. All of Ćulum's philosophical essays published between 1957 and 1966 were reprinted in

the collection Filozofske beleške (Philosophical Notes) (Beograd, 1967) which is my source here.

14. Theoria, no. 1–2, Beograd, 1976.

15. Vojislav Stojanović abandoned the calling of a professional philosopher in 1979 and in 1986 left the country to work as a journalist in Paris.

16. "Deontološka etika, sveci i heroji" (Deontological Ethics, Saints and Heroes), Filozofija no. 2, Beograd, 1972.

17. "Domet jednog argumenta" (The Range of an Argument), Filozofija, no. 4, Beograd, 1972.

18. "Etika i pozivanije na običan jezik" (Ethics and the appeal to the ordinary language", Theoria, no. 1–2, Beograd, 1976.

19. "Utilitaristička greška" (The Utilitarian Mistake). Gledišta, no. 7–8, Beograd, 1974.

20. Since 1982 Igor Primorac has been teaching philosophy at the Hebrew University in Jerusalem.

21. Published in Praxis, no. 3–5, Zagreb, 1974. The same issue of this journal carries an essay "Šta se dogodilo sa znakom asercije u modernoj logici?" (What Happened to the Assertion Sign in Modern Logic?) by Goran Švob the other analytic philosopher in Zagreb. These two essays were among the very few essays in analytic philosophy published in this well-known neomarxist journal.

22. Published in Theoria, posebno izdanje,Beograd, 1975.

23. The rejoinder and Sesardić's reply were published in 1976 and 1977 issues of the Beograd journal Theoria.

24. Published in Filozofske Studije, vol. 13. Beograd, 1981. A. Pavković has a rejoinder to this too in Theoria, 1984.

25. A collection of essays on the analytic philosophy of mind entitled Svest i saznanje : ogledi iz analitičke filosofije (Mind and Knowledge: Essays in Analytic Philosophy) appeared in Beograd in 1980 edited and introduced by Aleksandar Pavković. The editor preferred not to include essays by Davidson and Quine; among the included authors were Austin, Grice, Ryle, Pears, Deutscher, Anscombe, Hampshire and Hinton. The editor's introduction gives a short summary of arguments on each of the concepts discussed in the collection, (e.g, sense-perception, memory, thought, etc.). These concepts have not been the topic of much philosophical discussion in Yugoslavia.

26. For reasons of space I have not reviewed the work in philosophy of science and formal logic. Staniša Novaković, Svetozar Sindjelić and Neven Sesardić have published a number of works on various issues in the contemporary philosophy of science. Among philosophers, Aleksandar Kron and Kosta Došen have published, mostly in English, articles on various topic in logic including relevance logics, decision theory and the logic of quantum theory.

27. As I mentioned earlier my poor knowledge of Slovenian prevents me from attempting to review some of the work of the Slovenian colleagues among whom are Frane Jereman, Andrej Ule, Matjaž Potrč, Bojan Borstner and Marko Uršić. It is a very active group of philosophers who in 1986 started to publish a journal in English entitled Acta Analytica.

28. The document although internal was widely quoted in the polemics that ensued. My quotations are taken from the article "Analitička filozofija u očima njenih domaćih protivnika" (Analytical Philosophy as Seen by its Domestic Opponents) by Miloš Arsenijević and Rastko Jovanović, published in Književne novine, on December 1, 1985.

29. At the time of polemics, Životić held the office of the President of the Philosophy Society of Serbia.

30. One should mention a lone article by a non-philosopher Stojan Cerović who pointed out that the vicious attacks on analytic philosophy by philosophers may do the job which the authorities planned to carry out themselves.

31. Since 1985 no philosopher with an inclination to analytic philosophy has gained appointment to an academic or research post in Beograd.

32. I am grateful to Dušan Bošković, a philosopher and historian of ideas, for the help he has given me in writing this historical survey.

SOLUTION OF THE STACCATO VERSION OF THE ACHILLES PARADOX

MILOŠ ARSENIJEVIĆ

§1. PROBLEM

Let $\underline{P}_1$ and $\underline{P}_2$ be the following premises of a syllogism:

($\underline{P}_1$) After being in rest-state $\mathcal{A}_0$ (touching at A the area he is to move through), Achilles has moved along path AB according to the following pattern: a 1/2 h movement which covers one half of AB, followed by a 1/2 h rest (in rest-state $\mathcal{A}_1$, touching at the mid-point of AB the area in front of him); a 1/4 h movement which covers the distance between 1/2 AB and 3/4 AB, followed by a 1/4 h rest (in rest-state $\mathcal{A}_2$), and so on.., a $1/2^{n+1}$ h movement which covers the distance between $(2^n-1)/2^n$ AB and $(2^{n+1}-1)/2^{n+1}$ AB, followed by a $1/2^{n+1}$ h rest;... (n=0, 1, 2,...).

($\underline{P}_2$) 2 h after the beginning of his staccato motion (described in $\underline{P}_1$), Achilles is definitely at rest.

If $\underline{P}_1$ and $\underline{P}_2$ are true, what are we to conclude about the position of Achilles and the number of his rest-states after the 2 h have elapsed?

It seems obvious that Achilles cannot be anywhere between A and B, because for any such position we can demonstrate that Achilles has had to leave it before the 2 h have elapsed. It can be also shown that he could have reached no further than B.

A. Pavković (ed.), Contemporary Yugoslav Philosophy: The Analytic Approach, 27–55.

Concerning the number of rest-states, no finite number seems to be an adequate answer: for any k, however great, there is a point $(2^{k+1}-1)/2^{k+1}$ at which Achilles has had to be in rest-state $\mathcal{A}_{k+1}$.

So it seems that if Achilles has moved from A towards B in accordance with P_1, and if P_2 is true, we have to conclude that

(C) Achilles has reached B, after an infinite number of stops.

However, even though it seems that C is the conclusion which follows from premises P_1 and P_2 that can be true, unsurmountable difficulties are encountered with $\diamond C$.

For C to become true, Achilles has to perform an infinite number of strictly discrete successive acts of non-infinitesimal duration in a limited period of time. However,

a) the finiteness of the number of performed acts - continuous motions - is recursively preserved (where the fact that both the time interval for the performance of these acts and the duration of the rests decrease and converge to zero does not change anything). How could Achilles free himself from this property of the finiteness of the number of performed acts, and 'jump' into the situation of having performed an infinite number of them?

b) There is neither enough time within the 2 h period nor enough room within AB for an infinite number of movements defined in P_1: for every possible instant within the 2 h period (and every possible point within AB) it holds that up to that instant (and that point) Achilles has performed a finite number of movements. How could one single rest-state - the state of Achilles' being at B - change the situation so radically that it becomes true that Achilles has performed an infinite number of movements?

c) The geometric progression whose general member is

$$[(2^n-1)/2^n,\ (2^{n+1}-1)/2^{n+1}]$$

has no last member. How could Achilles find himself at B after having performed the series of strictly discrete acts of non-infinitesimal duration without having performed a last act?

This profound problem is typical of philosophy: the inevitable conclusion which seems untenable is drawn from the premises which seem as statements which evidently can be true.

If the problem arises from the fact that P_1 and P_2 seem possibly true - $\diamond P_1$ and $\diamond P_2$ - while at the same time conclusion C, which follows from P_1 and

$\underline{P}_2$ – $\underline{P}_1$, $\underline{P}_2$ $\vdash \underline{C}$ – seems necessarily false – $\neg\Diamond\underline{C}$ – then it seems that the only way out is to deny somehow $\Diamond\underline{P}_1$, or $\Diamond\underline{P}_2$, or $\Diamond\underline{P}_1$, $\Diamond\underline{P}_2 \vdash \Diamond\underline{C}$, or $\neg\Diamond\underline{C}$. The disjunction need not be exclusive.

We shall examine in detail the known proposals for the solution of The Achilles, bearing in mind that in some cases their authors did not tackle the problem in the form in which we have presented it, but were confronted with some related versions of it. The last fact is of importance in those cases where we do not obtain the answer to the problem presented in our version of it despite elaborate solutions of some other versions of it.

§2. PROPOSALS

The known proposals for the solution of The Achilles can be divided into four groups:

I. those which deny both $\Diamond\underline{P}_1$ and $\Diamond\underline{P}_2$;

II. those which deny $\Diamond\underline{P}_1$;

III. those which accept $\Diamond\underline{C}$;

IV. those which, at the same time, accept, or have no way to avoid, $\Diamond\underline{P}_1$ and $\Diamond\underline{P}_2$, and deny $\Diamond\underline{C}$, and yet do not show what may be wrong with $\Diamond\underline{P}_1$, $\Diamond\underline{P}_2 \vdash\underline{C}$.

I. $\neg\Diamond\underline{P}_1$ & $\neg\Diamond\underline{P}_2$

Negative Dialectic

In Plato's Parmenides, Socrates concludes – without Zeno's explicit agreement (Zeno being present in the discussion) that he, Socrates, is right at that point – that all of Zeno's arguments against plurality are aimed at the confirmation of Parmenides' monistic ontology[1] .

If we accept Socrates' reasoning, we can say that The Achilles is aimed at the confirmation of the non-existence of motion, and if we consider, in particular, our version of The Achilles, the rationale could be the following: if we accept that Achilles can move, we cannot deny that $\underline{P}_1$ is possible, and by adding $\underline{P}_2$, which seems evidently possible if $\underline{P}_1$ is possible, we get the paradoxical $\underline{C}$. Thus, in order to avoid the paradoxical consequence, we have to conclude that motion is impossible, with $\neg\Diamond\underline{P}_1$. In addition, in the world in which motion is impossible, time does not exist[2], thus also $\neg\Diamond \underline{P}_2$.

If we admit that if motion is at all possible there is no way to disallow the possibility of $\underline{P}_1$ and $\underline{P}_2$ to be true, and if, in addition, $\underline{C}$ is paradoxical, The Achilles, much as Zeno's dialectical proofs in general, results either in sceptical resignation or in the negation of the possibility of the existence of the world of change. This last result, expressed radically in Gorgias' teaching on ontological nihilism[3], explains the origin of label negative dialectic[4].

The idealist F.H. Bradley claimed that the acceptance of the non-existence of the world of change does not imply that nothing exists. Moreover, he asserted that the selfcontradictory world of change can be the appearance of the timeless Absolute[5].

Bradley does not explain 'how the appearances (in time and space) come to be, and again how without contradiction they can be real in the Absolute', but he insists 'that such knowledge is not necessary', because 'what we require to know is only that these appearances are not incompatible with our Absolute...Since it is possible that these appearances can be resolved into a harmony which both contains and transcends them; since again it is necessary, on our main principle, that this should be so - it therefore truly is real'[6].

Criticising this conclusion, Alfred Ayer asserted[7] that a self-contradictory description is not applicable to anything, and hence not to appearances as well. In spite of this, Bradley's point can be consistently interpreted as follows.

If something is red whenever we look at it, we may conclude that it is possibly red, but if it is red whenever we look at it with the left eye, green whenever we look at it with the right eye, and blue whenever we look at it with both eyes, we may conclude only that it looks red, green, and blue at the same time (different persons can simultaneously look at it in different ways), but we must not conclude that it is possibly red, green, and blue at the same time. Any particular description is consistent in itself, but all the descriptions taken together are inconsistent. The object we are describing may seem different, in accordance with the selected descriptions, but cannot be such per se. If we are really obliged - as Bradley believes we are - to describe motion in various ways which are inconsistent among themselves, we may say that though it appears as something real it cannot be real.

The main question is, however, whether, if we accept that motion exists, we are really obliged to accept the paradoxical $\underline{C}$. I am sure that most people would not like to proclaim the entire world of change a mere appearance, and this cannot satisfy us particularly in those instances - such as the presently considered - where we make that proclamation in view of our inability to find a better solution to our difficulties.

II. $\neg\Diamond \underline{P}_1$

Historically, there are two known strategies in denying $\Diamond \underline{P}_1$ without questioning the general hypothesis concerning the existence of plurality and motion (so also not denying $\Diamond \underline{P}_2$): atomism and radical empiricism.

1. Atomism

One of Zeno's arguments against plurality[8] is parallel to The Achilles in that it states that the hypothesis concerning the divisibility of bodies leads to the conclusion, which is assumed to be paradoxical for the reasons similar to those against $\underline{C}$ (§1), that a limited body can consist of an infinite number of constituent parts. Early Greek atomists used this argument of Zeno as an apagogic proof – not for the impossibility of any division, but for the necessity of the existence of atoms[9].

But, according to Leucippus and Democritus themselves, motion takes place in empty space[10], and hence there is nothing in the teaching of the physical atomists that would prevent Achilles from moving according to $\underline{P}_1$.

Beside atoms of the early Greek atomism, Epicurus introduced, however, by analogy (τῇ ἀναλογίᾳ[11]), minima which are not only impenetrable, but indivisible in every sense. He described them as 'that what is smallest within the atom', what is, as such, an absolute measure of everything larger[12]. A limited body as well as any limited part of empty space, contain necessarily a definite finite number of such minima[13], and this number is a priori fixed in each particular case.

These spatial minima we shall simply call topons, and this aspect of Epicurus' teaching – geometric atomism, because topons are not only physically but also geometrically indivisible. Epicurus introduced temporal minima too[14], which we shall call chronons, but we shall reserve the name kinematic atomism for later doctrines which contain only temporal, not physical and spatial minima.

Epicurean geometric atomism implies that the premise $\underline{P}_1$ in our version of The Archilles is not possible: according to $\underline{P}_1$ there is an $\underline{n}$ such that distance $[(2^n-1)/2^n, (2^{n+1}-1)/2^{n+1}]$ is smaller than a topon, which is absurd. This is sufficient, though we can also use the fact that according to Epicurus chronons exist, too, for the reductio ad absurdum of $\Diamond \underline{P}_1$.

There are many problems in geometric atomism. A physical atom can be of any shape, due to the fact that it is indivisible, and – according to early Greek atomism – ontologically characterized as an absolute unit by virtue of its impenetrability only. The geometrical atom, on the other hand, can hardly be of any shape, if it has to be absolutely, that is also geometrically, indivisible, and it is not at all clear – as it is not a geometrical point – how it can have size without being of any shape. It can be shown that the entirety of geometry breaks down if geometric atomism is true, but we shall not dwell on this point. I shall present here only the famous reducto ad absurdum of geometric atomism known as The French Interpretion of Zeno's fourth kinematic aporia[15].

Let all of the A's, just as all of the B's and all of the C's, occupy neighbouring topons, being in positions represented by the following diagram:

```
    A A A A
B B B B →
      ← C C C C
```

Let the B's and the C's be moving in directions indicated by the arrows, while the A's are at rest. The stationary A's and the moving B's give

```
  A A A A
B B B B
```

as the first possible next position. Similarly, the stationary A's and the moving C's give

```
A A A A
 C C C C
```

as the first possible next position. This means that the first possible next position of the A's, the B's, and the C's is

```
  A A A A
B B B B
   C C C C
```

which means that position

```
B B B B
   C C C C
```

of the B's and the D's has been skipped. But why cannot the B's and the C's find themselves in that position?

The French Interpretation stops at this point: the geometric atomists do not seem to have a satisfactory answer to the last question. But we can go a little bit further.

Suppose that the geometric atomists' reply is that the B's and the C's cannot be in position

```
B B B B
        C C C C
```

just because there are no topons making that position possible. What would, however, happen if some D's are moving parallel to the C's according to the following diagram?

```
      A A A A
B B B B→ ←D D D D
      ← C C C C
```

The B's and the D's will collide when B's and C's are in position

```
B B B B
      C C C C
```

that is, the B's and the D's will collide at the mid-point of the allegedly indivisible topon (after a half of the allegedly indivisible chronon).

Trying to solve Zenonean problems and, at the same time, avoiding the specific difficulties of physical and geometric atomisms, the twentieth century atomists rely solely on kinematic indivisibles, either based on the existence of time indivisibles, or on the necessity of the existence, in any particular case, of the shortest not actually divided and only a posteriori indivisible phase, or phases, of a given change[16].

Kinematic atomism's solution of our version of The Achilles which is based solely on time indivisibles states that after sufficiently great n Achilles can neither move from $(2^n-1)/2^n$ to $(2^{n+1}-1)/2^{n+1}$, nor rest for $(2^{n+1}-1)/2^{n+1}$ h, because the period of $(2^{n+1}-1)/2^{n+1}$ h is shorter than a chronon.

This version of kinematic atomism can be reduced to absurdity in a way similar to that used to reduce geometric atomism to absurdity.

Let the A's, the B's, and the D's be in the position presented in the following diagram:

```
      A A A A
B B B B→← D D D D
```

The A's are at rest, while the B's, and the D's are moving as indicated.

If the *B*'s need one *chronon* to reach position

```
  A A A A
B B B B
```

and the *D*'s also need one *chronon* to reach position

```
A A A A
    D D D D
```

the *B*'s and the *D*'s will collide after a half of the *chronon*, which is absurd.

If the solution of our version of *The Achilles* is to be based on the necessity of the existence of the shortest not actually divided phase, or phases, of a given change – but in such a way that a transition from

```
    A A A A
B B B B
        C C C C
```

to

```
  A A A A
B B B B
    C C C C
```

though not being *actually* mediated by

```
B B B B
      C C C C
```

does not imply that it *had to* happen during an indivisible period of time – this would not be a solution at all, because it does not exclude that the *B*'s and the *C*'s *too could have been* positioned thus:

```
B B B B
      C C C C
```

There is nothing to imply that Achilles *could not move* according to $\underline{P}_1$, even if his normal motion is actually undivided, or consists of a finite number of not in actuality divided phases.

Neither geometric, nor kinematic atomism *per se*, or any combination of the two, can make $\underline{P}_1$ impossible without exposing itself to unsurmountable difficulties.

2. Radical Empiricism

When dividing a given body, we shall reach a point where no further division is possible - if the parts we get by division are to be perceivable. Berkeley, the most radical empiricist, used this fact to destroy the idea of the infinite divisibility of matter - if its esse est percipii: the fact that there are perceptive minima necessarily puts limitations on any division.

Contemporary radical empiricists, David Hilbert and Max Black for instance, who do not want to rely on the esse est percipii principle, appeal to ordinary language and to scientific practice in order to derive a Berkeleian conclusion concerning infinity.

If we do not want to violate the limitation on the scale of applicability of the word 'jump', we shall never speak of a 'thousandth inch jump', just as we shall never speak of a multicoloured space of a billionth part of a pin, if we want not to violate the limitation imposed by scientific knowledge about the nature of colour[17].

How do we come to believe that it is possible for a physical body to consist of an infinite number of physically heterogeneous parts, and that it is possible for Achilles to move according to P_1? We discover that physical bodies have concealed parts with concealed properties that are not evident to direct inspection but are progressively revealed by other means. Then we generalize, we assume that what is true of a macroscopic body is true of any of its parts, namely, that every concealed part has concealed parts with concealed properties. By such a generalization, we in fact make an extrapolation from a given experience of macro-level phenomena to a possible experience of micro-level phenomena[18], overlooking, or ignoring, the fact that there are limitations of scale upon the applicability of the words we need to describe micro-level phenomena.

Hence, without using the esse est percipii principle, contemporary empiricists show that it is not certain that it is possible for a physical body to consist of an infinite number of physically heterogeneous parts, and for Achilles to move according to P_1.

However, although we are shown that it may not be possible for a physical body to consist of an infinite number of physically heterogeneous parts, and for a motion to consist of an infinite number of staccato movements, we are not shown that it must be impossible for any physical body to consist of an infinite number of physically heterogeneous parts, and for any runner to move according to P_1. Even if it is true that there are limitations of applicability of words referring to known physical properties, this does not imply that it also must be true of every concealed property which could possibly be revealed, and of every possible runner.

If we take Achilles to be Zeus, P_1 holds possible: in moving according to the rhythm defined in P_1, Zeus could be, by definition, not exposed to any alleged empirical limitation. In consequence, we have been offered no solution to our problem.

III. $\Diamond \underline{C}$

There are two variants of infinitism, the doctrine which takes $\underline{C}$ to be possible: infinitism without infinitesimals and infinitism with infinitesimals.

1. Infinitism Without Infinitesimals

The infinitists who do not use infinitesimals[19] proceed from the statement that according to standard analysis there is enough time and enough room for an infinite number of discrete successive acts, such as Achilles' movements described in P_1: according to standard analysis, Achilles' path AB metrically equals the sum of $\aleph_0$ space intervals $[(2^n-1)/2^n, (2^{n+1}-1)/2^{n+1}]$ which he covers by his movements, and, similarly, 2 h equal the corresponding sum of time intervals, $1 + 1/2 + \ldots + 1/2^n + \ldots$ ($n=0, 1, 2, \ldots$), during which Achilles makes his movements from $(2^n-1)/2^n$ to $(2^{n+1}-1)/2^{n+1}$ and rests after each of them.

After stating that according to standard analysis there is enough time and enough room for an infinite number of Achilles' movements described in P_1, the infinitists claim that for Achilles to perform an infinite number of acts described in P_1 it is enough to 'start and not stop after any finite number'[20]. In such a way Achilles will allegedly perform an infinite number of movements, in spite of the fact that the finiteness of the number of his movements is recursively preserved from movement to movement. Finally, the infinitists accept that Achilles can complete the series of an infinite number of strictly discrete successive acts of non-infinitesimal duration without having performed a last act.

To begin with the infinitists first statement: how is it possible that we find that according to standard analysis there is not enough time for an infinite number of Achilles' movements described in P_1, while the infinitists find that according to the same mathematical analysis there is enough time for an infinite number of these movements?

In our reasoning in §1, we stressed the fact that for any instant (any point) within the period of 2 h (within AB), it holds that up to that instant (that point) Achilles has performed a finite number of movements.

In their reasoning, the infinitists, however, stress the fact that no finite number can supply an adequate answer to the question of the number of

Achilles' rest-states when the 2 h have elapsed: after any finite number of performed movements, it holds that there is enough time within the 2 h for more movements to be made according to the rhythm defined in $\underline{P}_1$. In addition, they may say that after any finite number of rest-states Achilles is somewhere between A and B: $\forall n \exists \varepsilon ((2^n-1)/2^n < AB-\varepsilon)$ (where $\underline{n}$ is finite). From that the infinitists draw the conclusion that after the two hours have elapsed the statement that Achilles has performed an infinite number of movements must be true; thus $\diamond \underline{C}$.

The infinitists do not locate the possible mistake in the reasoning to the conclusion that there is neither enough time nor enough room for an infinite number of movements performed by Achilles in 2 h according to $\underline{P}_1$. They only show a thing which we did not deny in §1, that, namely, within the period of 2 h and within AB there is always enough time and enough room for more movements to be performed – independently of however great the number of the movements which have been already performed is. And from this they infer that the 2 h period and distance AB are sufficiently large for an infinite number of performed movements.

Where we encounter a problem the infinitists simply draw an 'infinitary' conclusion by stressing some facts and ignoring others. Our difficulties lie in the fact that no finite number seems to be an adequate answer to the question concerning the number of Achilles' rest-states when the 2 h have elapsed; yet at the same time, there is not enough time for an infinite number of them to have occurred. The infinitists take the fact that no finite number seems adequate as a sufficient ground for the conclusion that there is enough time and enough room for an infinite number of performed acts.

When dealing with the problem of the recursively preserved finiteness of the number of Achilles' movements, the infinitists use the same strategy; commencing from the fact that no finite number can supply an adequate answer to the question concerning the number of Achilles' rest-states when the 2 h have elapsed and the alleged truth that 2 h are enough time for an infinite number of performed movements, they simply infer that Achilles will have performed an infinite number of movements defined by $\underline{P}_1$ if he does not stop after any finite number; and they do not show how Achilles will free himself from the situation of having performed a finite number of movements, and how he 'jumps' into the situation of having performed an infinite number of them.

To illustrate how far the infinitists have to go in their strategy to derive conclusions while ignoring problems, we shall dwell a little bit longer on the discussion concerning the third difficulty from §1, namely, the problem of the 'last act' in the performing of the so-called super-task[21], the task of completing the series of an infinite number of acts.

Let us consider the 'game with a marble' played by Black's 'transferring machines' 'Beta' and 'Gamma', which Black 'invented' in order to show that it is impossible to perform a super-task[22]. Let there be a marble in a left-hand tray, which some device, named 'Beta', transfers to an empty right-hand tray in one minute; after that, 'Beta' rests for one minute, while 'Gamma', a similar device, transfers the same marble to the left-hand tray; 'Beta' then transfers the marble to the right-hand tray in half a minute, and rests for one half of a minute while 'Gamma' transfers the marble to the left; and so on, 'Beta' always moves the marble from the left to the right while 'Gamma' is at rest, and 'Gamma' moves the marble from the right to the left while 'Beta' is at rest, the periods of the transfer of the marble from the left to the right and back decreasing according to a geometric progression converging to zero. Where is the marble at the end of exactly four minutes, when the machines come to a halt?

A similar super-task is described by Thomson[23]. Thomson's lamp is a reading lamp equipped with a button which, if pressed, switches the lamp on when it is off and switches it off when it is on, where the button, in addition, may be pressed in such a way that the first jab requires 1/2 of a minute, the second 1/4 of a minute, and so on according to a geometric progression. If the lamp operates thus, is it on or off when one minute has elapsed and we have stopped pressing the button?

Paul Benacerraf argues[24] that according to the conditions given in the description of the operation of Thomson's lamp nothing implies what the state of the lamp is when one minute has elapsed: the lamp can be broken at the instant when the first minute elapses and the second minute begins, as well as it can be on or can be off after that; the given description only concerns the period of time delineated by the one minute. Similarly, the marble transferred from the right to the left and vice versa within the period of four minutes can be broken at the beginning of the fifth minute, just as it can fall out of the trays.

Adolf Grünbaum cites[25], however, the additional conditions which have to be met in order to enable us to say where the marble is, and whether the lamp is on or off - after the completion of the infinite processes described by Black and Thompson.

Let us assume that not only the time intervals of marble transfer but also the distances through which the marble transfers have to be effected decrease in proportion to the available successive times, converging to zero by the time the four minutes elapse. If the marble is at rest after that, it is, according to Grünbaum, on the line of contact of the left and the right trays, after being transferred an infinite number of times from one tray to another.

The additional conditions concerning the Thomson lamp super-task are the following. At time t_0, when the lamp is off, the initial vertical distance between the button base and the horizontal circuit-opening $E_1 E_2$ is 1/2 (see the diagram).

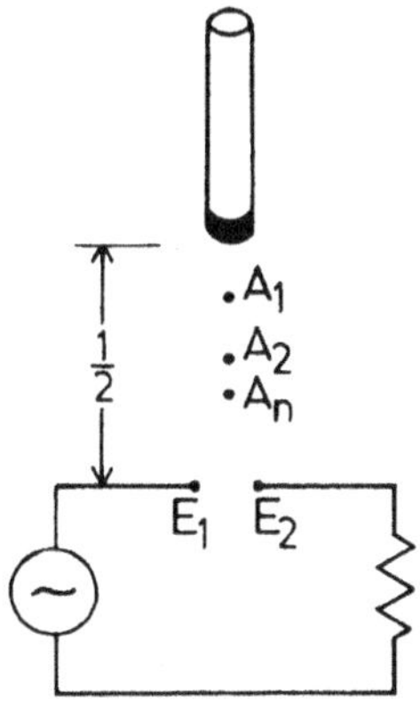

Then after the base of the button has been depressed once to close the circuit, let it be raised after each such depression not all the way to its initial position A but to intermediate points A_1, A_2, A_3, ... , A_n,... , whose respective distances from $E_1 E_2$ are 1/8, 1/32, 1/128, ..., $1/2^{2n+1}$, ..., corresponding to the same sequence of time intervals which the button needs to reach from $E_1 E_2$ to A_1, A_2,..., A_n,... respectively (and from A_1, A_2,..., A_n...to $E_1 E_2$). If these conditions are met, then, Grünbaum claims, after an infinite number of circuit-closing jabs, at the end of one minute, at time t_1, the state of the circuit is predictably closed, meaning that the lamp will be on thereafter, if the lamp circuit still exists intact at that and after that time.

One could say that, in order to reduce to absurdity Grünbaum's claim concerning the position of the marble after four minutes, it is enough to ask him whether the marble, being allegedly on the line of contact of the trays, has come to this position from the left or from the right. If it has come from the left, 'Beta' has moved it there, and 'Beta' needed a certain period of time to do that. If the marble has come to its last position from the right, 'Gamma' has moved it there, and 'Gamma' needed a certain period of time to do it. Whatever is the case, and however small the period of the last movement is, 'Beta' and 'Gamma' did not have enough time – from the beginning up to that last movement – to transfer the marble from the left to the right and back more than a finite number of times. But, Grünbaum would reply, no last transfer took place at all[26]. The marble has reached the last position without a last transfer from the right or from the left, that is, without arriving at its last position from anywhere!

Instead of explaining what this could possibly mean, to arrive at the last position from nowhere, Grünbaum uses a definitional fiat: in the sequence of marble transfers there is no last member.

Everything becomes even more curious when we turn to Thomson's process. Of course, Grünbaum claims that the jabbing motions issue in an on-state at time t_1, though that on-state is not the terminus of any continuously downward jabbing motion of positive duration (during which the lamp would be off)[27]. But, in a slightly different switching arrangement, concocted by Allen Janis[28], where nothing is essentially changed from a kinematical point of view, being moreover parallel to 'Beta"s and 'Gamma"s game with the marble, Grünbaum proclaims the corresponding Thomson process to be impossible only because the lamp is, by such an arrangement, neither predictably on nor predictably off at t_1 (again due to the fact that there is no last continuously jabbing motion) and the lamp cannot be in a neither-on-nor-off state after t_1 if the lamp circuit still exists intact at that time.

Janis' arrangement is the following. Let the switch button be movable through space intervals converging to zero and divided by $E_1 E_2$ as a mid-point (see the diagram above); the lamp is on or off depending on whether the button has arrived at $E_1 E_2$ from above or from below respectively: if the button base is at $E_1 E_2$ at time t, the existence of an on-state of the circuit at t requires that the button base has reached $E_1 E_2$ from above at or before t, and the existence of an off-state at t, while the base is present at $E_1 E_2$, requires the base to have reached $E_1 E_2$ from below at or before t.

Grünbaum's previous considerations enable us to assert that at time t_1 the button base cannot have reached $E_1 E_2$ either by a continuous approach from above or by a continuous approach from below. Thus, if the lamp circuit still exists intact at t_1 and thereafter, the lamp is then neither on nor off, in spite of the fact that the button base is at $E_1 E_2$.

Instead of adopting this conclusion concerning Janis' switching arrangement as something that sheds doubt on his previous considerations on infinite processes, Grünbaum, once more, ignores the problem, leaving his previous conclusions intact: Achilles can reach his goal after an infinite number of stops, without a last movement; the marble can reach the line of contact of the trays after an infinite number of transfers without a last transfer; the lamp can be on at the beginning of the second minute and thereafter after an infinite number of circuit-closing jabs, without that on-state being the result of any continuously downward jabbing motion of positive duration – under the previously defined conditions; but, the similar Thomson process is impossible under essentially the same conditions, viewed kinematically, only because by Janis' arrangement the non-existence of the last downward or upward jabbing motion implies the neither-on-nor-off state of the lamp by the end of one minute and thereafter. Thus, if there is no witness to the curiousness of the non-existence of the last act in completing the series of an infinite number of acts – as in the case of the marble transfers – the completion of the infinite process ought to be possible; if, however, such a witness does exist – the completion of essentially the same infinite process is impossible!

2. Infinitism With Infinitesimals

One of the difficulties with C (§1) consists of the conclusion that there is neither enough time within the period of 2 h nor enough room within AB for an infinite number of Achilles' movements defined in P_1. The reason for such a conclusion is based on the fact that for any instant, however close to the end-instant of the period of 2 h, just as for any point, however close to the right end-point of AB, it holds that up to that instant, and up to that point, Achilles has performed a finite number of movements.

According to non-standard analysis, however, where the so-called Archimedes' axiom does not hold[29], there would be an instant for which we could not demonstrate that if Achilles has moved according to P_1 up to that instant, then he has performed only a finite number of acts. Such an instant is any instant which is infinitely close to end-instant of the period of 2 h. Similarly, there is a point, and such a point is any point within AB infinitely close to B, for which we could not demonstrate that before reaching it, or passing through it, if Achilles has moved according to P_1, he has performed only a finite number of movements.

Now, according to one version of the formula dating from the middle-ages[30], a finite distance cannot be exhausted by an infinite number of subdistances if there are not infinitesimals among them, while according to an other version of the same formula, a finite distance cannot contain an infinite number of subdistances of non-infinitesimal magnitudes. The similar holds for a limited period of time and its subintervals.

Putting aside the additional problems which the theory of infinitesimals introduces, we shall point out that even in the formally coherent version of the theory[31] the introduction of infinitesimals does not succeed in shifting the original problem.

According to non-standard analysis, there is a 'gap' between the members of the sequence of Achilles' rest-states $\mathcal{A}_1$, $\mathcal{A}_2$, ..., $\mathcal{A}_n$, ... with finite subscripts and the members of that sequence with transfinite subscripts. There also exists no greatest finite subscript number in that sequence, and the number of members with finite subscript numbers is not finite. So, we are faced with the problem of Achilles' 'bridging the gap' between the positions with finite and the positions with transfinite subscripts, that is, with the problem of his 'jumping' into a position infinitely close to B. Independently of this problem, Achilles has to be in an infinite number of positions with finite subscripts, and the finiteness of the number of his rest-states is recursively preserved. And again, independently of the existence of infinitely small space distances, and infinitely short time intervals, there is neither enough room nor enough time for an infinite number of Achilles' rest-states $\mathcal{A}_1$, $\mathcal{A}_2$, ..., $\mathcal{A}_n$, ... with finite subscripts.

IV. $\neg\Diamond \underline{C}$, Without the Exclusion of $\Diamond \underline{P}_1$ and $\Diamond \underline{P}_2$ and With No Explanation as to What's Wrong With $\Diamond \underline{P}_1, \Diamond \underline{P}_2 \vdash \Diamond \underline{C}$

Under this heading, we shall consider two strategies of denying $\Diamond \underline{C}$, which are successful in dealing with some versions of The Achilles and some related aporias, but which strategies, in effect, do not exclude $\Diamond \underline{P}_1$ and $\Diamond \underline{P}_2$, yielding no answer to the question of the number of Achilles' rest-states when we assume $\underline{P}_1$ and $\underline{P}_2$ to be true.

1. Finitism

According to Geach, who subscribes to Frege's analysis of identity statements[32], the identity is always relative in the following sense: When one says "x is identical with y", this...is an incomplete expression; it is short for "x is the same A as y", where 'A' represents some count noun understood from the context of utterance...'[33] Thus x ≡ x is also meaningful only if in the given context x refers to a thing of a certain kind; for instance, x cannot refer to a part as such: 'part as such' cannot be a sortal.

On this reasoning, David Schwayder, and Max Black in his first analysis of Achilles and the Tortoise[34], tried to solve Zeno's metric and kinematic aporias: Achilles' path contains necessarily a finite number of parts properly identified, it is also finite measured by any means, and Achilles, in order to catch up with the tortoise, is not called upon to do the logically impossible, to complete the series of an infinite number of acts.

To count parts is possible only if one uses some count noun to single them out. If we, however, single them out by using proper sortals, we shall necessarily obtain a finite number of concerning parts: a finite number of pebbles, a finite number of blades of grass, a finite number of atoms, and so on [35].

Similarly, a limited body is metrically finite measured by any means. Any unit measure represents the size of something physical, however small. Be it a pebble, or anything smaller than pebble, it is contained a finite number of times in Achilles' path. This truth is untouched by the fact that the number of ways that may be used to define the unit measure is not limited[36].

In no case it is possible that a limited body have been measured by an infinitely changing measure unit. If Achilles' progress in chasing the tortoise were really such a measurement of the path that remains to be covered to the goal, he couldn't reach up with the tortoise, but, as Black put it,

'Achilles is not called upon to do the logically impossible', but we are those who create 'the illusion of the infinite task by the kind of mathematics that we use to describe space, time, and motion'[37].

It may be said that by accepting the finitists conclusions we can truly solve the version of The Achilles where Achilles moves legato. Moving normally, Achilles will reach the goal after a finite number of steps, heart beats, deep breaths, and so on, while the geometric progression which we use to describe his motion refers to nothing that he really does.

What, however, happens if Achilles moves staccato, according to $\underline{P}_1$? He then does move in accordance with a geometric progression, and its members do refer to something which Achilles actually does. By moving in this way, Achilles is really measuring his path with a unit which changes from step to step, being always recursively defined in terms of the previous step. If the number of members of that geometric progression is not finite, and if it is not possible to obtain the result of the measurement by using an infinitely changing unit, does this mean that Achilles will not reach his goal? And if he doesn't reach the goal, where will he be when the 2 h elapse? What is the answer to the question of the number of his actually differentiated movements?

There is nothing in the finitists' teaching which would make Achilles' movement according to $\underline{P}_1$ impossible. If he really does move according to $\underline{P}_1$, and if the number of members of the geometric progression is not finite, then it seems inevitable that the number of Achilles' movements, and rest-states, cannot be finite when the 2 h elapse; there is no explanation which would prevent us to derive $\diamond\underline{C}$ from $\diamond\underline{P}_1$ and $\diamond\underline{P}_2$, and, in particular, the question about the number of Achilles' rest-states if the 2 h have elapsed has not been answered.

2. Indefinitism

Whether something is one single thing, in spite of the fact that it has many parts, or a set consisting of many things, is, up to a certain point, a matter of linguistic convention. This is Aristotle's first reply to the so-called aporia concerning the whole and the parts raised in the struggle between the monists and the pluralists[38]. The existence, oneness, and multitude can be spoken of in multiple ways (πολλαχῶς λεγόμενα) and in certain cases we can alternately favour the oneness and the multitude.

However, there are cases where, in accordance with Aristotle's later teaching on primary meanings[39], we are dealing with a single thing that is only potentially something multiple.

A thing that has no heterogeneous parts, and also neither discrete or contiguously touching parts[40] – is one single thing in the primary sense, because it is only potentially multiple. Positively expressed, something that is homogeneous and continuous in itself[41] must be characterized as truly one single thing, because in reality it is in no sense multiple, due to the fact that its alleged parts would have been its parts only if they were actualized by an appropriate division; though the existence, oneness, and multitude can be spoken of in multiple ways, their primary meaning excludes the alternative of such a thing being truly multiple.

Achilles' path is really multiple insofar as it consists of a number of heterogeneous, discrete, or contiguously touching parts. Can the number of such parts be infinite?

Aristotle subscribes to atomists' point that Zeno's arguments against plurality show that we can never, by any division, successive or simultaneous, even in principle, actualize an infinite number of potential parts[42].

Aristotle, however, does not reach the atomistic conclusion that physical or spatial indivisibles must exist, he admits to nothing more than that by any division we cannot obtain an infinite number of parts[43].

If an infinite division is not possible, and at the same time there are no indivisibles, this means that the number of parts we could obtain by a division, successive or simultaneous, is a finite, but not a definite finite number. Though any particular successive division is carried out only up to a certain point, we always get a greater number of parts by continuing the division. Similarly, even though any particular simultaneous division can result in only a finite number of parts, a new simultaneous division that would yield a greater number of parts is always possible. In both cases – in the case of successive as well as in the case of simultaneous division – an infinite number of parts is never obtained, even though there is no greatest number of parts obtainable by division. This is the essence of indefinitism[44].

What holds for the path, also holds for Achilles' motion itself. Because 'oneness' has a variety of meanings, 'one single movement' is an ambiguous term[45]. A movement can be said to be one single movement beause it is continuous in itself, that is, because it is a legato motion. But this same movement can be said to be multiple if it consists of phases which are in form (κατ' εἶδος) different among themselves: the speed and the manner of its variation can change during the motion, in which case the motion is non-uniform (ἀνώμαλος)[46]. If, however, the speed and its variation are constant and uniform respectively during the whole legato motion, such a motion constitutes one single movement in the primary sense[47].

Is it possible for Achilles' motion within a finite time to consist of an infinite number of actually differing phases? According to Aristotle, this is not possible for the reasons analogous to the reasons why Achilles' path cannot consist of an infinite number of parts. Hence $\neg\Diamond\underline{C}$.

$\underline{P}_1$ is, however, possible, because there are neither space indivisibles nor time indivisibles. So, if we accept the indefinitist conclusions, we can solve the problems concerning the legato version of The Achilles, but this does not help us with our version of it. When the 2 h elapse, that is, when P_2 becomes true, then, on the one hand, we should not claim C true, because allegedly $\neg\diamond C$, and, on the other hand, we cannot cite a finite number as the number of Achilles' rest-states, because for any finite number it holds that Achilles has had to be in more rest-states according to P_1, and P_1 is possible. To state that the number of rest-states is finite but not definite makes, however, no sense when we speak of Achilles' actually performed differentiated movements.

§3. DISCUSSION

Bearing in mind our version of the problem (§1), we have seen that we cannot be satisfied by any of the considered proposed solutions of The Achilles (§2) for diverse reasons.

The price of the proposal put forward by the negative dialecticians is too high: be the world of change real or be it an appearance, we would like to speak of it in a consistent way. However, the essence of their proposal is that this is impossible. We shall accept their proposal only as a last resort.

The atomistic versions which allegedly solve the problem by making P_1 impossible have been reduced to absurdity - not to mention other difficulties. True, the reductio ad absurdum can be avoided by additional restrictions, but they are inadmissibly ad hoc: BBBB and DDDD (see diagrams above) shouldn't start moving at all, if they are going to meet at half topon, or after half a chronon.

The restrictions imposed by radical empiricists seem unjustifiably limiting: nothing we know about the empirical world justifies the conclusion that it must be impossible for anybody to move according to P_1. We must also have a solution for the case that such a movement is possible.

The infinitists' strategy of using a definitional fiat to 'kill' the problem wherever it arises can hardly convince us that $\diamond C$. Even if we accept that when viewed mathematically it is meaningful to speak of $\aleph_0$ finite distances $[(2^n-1)/2^n, (2^{n+1}-1)/2^{n+1}]$ which cover the whole distance between zero and 1 without any such distance being the last one on the right, we can hardly accept that a marble can reach its final position after being transferred an infinite number of times from the left to the right and vice versa, arriving there from nowhere. The necessity of 'killing' the witness introduced in a Thomsonean process that is quite analogous to Black's process in order to make that Thomsonean process possible, shows us most characteristically how far the infinitists are forced to go in using the method of definitional fiat. - The infinitists who use the infinitesimals do not shift the original problem at all - let aside the additional difficulties.

The finitist and the indefinitist teachings which stress the importance of the manner of differentiating parts and distinguishing between real and potential parts seem quite acceptable in implying the possibility of $\underline{P}_1$ and $\underline{P}_2$ to be true while asserting $\underline{C}$ to be impossible, but, not faced with, or ignoring, the staccato version of The Achilles, neither the finitists nor the indefinitists offer an adequate answer to the question about the number of Achilles' rest-states in the case that $\underline{P}_1$ and $\underline{P}_2$ are true.

Looking at all these deficiencies of the offered proposals, we may try to intuit the 'ideal' solution to our problem. The 'ideal' solution would be a solution which would

a) preserve the consistency of the world of change;

b) introduce no indivisibles of any kind;

c) impose no radical empiricist restrictions which would make $\underline{P}_1$ impossible;

d) justifiably accept $\neg\Diamond\underline{C}$ after explaining why $\Diamond\underline{P}_1$ with $\Diamond\underline{P}_2$ do not entail $\Diamond\underline{C}$.

In order to formulate such a solution, we shall now consider the acceptable aspects of the forwarded proposals.

The merit of the negative dialectic, from ancient times to today, lies in indicating the difficulties and deficiencies of the forwarded or possible proposals of the solution to Zenonean paradoxes: if the observed deficiencies cannot be resolved within the proposed solutions, the result of Zeno's dialectic should be accepted at least in its sceptical aspects: we have to admit that we are unable to speak of the world of change in a consistent way. It may be that in order to preserve that world we have to make peace with the inevitable incompatibilities between the world described in the Zenonean premises that make the goal unreachable and the everyday world in which Achilles has no difficulties in catching up with the tortoise.

The merit of atomism, i.e. of its variants, lies in the insistance that by the acceptance of the infinite division of Achilles' movement his goal remains unreachable.

The unacceptable conclusion based on $\underline{P}_1$ and $\underline{P}_2$ asserted in infinitism without infinitesimals necessitate our avoiding the admission that the conjunction of $\underline{P}_1$ and $\underline{P}_2$ can ever be true, and this represents the foremost merit of that variant of infinitism.

The merit of the infinitism with infinitesimals lies in its negative implication, which states that a finite distance, or a finite period of time, cannot contain an infinite number of subintervals if all of them are non-infinitesimals.

The merit of finitism lies in its stressing that measured by any appropriately predefined unit a limited body, or a limited period of time, always renders a finite result.

The merit of indefinitism consists in its restrictions imposed on the formulation of the problem itself, related to the difference between real and potential parts: we are faced with a real problem only when it is defined in terms of Achilles' actually differentiated movements, such as defined by the description of his staccato motion.

Armed with the acceptable insights of the proposals as well as with our experiences of their deficiencies, we shall now sketch a new solution of the problem.

§4. SOLUTION

Looking formally at our classification of the proposals for the solution to The Achilles, one may find that it is of note that – with the sole exception of the negative dialecticians, who deny the very possibility of the existence of the world of change – nobody has questioned the possibility of $\underline{P}_2$. Of course, considering the content of $\underline{P}_2$, one can easily discover why this is so: $\underline{P}_2$ only implies that the 2 h of Achilles' movement will elapse and says that when 2 h elapse Achilles is definitely at rest – all this seeming trivially possible. The fact, however, that not only has nobody questioned $\Diamond\underline{P}_2$, but that nobody has even considered the possibility of $\neg\Diamond\underline{P}_2$, suggests that everybody – pace negative dialecticians – has taken the unconditional possibility of $\underline{P}_2$ – $\Box\Diamond\underline{P}_2$ – as an evident truth. But, is what $\underline{P}_2$ implies really unconditionally possible?

Assuming that Achilles' staccato motion is a way to measuring distance AB, accepting also that after any finite number of movements Achilles can move once more in accordance with $\underline{P}_1$ without reaching B, and taking as true that it is impossible for a finite distance to have been measured by an infinitely changing unit, we are unexpectedly led to conclude that, though generally possible, $\underline{P}_2$ is not possible if $\underline{P}_1$ is true; $\underline{P}_1$ being true, the 2 h cannot elapse, and they will not elapse! There is a world, and that is exactly the world described by $\underline{P}_1$, in which $\underline{P}_2$ is impossible; so $\neg\Box\Diamond\underline{P}_2$.

The conclusion that it is not unconditionally true that 2 h can elapse may look very odd, because it seems that *by simply waiting* we can show that 2 h must elapse. But would this really be a proof that 2 h *must* elapse? No! That would only prove that the 2 h *have elapsed*. Would 2 h elapse in Achilles' world when he moves according to P_1 ?

The lesson we have learnt through the acceptable aspect of radical empiricism can help us here. For *us*, who *do live* in the world where there are minima everywhere, and whose perceptive and apperceptive thresholds are comparatively unchangable, 2 h *must* elapse. But, in Achilles' world where movement takes place according to P_1 there are, *ex hypothesi*, no minima. Without minima, his 2 h is not a finite time.

The fact that 2 h have *de facto* elapsed can hardly *prove* that the world of Achilles' movement according to P_1 is impossible. But this *implies* that in the *particular* case, if 2 h *have* elapsed, Achilles *couldn't have* moved in accordance with P_1 during the *whole* of the 2 h. If those 2 h have elapsed, Achilles has had to stop moving according to P_1 somewhere.

If P_1 has been true, then, necessarily P_2 has not become true, and if P_2 is true, P_1 couldn't have been true. If the description of Achilles' movement given in P_1 has not become false, the 2 h haven't elapsed, and Achilles has performed a finite number of movements. If, however, P_2 has become true, this entails that Achilles has had to *stop moving* according to P_1 *somewhere* – where exactly is a matter of *empirical* fact.

If P_1 has not become false there is no question of the *last* movement by which Achilles has reached B, simply because Achilles cannot reach B until P_1 becomes false. If, on the other hand, Achilles has reached B, he has done it, necessarily, by performing a *last* movement in a series of a necessarily *finite* number of movements, since in order to reach B Achilles has had to stop moving according to P_1 (somewhere).

Let $F(\mathcal{A}_k)$ mean that Achilles has been in k rest-states. If P_1 is really possible – and we assume it is – then $\forall k \Diamond F(\mathcal{A}_k)$. But, $\forall k \Diamond F(\mathcal{A}_k)$ does not imply $\Diamond \forall k F(\mathcal{A}_k)$. One is led to the transition from the former to the latter by the addition of P_2, which is possible, to P_1, which is also possible. P_2 being assumed, however, *incotenable* with P_1, $\Diamond \forall k F(\mathcal{A}_k)$ can be avoided.

This resolution of the problem cannot be formally expressed in any modal logic system with the strict implication reducible to the material one. We need to state $\Diamond P_1$, $\Diamond P_2$, $\Box(P_1 \Rightarrow \neg P_2)$, and $\Box(P_2 \Rightarrow \neg P_1)$, but then, supposing P_1 and P_2 (being possible) to be true, we can't avoid a contradiction.

The solution can easily be expressed, however, with the aid of the *relevant* implication. In relevant logics, it doesn't hold that by the addition of a new hypothesis the hypothesis previously introduced must remain deducible[48]: $A, B \vdash A$ does not always hold, and, in particular, we mustn't continue to assert $\underline{P}_1$ – $\underline{P}_1$, $\underline{P}_2 \vdash \underline{P}_1$ – if added hypothesis $\underline{P}_2$ is incotenable with $\underline{P}_1$, the incotenability of $\underline{P}_1$ and $\underline{P}_2$ being defined as

$$\neg(\underline{P}_1 \circ \underline{P}_2) \overset{def}{=} (\underline{P}_1 \rightarrow \neg\underline{P}_2)$$

(where '$\rightarrow$' is relevant implication and '$\circ$' intensional conjunction reducible to it with the aid of the ordinary negation[49]).

If we adopt $\underline{P}_1$ and $\underline{P}_2$ as possible but incotenable premises, we can save the world of change without any of the deficiencies of the proposals offered so far, conserving all their acceptable aspects.

In principle, there are no indivisibles, but in both the everyday and scientific worlds the empirical constraints exclude situations in which $\underline{P}_2$ would not be possible, and so, under these constraints, $\neg\Diamond\underline{P}_1$. In a possible world where these constraints are removed, $\underline{P}_1$ is, however, possible, but, if movement is governed by $\underline{P}_1$, $\underline{P}_2$ cannot ever be true; thus, nevertheless $\neg\Diamond\underline{C}$.

In principle, the greatest number of Achilles' movements defined by $\underline{P}_1$ does not exist, but the fact that, if Achilles has moved according to $\underline{P}_1$, the 2 h could not have elapsed, excludes a situation where no finite number could supply an adequate answer to the question of the number of Achilles' rest-states, and that is in accordance with the fact that the finiteness of the number of Achilles' movements is recursively preserved. The fact that within the finite distance AB and the period of 2 h there is always enough room and enough time for a number of Achilles' movements greater than any given finite number is reconcilable with the fact that within that distance and that period of time there is neither enough room nor enough time for an infinite number of performed movements – if $\underline{P}_1$ and $\underline{P}_2$ are incotenable.

The fact that when we speak of subdistances and subintervals in a general, non-specified way, we cannot cite a definite finite number of subdistances of AB, and subintervals of the 2 h period, doesn't mean that the number of parts of a limited body and the number of phases of a change performed in a limited period of time are not definite finite numbers when referring to real parts and real phases: Achilles' path does consist of a definite finite number of physically heterogeneous parts, and Achilles' motion during the whole period of 2 h has had to consist of a definite finite number of performed differentiated movements, as those in Achilles' staccato run; the number of movements has not been fixed only within the time when it has been becoming greater and greater, namely within the 2 h, but all the same, at any given moment, it has been definite and finite.

One could try to reduce our claim that $\underline{P}_1$ and $\underline{P}_2$ are incotenable to absurdity by a thought experiment which would 'mix' the world of $\underline{P}_1$ and the everyday world in which $\underline{P}_2$ is to become true.

Let Achilles move not staccato but in a normal manner, legato, and at a constant speed with which he will reach B in 2 h – this being a more appropriate way of movement with regard to the fact that he is, after all, a man. Let Zeus, however, move parallel with him at double speed but staccato, covering the distances of Achilles' path and resting after each of them in the way described by P_1. On one hand, Achilles will reach B just as all of us would, which means that the 2 h will have elapsed. On the other, if movement according to P_1 is at all possible, Zeus, the immortal, will be able to move in this manner. He may be capricious, not wanting ever to stop moving in such a way. Where will Zeus be,and in how many rest-states will he have been, when Achilles reaches the goal?

According to our solution, the essential difference between this situation and the original one (from §1) lies in the fact that now there are two runners, the one who belongs to the everyday world and for whom 2 h must elapse, and the other for whom, belonging to the world of P_1, 2 h cannot elapse. However, the two runners move in such a way that Zeus is never behind Achilles. Thus, it seems, either Zeus' 2 h must elapse if Achilles' 2 h must elapse, or Achilles' 2 h cannot elapse if Zeus' 2 h cannot elapse.

It may seem that the situation with two runners indicates that our solution is implausible.

However, first, formally speaking, there need not be a contradiction in the set of the following statements:

- it is possible that Zeus has moved according to the rhythm defined in P_1 (Zp_1);
- if it hasn't become false that Zeus has moved according to the rhythm defined in P_1 (Zp_1), then, necessarily, his 2 h have not elapsed ($\neg Z_{2\,h}$);
- if Zeus' 2 h have not elapsed ($\neg Z_{2\,h}$), then, necessarily, Achilles 2 h have not elapsed ($\neg Ach_{2\,h}$);
- if Achilles 2 h have elapsed ($Ach_{2\,h}$), then, necessarily, Zeus' 2 h have also elapsed ($Z_{2\,h}$);
- Achilles 2 h have elapsed ($Ach_{2\,h}$).

A contradiction would result only if by defining necessary implications, which occur in these statements, the relevancy conditions are ignored or mistakenly treated. If we, however, assume Zp_1 and $Ach_{2\,h}$ to be relevantly inconsistent, that is, incotenable [50], no contradiction can be derived from the set of the above statements.

And, second, we have found out why we are prone to believe that one of the two statements which we take as incotenable must unconditionally become true: we do live in Achilles' world where he moves legato, while we do not live in the world of Zeus who moves staccato. Living in Achilles' world, we can always easily wait for him to reach the goal, while we can never follow all of Zeus' movements in order to see whether Achilles will reach the goal.

This means that we can always easily witness Zeus' giving up his staccato motion, while we can never witness the situation in which the statement about Achilles' reaching the goal cannot become true because of Zeus' following the law of his staccato motion.

Let us suppose that Hera, the godess, waiting for her husband at B, spies what he is doing with Achilles. Following every one of Zeus' movements, she will, contrary to us, embrace neither Zeus nor Achilles. In order to meet her husband, she would have to stop spying on him: she would have to 'jump' into a world incompatible with Zeus' permanent staccato motion, leaving Achilles to reach the goal, because this fact would force her husband, by logical necessity, to give up his staccato dance with Achilles. Where exactly would he do it, is not a priori determinable. Where he gave up the dance, if Achilles has reached the goal, is a matter of empirical fact.

Achilles' world encompassing his reaching the goal is inaccessible from Zeus' staccato world, the two worlds being incomposable.

It may seem incorrect to say that Zeus, moving staccato, will never reach the goal, because in ordinary language 'x will never happen' entails 'x will not happen even if you wait more than 2 h'. But this entailment holds only in ordinary language, which is a suitable language for ordinary experiences. The experience of Hera who is spying on her husband is not such an experience, and by doing this, her 2 h will not elapse; two hours are not necessarily a finite time.

NOTES

1. See Plato, Parmenides, 128 A.
2. Cf. Plato, Timaeus, 38 A and Aristotle, Physics, 221 b 3–7ff.
3. See DK 82 B 3.
4. Cf. G.W.F. Hegel, Vorlesungen über die Geschichte der Philosophie II (Werke 19), p. 62, Suhrkamp Verlag, Frankfurt am Main, 1971.
5. See F.H. Bradley, Appearance and Reality – A Metaphysical Essay, ch. XVIII, Clarendon Press, Oxford, 1966.
6. Ibid., p. 181.
7. See A.J. Ayer, Metaphysics and Common Sense, p. 66, MacMillan, 1969.

8. DK 29 B 1.

9. Cf. Aristotle, On Coming-To-Be and Passing-Away, 316 a 15ff.

10. See ibid., 325 a 31.

11. Epicurus, To Herodotus 58

12. Ibid., 59
13. Ibid., 56-57

14. Ibid., 61-62

15. Cf. P. Tannery, Pour l'histoire de la science Hellène, Paris, 1887. F. Evelin, 'Le mouvement et les partisans des indivisibles', Revue de metaphysique et morale 1,1893, G. Noël, 'Le mouvement et les arguments de Zénon d'Elée', Revue de metaphysique et morale 1, 1893. - I don't think that this interpretation, followed by many scholars, is historically correct (see M. Arsenijević, Prostor, vreme, Zenon (Space, Time, Zeno), pp.91-92, Serbian Philosophical Society and Graphic Institute of Croatia, Belgrade-Zagreb, 1986), but, in any case, it can be used as an argument against geometric atomism.

16. I shall not discuss here Grunbaum's interpretation, according to which all of the famous kinematic atomists, W. James, A.N. Whitehead, and P. Weiss, allegedly side the kinematic atomism of the former kind (cf. A. Grünbaum, 'Relativity and the Atomicity of Becoming', Review of Metaphysics 2, band 4, 1950 and Modern Science and Zeno's Paradoxes, p. 48, George Allen and Unwin Ltd., London, 1968).

17. M. Black, Problems of Analysis, pp. 116-118, Cornell University Press, Ithaca, New York, 1954.

18. Berkeley was the first to claim that our faith in infinite divisibility has its origin in this extrapolation, based on the way in which the mathematicians proceed: whenever the geometer speaks of particular lines and figures, he 'considers them abstracting from their magnitude', which implies 'that he cares not what the particular magnitude is, whether great or small, but looks on that as a thing indifferent to the demonstration' (Berkeley, The Principles of Human Knowledge, §126, p. 154, George Routledge and Sons, London and New York, 1878). David Hilbert, a mathematician, was of the same opinion, and he tried to solve The Achilles on that basis (see D. Hilbert and P. Bernays, Grundlagen der Mathematik I, § 1 c 2, p. 16, Springer Verlag, Berlin, Heidelberg, New York, 1968).

19. Infinitism is the religion of mathematicians. Among philosophers who used this strategy in trying to solve Zeno's paradoxes the most influential are :
B. Russell (see Our Knowledge of the External World, lectures VI, VII, The Open Court Publishing Company, Chicago and London, 1914),
R. Taylor (see 'Mr. Black on Temporal Paradoxes', Analysis 12, 1951, and 'Mr. Wisdom on Temporal Paradoxes', Analysis 13, 1952),
A. Grünbaum (see Modern Science and Zeno's Paradoxes, ch.II, George Allen and Unwin Ltd., London,1968), and
W.C. Salmon (see Space, Time, and Motion, ch. II, Dickenson Publishing Co., Inc., Encino and Belmont, California, 1975).

20. J. Watling, 'The Sum of an Infinite Series', Analysis 13, 1952, p. 48.

21. Cf. J.F. Thomson, 'Tasks and Super-Tasks', Analysis 15, 1954.

22. See M. Black, 'Achilles and the Tortoise', Analysis 11, 1951, p. 97.

23. See above, n. 21.

24. P. Benacerraf, 'Tasks, Super-Tasks and Modern Eleatics', The Journal of Philosophy vol. LIX 24.

25. A. Grünbaum, Modern Science and Zeno's Paradoxes, pp. 94ff.

26. Ibid., p. 105.

27. Ibid., p. 100.

28. See ibid., p. 101.

29. Cf. A. Robinson, Non-Standard Analysis, pp. 55ff. and pp. 266-7, North-Holland Publishing Company, Amsterdam, 1970.

30. Cf. C.B. Boyer, The Concepts of the Calculus: A Critical and Historical Discussion of the Derivative and the Integral, ch. III, Dover Publications, New York and London, 1939.

31. The solution of The Achilles proposed by G.J. Whitrow in 1961 is in accordance with non-standard analysis, which came into being a few years later (cf. G.F. Whitrow, The Natural Philosophy of Time, p. 151, Thomas Nelson and Sons Ltd., 1961 and A. Robinson, Non-Standard Analysis, p. VIII and pp. 58ff.).

32. See P.T. Geach, Logic Matters, p. 238, Basil Blackwell, Oxford, 1972.

33. Ibid., loc. cit.

34. See M. Black, 'Achilles and the Tortoise', *Analysis* 11, 1951.

35. Cf. *ibid.*, pp. 100–101.

36. Cf. D.S. Schwayder, 'Achilles Unbound', *The Journal of Philosophy* vol. LII 17, pp. 454ff.

37. M. Black, 'Achilles and the Tortoise', *Analysis* 11, 1951, p. 101.

38. See Aristotle, *Physics*, 185 b10ff.

39. Cf. Aristotle, *The Nicomachean Ethics*, 1096 b 27, and *Metaphysics*, 1003 a 21.

40. Two parts are contiguously touching if they are physically heterogeneous or if they are not physically grown together. For Aristotle's definition of contiguity see *Physics*, 227 a 6.

41. For Aristotle's definition of continuity see *Physics*, 227 a 10. That the distinction between contiguous and continuous touchings can be inclusive see M. Arsenijević, 'Dodirivanje' ('Touching'), *Filozofske studije (Philosophical Studies)* VI, 1975.

42. See Aristotle, *On Coming-to-Be and Passing-Away*, 317 a 8ff.

43. See Aristotle, *Physics*, 231 b 15.

44. For the justification of my use of name *indefinitism* for the theory under consideration cf. G.W. Leibniz, 'The Theory of Abstract Motion: Fundamental Principles', *Philosophical Papers and Letters*, p. 139, D. Reidel Publishing Company, Dordrecht-Holland and Boston-U.S.A., 2nd edition, 2nd printing, 1976, and Kant, *Kritik der reinen Vernunft* (Werke III), pp. 351ff., Druck und Verlag von Georg Reimer, Berlin, 1911.

45. Aristotle, *Physics*, 227 b 3.

46. Cf. *ibid.*, 228 b 15ff.

47. In our century, H. Bergson proposed a solution to Zeno's *Arrow* completely based on Aristotle's analysis of the nature of the oneness of a motion (see H. Bergson, *L'évolution creatrice*, pp. 334ff., Librairie Felix Alcan, Paris, 1932).

48. See A.R. Anderson and N.D. Belnap, Jr., Entailment - The Logic of Relevance and Necessity, §§3, 4, Princeton University Press, Princeton and London, 1975.

49. Intensional conjuction was introduced by R.K. Meyer in 'Some Problems No Longer Open for E and Related Ligics', Journal of Symbolic Logic 35, 1970.

50. De facto, the term 'relevant consistency', used in relevant logics, and Nelson Goodman's neologism 'cotenability' are being used synonymously (cf. A.R. Anderson and N.D. Belnap, Jr., op. cit., pp. 345-6).

PACIFISM : IS ITS MORAL FOUNDATION POSSIBLE OR NEEDED?*

JOVAN BABIĆ

The gods of war have always occupied a higher social position than the gods of peace. However, the value of social peace has always been the main source of pacifistic feelings and movements which express moral condemnation of war and violence. Although individuals, nations and humankind as a whole have, as W. James noted, acquired a substantial part of their identity through the risks, glory and agony of their wars, it has nevertheless been established as something obvious, for instance, explicitly by Hesiod, that peaceful and constructive work, as a crucial presupposition of human happiness, is possible only in the state of peace. However, that state does not come of itself. One has to struggle for its establishment and preservation. Difficulties in giving a general characterisation of pacifism are not hard to conceive, for the pacifist standpoint includes not only various doctrines, as Narveson has observed,[1] but also different social movements and individual standpoints. Yet, there is one common characteristic of pacifism, that can be ascribed to it with certainty: it is in a way the standpoint which both involves and denies being in counterposition. Pacifism is the standpoint of protest, opposition and resistance that paradoxically aims at abstinence from physical violence. Pacifist standpoint is an attitude towards violence, in particular war-violence. That attitude is expressed in the refusal to respond to violence with (physical) force, which in the last resort implies the denial of the validity of those principles which justify (self)-defence, punishment and prevention of crime. At the level of the individual, pacifism, as a moral standpoint, is based on a specific Weltanschauung which includes not only the principle that violence has an intrinsic negative value but also the principle that attributes an absolute value to human life. This characterisation may have a much more stringent form, that regards even ordinary unkindness and the use of relatively harmless violence as absolutely forbidden. At the social level, the pacifist standpoint generates collective feelings with a tendency towards social institutionalisation, in the form of various movements, sects, etc. Nowadays, at this level, there is a specific narrowing of the meaning of the term 'pacifism' so that it denotes various forms of peace-activities which aim at disarmament, activities which specifically oppose war-violence and even particular types of war, e.g., nuclear, chemical or biological.

However, many contemporary movements, such as movements against nuclear arms, that are inappropriatly labelled 'pacifist', for those movements are not always against violence as such, and often are not peace-movements at all.

A. Pavković (ed.), Contemporary Yugoslav Philosophy: The Analytic Approach, 57–70.

Yet, among the positions that can be legitimately called pacifistic, special significance must be given to those which are specifically war-oriented, i.e., to the peace-movements. Although a significant amount of their activities is more or less tactical, it cannot be denied that these social movements sincerely accept, at least in part, the pacifistic Weltanschauung, from whence comes the claim that pacifism is a genuine moral standpoint. This claim seems dubious or problematic, and my aim in this article will be to show that the pretension of pacifism to be a genuine moral standpoint is unnecessary for the justification of the pacifists' political ends and activities.

Most people regard not only defence, but to some extent revenge (as well as punishment in the retributive sense), as morally justified. Many will plead for prevention not only of violence, but of all forbidden actions. What is common to all these positions is that violence is regarded as evil, but on this point they are not in accord with pacifism: according to them the use of violence under certain conditions is indeed justified. The viewpoint according to which prevention is justified is sometimes conceived in such a way that the aim of final elimination of violence is transformed into an almost unlimited justification of the use of violence; for example, in some revolutionary doctrines any violence is justified if it aims at the final elimination of the conditions for the existence of violence. Punishment is justified by similar arguments; partly by the same justification regarding its purpose, and partly by reference to the assumption that the negative value of two acts of violence (crime and punishment as one whole) is lesser than the negative value of either of them taken by itself (e.g. crime without punishment). Finally, defence (or self-defence) must not be confused with either punishment or prevention: it is an immediate reaction to violence which it is supposed to prevent, but its immediate purpose is not the prevention of violence as such, but preservation of some particular value which is threatened (e.g. life, dignity, physical integrity). It is not an accident that pacifism is characterised in terms of rejection and abandonment of defence. One who rejects defence, to be consistent, must reject punishment as well, and above all prevention. Hence, a pacifist is one who holds violence to be such an evil that he even ought to refrain from defence. The contrary, however, does not hold: one who accepts defence, but rejects prevention (and/or punishment) cannot be called a pacifist. Namely, he cannot develop his position simply on the basis of the attitude towards violence, but must (and, as we shall see, can) provide an independent reason (or reasons) for his different treatment of violence as it appears in defence and as it appears in prevention (and/or punishment).

Now, the question is whether it is possible to defend the position of abstaining from defence. What does such an abstinence involve? Well, it certainly involves the decision to abstain from defending oneself (in the circumstances when defence may seem appropriate). But this decision may be made on the basis of at least two kinds of considerations. First, it can be the decision to give up defence in some specific situation when one is really under attack. This decision, for example, may have various causes. Its cause may be fear. Or it may be the opinion that showing one's firm intention to abstain from defending himself will make an impression on the attacker, make the attacker feel ashamed or in some other way be too inhibited to pursue his malicious goals; and, as we all know, this strategy is the best defence in some cases. Is this behaviour in any relevant sense pacifist? One can mistakenly estimate what the best tactic in fact is, and, as it often happens, contrary to one's intentions, encourage the attack. What should one do next time he finds himself in a similar situation? Next time the situation, circumstances or one's mood may be different, and so the estimate of the best tactic as well. It seems obvious that the decision to abstain from defence which is a result of tactical estimates is not compatible with pacifism. A decision compatible with pacifism cannot be instantaneous, but must be based on some principle. But if this decision is to be based on some principle, what must be the nature of that principle? One might, of course, decide never to defend oneself. A smoker or a kleptomaniac can in the same manner firmly decide to end his regrettable practice. Some other person, again, can decide to fast every Wednesday or not to eat meat at all. Let us grant that the four of them live up to their decisions. This is not enough, however, to regard these practices as having moral justification. For these practices to be morally justified, it is necessary that they may be universalized, and this means that it can be rightfully expected that they may be performed by everyone: in this respect, one who decides to be truthful or honest is in a much better position. In our last example, the man who chooses to fast every Wednesday cannot expect nor require others to conform with his practice; similarly, if I decide never to defend myself whatever the circumstances I do not have the right to expect, and even less to require, from others to follow me and abstain from defending themselves when they need it.

The decisions of the smoker, the kleptomaniac, the man who would not eat on Wednesdays and the like <u>may</u> lack moral justification; they are vows in Gandhi's sense, hence some sort of oath to oneself. I believe that precisely the non-universalizability of the pacifistic claim is the major source of the tendency to institutionalize pacifism as a social movement (sects, societies, associations): non-universalizability makes of pacifism, as it were, a private enterprise and thus the validity of its claims necessarily discriminatory - they are binding only for those who already accept pacifism. Hence,its acceptance is a matter of private choice. This, in turn, means that the moral right, i.e., the right to defence, has a primacy with regard to the private character of any involvement with a pacifist movement. The general character of the relation between private and universal claims means that at best the pacifist position is permissible. Of course, axiologically this is very weak: permissibility has its limitations (if you are not a pacifist you have a right to blame your pacifist friend - and even to end the friendship, e.g., because he failed to defend you from a rapist); furthermore, it is often morally indifferent; and finally, even supererogatory acts (e.g., heroic ones) are <u>only</u> permissible.

The right to defence, however, is not just permissible: it is everyone's right and not just of some who carry a special vow. But, the right to decline or simply to fail to use the right to defence must also be universally acknowledged; and this does not imply any particular attitude towards pacifism, (except, of course, that it is permissible) for the right to decline or wave a right follows from the concept of a right itself. There is no obligation to exercise rights. They are rather unlike duties which, unless they are annuled must be fulfilled. There are cases, I believe, in which a person is in some sense under the obligation to defend himself, although this would not follow from his right to self-defence, but rather from the character of the content of the value which is put in danger, e.g., some absolute value such as dignity.

From the fact that the moral right to defence has such a primacy it follows, then, that it may not be wrong[2] to respond to violence with force, if that is the only or the best possible defence. An argument for such a view is advanced by Narveson who examines the question of how much force one may be justified in exercising in defence, and the answer he gives is both simple and precise: 'Enough'.[3] On first sight this may seem overpermissive, but it is not. Independently, from all other consequences, which are accidental to violence, e.g., pain, what is morally contestable in violence is coercion, that is the restriction of freedom. However, an act of restricting one's freedom, even if it is coercive, does not have to be a violent act: to be violent an act must be, at least in part, directed to the restriction of one's freedom - i.e., have the latter as a part of its objective - while a non-violent act of restricting one's freedom, even if it is coercive, produces the latter only as a side-effect. Natural laws or unintentional acts are not the origin of violence. To have a right implies the right to do whatever is necessary to prevent the infringement of that right.[4] But if rational arguments or simple appeal to a right were always sufficient to prevent infringements of rights, i.e., if rights were violated only as a result of ignorance or misunderstanding, then violence would not exist and the pacifist way of speaking of rights would, therefore, loose its meaning. If this were true, that is, if rights were violated only as a result of ignorance or misunderstanding, any infringement of a right of mine could not be a consequence of a violent act; namely, if I have been in the position to present rational arguments or appeal to that right, the infringement of it would not even have had occurred. It is because this counterfactual is true (under the specified condition) that the act of infringement of the right in question cannot qualify as an act of violence. Since there is no violence, what then would be the point of the pacifist's rejecting one's right to defend himself from violence? Consequently, it is not only empirically probable that what is necessary to prevent the infringement of some right requires the employment of force, but this is a logical consequence of a conceptual analysis of the 'infringement of a right': precisely this infringement involves violence.[5] Whether the force needed to prevent violence is actual physical force depends, of course, on empirical conditions.

Here we may invoke Hobbes. In the natural state as described by Hobbes there is a war of all against all; this fluid state of 'war' is not, however, a real war, but only a metaphor for some state which is not a social state, which does not possess social presuppositions for the minimum of confidence, and hence here the state of peace is not only empirically impossible, but, what is more, inconceivable. So it follows that the concept of peace is meaningful only in the civil state, the state which is defined in terms of rights. This in turn means that peace, and hence the pacifist attitude as well, require a number of normative conditions, but this, of course, is not by itself an argument against pacifism. The existance of state, marriage, time-tables, etc., presupposes a series of normative or conventional conditions, but this is not an argument either for nor against those institutions. However, this does go against an argument which is sometimes offered in support of pacifism. According to the argument, the pacifist position is a matter of natural law or an expression of some kind of natural right to life; hence the resulting prohibition on taking another's life. Murder or rape, as well as fraudulence or malice are not unnatural but immoral.

One of the possible explanations for pacifist's belief that the use of physical force is always morally wrong is perhaps the fact that it presupposes action (presumably evil action), therefore an active, i.e., blameful, position, while inaction (as sufferance of evil), as a passive position, is presumably always blameless. Two objections may be advanced against such a view of 'clean hands and clear conscience', according to which evil must never –'even if the heavens fall' – be done. The first has already been mentioned the impossibility of it becoming the subject of a universal moral obligation. Its maximalistic requirements make its universalization impossible. However, since the pacifist claim is permissible, and hence at least _prima facie_ legitimate, the above objection must not be understood as a condemanation of political or religious movements guided by the pacifist ideal: for the constitution of such movements and their goals permissibility is quite sufficient. The objection concerns only the moral aspects of the pacifist claim. The second objection concerns the definition of 'evil' adopted by pacifists. They indentify 'evil' with the use of physical force, and although this definition logically commits them to regarding the latter as the _sole_ evil in this world, they typically perceive the use of physical force only as the _worst_ of all evils. Pacifistic claims oscillate between these two theses. The second, weaker thesis presupposes the belief in a hierarchy of evils such that the use of physical force is the worst of all of them; that which would count as a conclusive rejection of this thesis will, _ipso facto_, rule out the stronger one. The following two objections are perhaps serious enough to refute the weaker thesis. First, it is not clear how such a hierarchy of values (evils) with respect to the content is to be constructed.[6] The second objection questions the thesis that the use of physical force may be determined on _a priori_ grounds as the worst evil. Besides the fact that coercion must not be physical and that physical force may not be the most effective sort of coercion, it is not, in this context, entirely clear why it should be possible to determine on a priori grounds that acts of coercion are necessarily more evil than acts of deprivation. The difference between actions and omissions can be of no use here, for to refer to any characterization of the content of coercion, e.g., that it is physical, is to transform the question of general evaluation into the question of evaluating specific types of acts (i.e., those involving physical force), that is, into the question of judging the lesser evil. And on this level, the level of specific evaluation, the difference between coercion and deprivation is lost. Deprivation is not just an omission, it is much more than that, it requires the agent to take, as it were, an active role; for this reason deprivation belongs to the category of actions rather than omissions. Furthermore, in drawing the distinction, within the concept of action itself, between coercion and deprivation, physical force is only a part of one side of this dichotomy; that some cases of deprivation are wrong in the same sense in which physical force may be wrong should be obvious. Moreover, there are instances of coercion which do not have to involve physical force, for example, different types of manipulation, and they may be wrong in just the same way that physical force is. The content of the notion of manipulation clearly suggests that, in principle, there is no reason to distinguish between coercion and deprivation solely on the basis of their degrees of evil: manipulation may equally consist of coercion or deprivation, both in its mechanism and in its final result. From all this it follows that to place physical force at the top of the pyramid of evils, as the pacifist does, is arbitrary and is a matter of his subjective choice, i.e., his private matter.[7]

Now we can consider the question of whether accepting pacifistic principles could perhaps be regarded as an act of supererogation. The value of supererogatory actions is precisely in that they exceed that which may be universally required. But an action of that kind has its value only as an act of an individual; it is impossible, and even if it were possible it would be morally impermissible, to single out beforehand a specific class of people which would then be, as a result of accepting a particular set of normative principles, under the obligation to perform supererogatory actions. Apart from the moral impermissibility of prescribing supererogatory actions as such which would conflict with the right of one to have the scope of his duties determined with a significant degree of precision, which is in turn a necessary condition for the very possibility of supererogatory actions. This has an axiologically strange consequence: can we imagine a brotherhood of heroes the member of which you become by making a vow that you will perform heroic deeds. It is hypocritical and morally paradoxical to set such a goal for oneself: in the very act of setting such a goal this goal is degraded.

* * *

At the social and political level pacifism emerges as a practical idea which then requires from anyone to make one's mind with respect to it. Because pacifism excludes precisely the most effective means of achieving standard political aims, this means that majority of political systems and doctrines are in the position that they must state and defend their negative attitude towards it. Thus, for instance, Marxists have regarded pacifism as a part of an antirevolutionary, bourgeois ideology and even as 'the twin of fascism' (Stalin). On the other hand, despite their tendency to project the image of their religion as a religion of peace, Christians had supported the abstinence from the use of force only for a short period; immediately after its official recognition, in the year 313, the church changed its mind and rejected the pacifistic doctrine, introducing at the same time, the doctrine of just and unjust wars.

The criticism of pacifism at this, social, level depends on two major factors: the critic's attitude towards society and social questions in general and the definition of pacifism itself. The first is not theoretically interesting in the context of the present paper; the critic's moral views may, of course, be relevant, but not his political views, except to the extent that they are implications of his moral views. But the different meanings associated with the concept of pacifism and the difference in political programs following from them may provide theoretical reasons for criticism and practical reasons for adopting specific attitudes towards a movement which defines itself as pacifistic or is perceived as such. Almost everyone would agree with the condemnation of violence and war, if this condemnation means the preference of peace to war. But hardly anyone would unconditionally accept the rejection of defence. In the case of war, a consistent pacifist, given his condemnation and rejection of any but non-violent involvement in war, is committed to rejecting any ethics of war and, just like the extreme war-apologist, will not be in the position to set any moral limits in conducting wars. The pacifist will simply not be able to separate various acts in war according to different degrees of their moral wrongness. For, according to him, all such acts are absolutely impermissible. This, furthermore, means that a consistent pacifist will not be able, in a concrete situation, to make distinctions between things that ought to be distinguished and that he will reject them in the same manner and with the same rigour and unflexibility. In the case of an organized pacifist movement its members will oppose, without discrimination, different phenomena which, although perhaps all wrong, do not have to be equally wrong and hence intolerable to the same degree.

Moral puritanism exhibited in such an extreme attitude has as its consequence the impossibility of distinguishing between the acts of torture, war-crimes or genocide, on the one hand, and taking part in a war, killing in self-defence or even taking weapons in one's hands, on the other. If we were to evaluate this world on the model of some ideal perfect world, then, of course,none of these acts could count as good or desirable. However, such an idealisation is morally entirely fruitless unless it offers not simply a way of determining whether an act or state of affairs conforms to its ideal content, but also a practical mechanism which will make it possible to see which act, among those which do not conform to their ideal content, is better or worse. Basing one's position on the conditions which do not exist and perhaps are only logically possible, but in fact will never actually exist in our imperfect world, is not only morally futile but at the same time methodologically untenable. Social movements which adopt and essentially depend on some such idealisation, and as a result have set maximalistic demands, despite their likely militancy, will never be able to impose themselves on to the totality of a society for good. These movements may exist only as isolated organisations, sects, etc., in the midst of the rest of the world (we may perhaps say, against a backdrop of 'normal' people) and if it happens that they succeed in imposing their views in the form of a prevailing religion or ideology, very soon they have to introduce differences in the scope of their demands, which means that their initial maximalistic demand becomes void: they now abandon ordinary people and preserve their grand demands only for the 'chosen' few. Bogomiles are a good example, but, of course, not the only one.

Another general problem which has immediate, although not necessarily negative, consequences is the problem of predicting or calculating the chances of a social project's success. This problem is much more complex than the similar problem concerning individual projects. It has been claimed earlier that such calculations represent a tactical rather than a moral viewpoint. But at the social level this is not necessarily so. If the prediction is that a certain action (e.g., a widespread and organised civil disobedience, boycott, strike, etc.) is likely to bring about the desired result then the decision to carry it out ceases to be (just) a tactical and often becomes a moral question as well. However, a question arises of why the decision would have moral significance if the likelihood for the success of the given action is high and would have no moral significance if the likelihood is low or indeterminate. This clearly shows the difference between the individual and social level: something which at the individual level may be wrong (degrading or something that cannot be recommended as a general practice, for instance, to decline the demand to have one's rights respected), may at the social and political level, where the question regarding the consequences of one's decisions has different import, not only be permitted (as at the individual level) but morally recommendable and in some circumstances even obligatory. What will count as permissible or recommendable at the political or social level - regardless of what it may represent at the individual level - is an empirical question. To fail to do away with some evil at the social level may be morally wrong, yet its elimination may require disregard for some legitimate individual interests and itself be a source of coercion. What form this will take in a concrete situation will depend largely on the situation itself and, of course, on the corresponding estimates. The consequence of all this for the subject of this paper is that possible justification or desirability of pacifistic practice does not follow from the pacifistic principle as such, but is the consequence of the structure of a particular social and political situation wherein pacifism just happens to have a role corresponding to its goals.

As a political movement, pacifism often relies on such calculations. Since its goal is peace, it can not, and has not excluded considerations of determining the methods and estimating the likelihood of attaining peace and in particular, world-peace. Ideas of how to achieve this goal have been various. One of them was an optimistic or quasi-optimistic vision that the development of weaponry and destructive techniques will gradually lead to dissolution of war. The second, more interesting and for mankind more challenging one, was the idea of the world becoming one state. So Bertrand Russell, who in 1918, spent six months in jail because he propagated resistance to conscription, in the midst of World War I (in 1916) wondered whether it had been better to ensure universal peace even at the price of the supremacy of one power, that is Germany.[8] Had Napoleon managed to conquer Europe and Great Britain, thought Russell, the world would have been happier, more civilized and the peace would have reigned. National freedom, the freedom of states, would have been indeed sacrificed, but freedom of the individual would thus have had its chance, for exactly the existence of national states and economic nationalism which follows from it (and in Russell's opinion, this is the dominant force of our time), represent the main cause of war, and war eliminates the very conditions for the existence of human freedom. This is not, of course, the place for a detailed discussion of the concept of the World-state, hence I shall only pose a few obvious questions. Is it possible to accomplish the unification of the world in one state on the voluntary basis or is it, on the other hand, necessary that there be a supremacy of one (or perhaps two) superpowers? In this latter case, the question arises of whether such a union based on subjugation or on a balance of deterrence would have any value. The answer to this question would also provide the answer to the question of whether this would be the way to obtain the pacifistic ideal of world-peace.

To elaborate on this point, let us consider an example of how some singularly valuable things may be obtained just as a side-effect of some activity the purpose of which was something entirely different. Hegemonistic policy may in fact lead to the realisation of peace, or even universal peace, but this desideratum, although it may appear on the list of proclaimed goals, is never that which motivates the launching of that policy. Suppose that some nation believed in its mission to rule the world, then the world-peace would have been just a side-effect, were this mission to be fulfilled.

Now, the essential question regarding the idea of world-state is whether such a product would warrant more freedom and creativity, and less disrespect for laws, or whether it would be a large-scale tyranny. There can be no a priori answer to this question, but there is no obvious reason why the world-state should be in any way special with regard to existing national states: whether it be an empire or a federation[9] the world-state can just as well turn out to be a tyranny - so the same patterns of social and political organization are open to the world-state as to the existing national states. This means that the universal peace, even if it is generally a better alternative, is not necessarily a good alternative. We may imagine a scenario wherein of two power blocks one was disposed to such a degree of fanaticism that it would prefer the destruction of the world to a compromise; then a pacifist, like Russell, would have to categorically demand that the other block yield and capitulate. Suppose, furthermore, that those before whom the capitulation is demanded are the Nazis. Is yielding or capitulating before a blackmail really the right choice, on the assumption that the destruction of the world is not just among the preferences of the fanatics but also in the scope of outcomes they are actually capable to bring about?

It is not, however, certain that indulgence and passive resistance will make the impression on the aggressor. The answer that it may, besides the fact that it may be given by anyone and not just by the pacifist, is inadequate; on the other hand, the pacifist's answer that it will may be a mere expression of his craving that it be that way. This latter answer, just as the widespread belief (or prejudice) that the balance of power and deterrence implicitly assumes that neither side has aggressive intentions for the role of increasing armament is not aggression but deterrence; this means in the case of pacifism that every assault is somehow provoked. However, if this were true then what appears perfectly rational in the case of nuclear weapons, i.e., unilateral disarmament,[10] would be so in the case of any other weapons too. This is where the weakness of pacifism lies. Pacifism, simply put, does not always work; pacifist position becomes meaningless in the case when the oppressor is ruthless and fanatic. Thus, even when resistance is most urgent it will be condemned by the pacifist. How is it possible for pacifists to prepare to refrain from resisting even when resistance is most urgent? It seems that pacifists are so psychologically constituted that for them any violence,[11] is absolutely unbearable. In order to see what it is exactly that is unbearable for the pacifist we must understand what he means by 'violence'. There are only two possibilities consistent with his position. He may mean 'violence in which he takes an active part' or he may mean that violence by definition is something that includes resistance; in the latter case it would be sufficient to refrain from resistance to make a given act non-violent. The second possibility is clearly unsatisfactory because as a definition it seems counter-intuitive, while the first is just the consequence of a certain psychological state. (This does not strike me as a promising line either, but may deserve further investigation). In neither case have we been given sufficient grounds for some moral or philosophical viewpoint; hence, we are left with nothing which could require moral (or philosophical) justification.

Concerning the first meaning of 'violence' one can also ask how it is possible that the pacifist can adopt such a partial and selfish viewpoint so that something is violence only if he takes part in it? The reason may be his refusal to share responsibility for some painful events that have happened. After horrid experiences, e.g., the World War II, moral recovery may appear difficult to achieve, but to the pacifist it may be even unthinkable, so that, according to him, in the light of what has happened, peaceful, non-resistant surrender would have been better than resistance and war. So, contrary to what may be expected, the pacifist is here more alike to those who 'remember' rather than those who are 'forward oriented'. It is as if pacifists, with their limitless faith in the rightfulness and purposefulness of 'turning the other cheek' and the possibility of prevention of 'the worst', tacitly fear that the faculty on which this possibility is based is not all that well ingrained in human nature. Incapable of admitting this, they supress this problem and cover it with a surplus of eagerness. This, of course, does not mean that pacifists are insincere or cowards, nor that the evil ought not to be fought at all. Evil ought to be confronted, if for no other reason, in order not to be ashamed for what other people have done and to spare ourselves the hatred and will for vengeance. All such reasons are perfectly legitimate, just as selfish (on the condition that they do not harm others) or supererogatory motives are. But unless it is highly probable that pacifist's course of action will produce the desired effect on the intruder, his procedure becomes meaningless.

It may be replied, however, that should the pacifist's attitude in a given circumstance prove insufficient as an assault deterrent, this nevertheless, would not mean that it is meaningless, for the purpose of this attitude is, as it were, limited to the pacifist himself: his respect for, e.g., the principle that it is better to die than to kill a man, or the principle (which Gandhi adopted) that the opposition to evil is to be accompanied by good-will rather than by hatred and revenge. But these are all two-place relations: 'taming' of a thunderboldt by lightning-rod or of a river by embankment is only a metaphor which does not involve either revenge or good-will, for there is no one towards whom these attitudes would be expressed; if one kills an infuriated bear which attacked him, we shall not call this murder or revenge. Moreover, Gandhi's kindheartedness towards his enemy has its purpose precisely in making him ashamed, that is, in a successful demonstration of the peculiar moral strength inherent in passive resistance; and, therefore, in issuing a warning the role of which is to remind the assailant of his potential humanity. But this means that the goal of pacifistic action is not limited solely to the pacifist himself, but on the contrary, that it is primarily directed to the assailant. However, if the assailant is not to be treated as an end in itself, or perhaps ought not to be treated so because of his unscrupulousness and fanaticism (which reveal his inability to show any respect for that which is universally human), then the pacifistic attitude gets divorced from the sole condition which would give it a (possible) purpose, and becomes pointless. In such a case, then, resistance is not only justified in the form of defence, but a counterattack (and sometimes perhaps even the preventive attack) is also justified. Readiness to defend oneself functions as a potential warning to the aggressor as to what is at stake. But, of course, resistance is justified primarily as defence: not only that there is nothing wrong in not suffering injustice, but injustice ought not to be suffered or tolerated. To advocate patience (as a form of tolerance), which is a standard way of expressing the pacifist attitude, can be justified in this context only conditionally, and on the assumption that patience will not have to last for ever, that is, for an unreasonably long time.

Now there is a practical question as to whether pacifists, as people with special duties, have perhaps some special rights with respect to war?[12] I have tried to demonstrate earlier that there cannot be special duties, i.e., duties which, disguised as moral duties, would pertain only to members of a certain organisation who gave a special oath. But some organization may prescribe to its members some special obligations, simply because they are members and on the basis of the oath they have given; and in the case that nothing is required by the oath which is in conflict with morality such obligations may be perfectly legitimate and even stimulating. Thus, it is a fact that various beneficial organizations or institutions have the capacity to motivate their members to great efforts in trying to achieve the ends which must be regarded by everyone as honorable and good, hence increasing significantly the likelihood of those ends being in fact achieved. It is not sufficient to be a missionary to be Mother Theresa, but it is not an accident that Mother Theresa is a missionary. Opposition to war, especially if it is massive, can surely give significant results. Moreover, there is no need for a special social organization to make it obligatory for one to oppose war, it is sufficient that one agrees with Tolstoy that the duty of conscience is higher and more important than other duties; and if one's conscience forbids one's participation in a war, then this cannot be condemned. In some countries here there are even laws concerning this. But there arises an almost technical question: how are we to know whether someone opposes his participation in a war because of his conscience or some other reason? Of course, this question is of no great moral importance, for other reasons may be equally legitimate (for instance, why should one whose conscience, whatever it is, forbids him to participate in a war be released from the obligation to participate, while one who must care for a sick wife and six children be treated as a deserter? But where to place a limit?). Given that what is here at stake is life itself – that which is a necessary condition for the realization of any other values –it is not at all necessary to advocate pacifist views on life as an absolute and ultimate value; it is sufficient that one has more important goals than the ones for which the war is waging, to be in the position to employ rational arguments (that is without becoming a pacifist and without giving an oath) against the view that war is the way to solve human conflicts. This, however, requires further explanation; for a person need <u>not</u> have any concrete, materially determined aims in order to reject this particular <u>way</u> of solving conflicts. Moreover, it may be said that there is, at least in a <u>prima facie</u> sense, a moral duty to reject war as <u>the way</u> of solving conflicts; it cannot be a part of a moral purpose of a man or a nation to put himself or itself in such a contingent and arbitrary situation. But as with any other <u>prima facie</u> obligation this obligation may be overridden. However, a complication crops up here: the other obligation does not have to be the overriding one, it may simply be such that it (as a value) justifies the use of the given, i.e., violent, way of solving conflicts (as, e.g., in self-defence). In that case, the agent, although he prima facie condemns such a way of solving conflicts, will approve of war (i.e., he will not recommend capitulation), but this does not mean that by doing so he automatically ought to sacrifice his other goals and report himself as a volunteer. The issue is complex and it implies that not only in the case of a war of which I do not approve, but even in the case of war of which I do approve, I may still claim that the war does not concern me so much that it would be my duty to take part in it. Of course, the question of the legitimacy of decisions made on the basis of such attitudes is a new, complex question which implies or presupposes answers to a host of other questions; above all, the question of moral foundation of political obligations, and related to this, the question regarding possible rights and obligations concerning loyalty, civil disobedience and uprising, rebellion or revolution. What seems certain, however, is that pacifism has no moral foundation, either at the individual level or as a social movement. That is, it is neither morally obligatory nor morally impermissible. So its possible value (or negative value) derives from contingent features of a specific social situation, which in turn, are to a large extent a matter of personal or political assessment. Peace, although a universal human end, and so the subject of everyone's moral obligation, would have to be the absolute end in itself, the ultimate end, for pacifism to become an obligatory moral viewpoint. As it stands, it is not necessary to be a pacifist in order for one to plead for peace.

NOTES

* I am indebted to Nikola Grahek and Aleksandar Jokić, the latter not only for his help in translation but also for many helpful comments.

1. Cf. J. Narveson, 'Pacifism: A Philosophical Analysis', Ethics 75 (1964/5) 4.

2. I say 'may' because I allow for such cases wherein it is not appropriate to answer violence with force because of the presence of some overriding reason; for example, in parent-child relation, when the child is unjustly inflicted with physical pain, e.g., as a result of a mistaken judgment, it would be inappropriate for the child to respond with force to this act of violence, because of the heirarchical nature of the parent-child relation; that is, because of an external and independent reason. Similarly, when someone takes part in some cult ritual (e.g., religious one) consisting in whipping, without sincerely accepting the whole institution and hence disbelieving the supposed sacred nature of the ritual, but nevertheless decides to take part in it for, say,

careeristic, patriotic, or other tactical reasons; because of the insincere acceptance of the whipping ceremony, the whipping becomes an act of violence, but because of the voluntary participation in the ceremony it would be inappropriate to answer whipping with force.

3. Ct. Narveson, op.c.,p. 268.

4. Ibid., p. 266.

5. Note that a restriction on one's freedom does not presuppose violence whereas an infringement of a right does, i.e., a non-violent infringement of a right is impossible.

6. For an argument that such a construction is impossible see my Kant i Scheler, ICC, Beograd 1986, especially chs. III and V.

7. Cf. supra, pp. 6-7.

8. Cf. B. Russell, Why Men Fight: A Method of Abolishing the International Duel, New York, The Century Company 1917, p. 98.

9. On an idea of world-federation see I. Kant, On Eternal Peace.

10. Nuclear weapons are of no use for purposes of conquest.

11. But recall that the pacifist's claim, most likely, is that violence is the sole evil, and not just the worst of all evils. See the distinction between the strong and weak thesis, supra, p.

12. Cf. e.g. C. Ihara, 'In Defence of a Version of Pacifism', Ethics 88 (1978) 4.

THE ROLE OF GENERAL TERMS IN SINGLING OUT THE REFERENT OF A DEMONSTRATIVE

VOJISLAV BOŽIČKOVIĆ

In this paper I will primarily deal with some aspects of the reference of a demonstrative within the framework of the Fregean sense theory. Due to limitations of space I cannot go into discussions regarding my acceptance of this framework. I will simply assume it. I believe, however, that the topics and the arguments of this paper are relevant not only within the sense theory.

The Principle of Parity of General Terms – Introduction

Suppose that there is a situation in which the speaker attempts to convey information uttering the demonstrative 'this' in such a way that more is intended than just the communication of its linguistic meaning[1], i.e., more than just letting the hearer know that a word of a certain linguistic type is used. We may also suppose that the speaker, at the same time, points towards a spatio-temporal region in such a way that his utterance is not relevantly shared with the hearer on the basis of their previous discourse[2] The hearer will, we may suppose, either single out some object – due to various reasons that he may have – believing it to be the thing the speaker is referring to, or, lacking what he takes to be 'sufficient information', he will single out no particular object. Given that the latter case is irrelevant for the topics I am discussing I will consider only the former one. (To avoid complications I will suppose that the object which the hearer singles out corresponds to the object the speaker refers to and intends to refer to).

Managing to pick out an object which he takes to be the referent of 'this', uttered by the speaker on the occasion described, the hearer makes use of some extra-linguistic factors involved, e.g., spatial positions, distances, and the like.

Suppose now that in the circumstances described the speaker utters 'this horse' instead of 'this', and suppose that as before the hearer singles out an object taking it to be the referent of 'this horse'. For the sake of simplicity I will also suppose that this object is a horse and that it is the only horse in the hearer's vicinity. In this case the hearer's singling out of the object in question is partly directed by the linguistic meaning of the general term 'horse' as well as by some extra-linguistic factors.

A. Pavković (ed.), Contemporary Yugoslav Philosophy: The Analytic Approach, 71–83.

Comparison of these two cases suggests that – besides on the basis of the linguistic meaning of 'this' – the role of determination of the referent in question, played in the latter case by the linguistic meaning of the general term 'horse', is in the former case accomplished by the extra-linguistic features. (There is no doubt that some extra-linguistic features are also present in the latter case: circumstances in which the hearer understands the linguistic meaning of 'this horse', and yet fails to single out any object as its referent, should remind us of the presence of these features).

Demonstrative phrases containing more than one general term can be, prima facie, accounted for in a similar way. This also seems to be the case with demonstrative sentences. Suppose that the speaker utters the sentence

(1) This horse is a steed

which contains two general terms. We may suppose that the hearer singles out the relevant object as before. Given that the term 'steed' is more specific than the term 'horse' it can enable the hearer to single out the relevant object more easily than in the case where only 'this horse' was uttered.

There might, of course, be some considerable differences as regards the psychological significance of each of these general terms for the hearer. It is, however, not obvious, on its face value, that there is a difference between them at the semantical level, i.e., it does not seem that due to grammatical or some other role in the sentence one of the two general terms plays a 'decisive' role in comparison with the other, as regards the hearer's singling out of the relevant object. (The relation of this to the question of truth-value of a sentence will become clearer in the next section). Moreover, it seems that the general term attached to the demonstrative as a part of the grammatical subject is not semantically 'privileged' in comparison with the one playing the role of grammatical predicate. We are tempted to generalize this further and establish a principle stating that all general terms in a demonstrative sentence are on a par as regards their semantical role in the hearer's singling out of the referent of the demonstrative in question. We might call this principle 'the Principle of Parity of General Terms'. In the remaining part of this paper I will examine whether this principle is valid.

The first hint towards the justification of the principle is the so-called predicative analysis of all general terms. Namely, it seems that (1) can be analysed like

(1a) This x (Horse x & Steed x),

where 'this' is a variable-binding, formula-forming operator on formulae (of the same logical type as the quantifiers).

According to this syntactical analysis both general terms play the predicative role in the sentence. It might be claimed that it also displays the equality of semantical roles of the terms in question – in accordance with 'the Principle of Parity of General Terms'.

The objection, however, might be raised that this kind of analysis is artificial, i.e., that it obliterates the difference in the semantical significance of the general terms in (1) as the sentence of a natural language. In light of this possible objection I will take a different strategy: I shall try to show that two demonstrative sentences containing the same demonstrative (type) and the same general terms, but in reverse order, express the same Fregean thought. By virtue of this I shall derive the conclusion that in the chosen pair of sentences the general terms in question are semantically on a par. In other words, the validity of 'the Principle of Parity of General Terms' will be implied regarding the chosen pair of sentences. In light of this, I will carry on the investigation of 'the Principle' as regards its generalization and accept its limited generalization.

Consider

(1) This horse is a steed

and

(2) This steed is a horse.

As the identity of thoughts according to the Fregean theory implies the extensional equivalence of sentences expressing them, showing that (1) and (2) express the same thought implies their extensional equivalence; there is thus no need to argue for the extensional equivalence on separate grounds.

<u>The Intuitive Criterion of Difference</u>

As a basis of estimation of whether (1) and (2) express the same thought I shall accept a self-evident criterion present in Frege's writings which Evans calls 'the Intuitive Criterion of Difference', and formulates it in the following way: The thought associated with one sentence S as its sense must be different from the thought associated with another sentence S′ as its sense, if it is possible for someone (not anyone) to understand both sentences at a given time while coherently taking different attitudes towards them, i.e., accepting (rejecting) one while rejecting (accepting), or being agnostic about, the other. (/3/ 18–19).

We should bear in mind that this criterion is designed primarily for consideration of sentences containing co-referring proper names. If, on the other hand, a relevant difference is admitted as existing between proper names and demonstratives, 'the Criterion' should be adjusted accordingly. I will leave aside the question of difference between these two kinds of expressions, pointing out just one phenomenon as regards demonstratives: There might be some hesitation in accepting that the demonstrative 'this' refers to the same object as used in (1) and as used in (2) even if (1) and (2) are uttered on the same occasion. If 'the Criterion' is to be efficient, it must be made clear whether the same object is being referred to or not. If the former is the case, 'the Criterion' suggests that (1) and (2) express the same thought. If this is not the case, i.e., if two different objects are being referred to, (1) and (2) don't express the same thought, for the difference in reference, according to the Fregean theory, implies the difference in senses of the corresponding expressions.

There is, however, some reluctance to accept the judgement that the pair of sentences (1) and (2) express the same thought even in the case where the same object is being referred to by the demonstrative 'this' in each of them. Namely, there is the asymmetry, pointed out by Barry Taylor (/6/), one feels in the semantic roles of F and G in that F is G [3] , which has a number of manifestations. One is a hesitation to accept the equivalence of that F is G with that G is F . Another is that the Donnellan-like claim that that F is G can be true in some contexts when the demonstrated object is not F, though never when it is not G, is not self-evidently absurd.

As this might suggest that a deeper analysis reveals a possibility of taking different attitudes towards (1) and (2), it is worth examining the source of the 'feeling' of the asymmetry in question.

It is natural to suppose that the hearer will, after hearing the speaker's utterance of a demonstrative sentence, wishing to pick out the referent of the demonstrative, 'reach out' to single it out according to 'instructions' provided by the speaker's utterance of the sentence. Now, it can happen that - due to various psychological factors - the hearer picks out one object as the referent of the demonstrative 'this' in (1) and another as its referent in (2), though the speaker intended to refer to the same object and uttered (1) and (2) in the same situation. Suppose that in the speaker's and the hearer's visual field there are two horses, one equipped with conventional gear for a steed, the other not. Hearing

(1) This horse is a steed,

uttered in such a way that, say, the speaker's pointing is imprecise so that it could be taken to be directed to either of the horses, the hearer may take it as too obvious that the equipped horse is a steed and so think that it is not worth mentioning this too obvious fact. He may take the speaker to be saying that the horse who at the first sight does not look like a steed is nonetheless, a steed. On the other hand, he may believe that by uttering

(2) This steed is a horse,

the speaker intends to convey that the horse with gear is a real horse and not, say, a life-size dummy horse.

Cases of this kind give rise to the hesitation to accept the equivalence of that F is G with that G is F mentioned above. But, what happens here is that the hearer attaches different thoughts to (1) and (2) as a result of his belief that the demonstrative 'this' in (1) and (2) refers to different objects. On the other hand, in the cases in which he takes them to refer to the same object, it is implausible - in accordance with 'the Intuitive Criterion of Difference' - to ascribe to him different attitudes towards (1) and (2). An examination of whether the hearer will take the demonstrative's occurrences in (1) and (2) to refer to the same object or not can discover some important psychological distinctions, but the resolution of the 'puzzle' raised by the aforementioned asymmetry at a purely semantical level depends on which hypothesis we accept. If we assume that the objects referred to by 'this' in (1) and (2) are different, the possibility that (1) and (2) express the same thought is ex hypothesi precluded, for the sentences whose corresponding referring terms refer to different objects cannot express the same thought. If, on the other hand, we assume that the object referred to is the same, and moreover that the hearer takes it to be the same, then, according to 'the Intuitive Criterion of Difference', he cannot coherently take different attitudes towards (1) and (2) at the same time. If the object referred to by 'this' in (1) and (2) is assumed to be the same but the hearer believes that two objects are in question, we will say that he is mistaken, not that there is a semantic ambiguity. (In the case when two objects are assumed to be referred to by 'this' in (1) and in (2), but the hearer believes that one object is in question, we will again ascribe a mistake on his part).

Now, assuming that one object is in question, we might hesitate to accept the foregoing verdict that (1) and (2) express the same thought, in the case where the demonstrated object is a horse but not a steed. It seems that (remember Donnellan - like the claim mentioned above)

(2) This steed is a horse

can be somehow true in this case, though

(1) This horse is a steed

is evidently false. Envisaging some of the possible circumstances of linguistic communication, however, suggests that this difference is of pragmatic, not of a semantic, kind. Suppose that the speaker uttered (1) and/or (2) believing that both general terms apply to the given object. The hearer, on the other hand, in accordance with our hypothesis, believes that the object in question is only a horse, not a steed. He may now point out to the speaker that he is mistaken as regards the information presented by the way of (1), but not by the way of (2). This, however, depends on the overall purpose of their conversation (etc.), and, if the purpose at hand requires, the hearer will also point out to the speaker that he is mistaken with respect to (2).

The Criterion of Substitutivity

The significance of 'the Intuitive Criterion of Difference' becomes clearer in light of discussion of the question of whether two sentences express the same thought in terms of the so-called 'Criterion of Substitutivity'. A deeper analysis shows that answering this question on the basis of the latter criterion is in fact done by presupposing the former criterion. My analysis will deal with belief contexts only, but it seems, mutatis mutandis, to be applicable to other 'intensional' contexts.

As two sentences may be extensionally equivalent and yet express different thoughts, their intechangeability salva veritate is not sufficient to ensure that they express the same thought. We need, therefore, a use of a belief sentence where its truth-value is affected when its dependent clause is replaced with another extensionally equivalent but not intensionally equivalent with it. Contexts in which the truth-value is so affected are so-called oblique contexts. But, in their typical - deictical - use, demonstratives like 'this' and 'that' will make the context transparent, not oblique, i.e., the belief sentences

(3) A believes that this horse is a steed

and

(4) A believes that this steed is a horse

are de re. This, given standard accounts of de re belief, makes the context transparent, not oblique. According to Burge (/1/), a de re belief is a belief whose correct ascription places the believer in an appropriate nonconceptual, contextual relation to objects the belief is about. (The term 'nonconceptual', he says, does not imply that no concepts or other mental notions enter into a full statement of the relation). Attributing a de re belief about a given object one is free to substitute any co-referring designator of the relevant object.

A belief ascription de dicto, on the other hand, relates the believer to the 'content' of the embedded clause; in accordance with the framework within which I am proceeding, the 'content' is to be taken as the Fregean thought as expressed by this clause in its 'original' context, or, as Frege would say (/4/) as its customary sense.[4]

A belief de dicto involves the believer's conception of the object his belief is about, and the substitution of one referring term with another, co-referring with it, does not guarantee that the believer will believe in both resulting sentences, e.g., the believer may believe that

(5) Nine is greater than seven

and at the same time disbelieve that

(6) Smith's favourite number is greater than seven,

given that nine is Smith's favourite number. In other words,

(7) A believes that nine is greater than seven

need not be equivalent to

(8) A believes that Smith's favourite number is greater than seven.

As these contexts are oblique, the equivalence of (7) and (8) would be forthcoming if and only if the embedded clauses, i.e., (5) and (6) had the same sense.

The judgement that the customary senses of (5) and (6) differ is based on the insight that 'nine' and 'Smith's favourite number' don't have the same sense, which in turn implies the difference of senses of (5) and (6). In the Fregean theory this judgement is represented by 'the Intuitive Criterion of Difference'. (The accounts not based on the sense theory, on the other hand, would interpret the corresponding insight in accordance with their apparatus, and the difference in attitudes towards (5) and (6) would be accounted for in terms of the corresponding conception of 'content'.[5] When the 'content' is the Fregean sense, 'the Intuitive Criterion of Difference' is a self-evident representation of the judgement in question).

Two things mentioned above, i.e.,

(i) that 'the Criterion of Substitutivity' allows for judging whether two sentences express the same thought, or not, only if the contexts are oblique, i.e., the beliefs are de dicto;

and

(ii) that the belief contexts are de re when demonstratives in demonstrative embedded clauses are used deictically,

suggest that our examination regarding whether two belief sentences containing demonstrative clauses are equivalent or not should take into account the cases in which demonstratives are used de dicto. Burge has pointed out that demonstratives have de dicto uses when they occur anaphorically as pronouns of laziness. It seems to me that de dicto usage of a demonstrative is also provided in some other 'parasitic' cases. Given that the cases of anaphorical usage might raise some problems which I cannot discuss here, I will envisage a non-anaphorical 'parasitic' case. Suppose that the person A hears the person B saying

(1) This horse is a steed,

being situated in such a way that he does not face the object B is referring to uttering (1), but accepts (1) as true and utters it. Seen in this light, the ascription of belief to A, i.e.,

(3) A believes that this horse is a steed

may be regarded as de dicto. A's conception of the relevant object plays a crucial role in his thinking about the referent. (Very roughly, we might think of the conception in question as 'the object referred to by B's last utterance of the demonstrative 'this' ').

A's lack of knowledge of, say, the fact that the demonstrative 'this' as used by B in the given situation is co-referential with 'Bucephalus', supposing that this is the horse's name, will make it the case that in its de dicto reading (3) need not be equivalent to

(9) A believes that Bucephalus is a steed.

They need not be equivalent since A need not believe that the demonstrative 'this', as used in the given situation, and the name 'Bucephalus' refer to the same object. In other words, it is possible for someone to understand both embedded sentences at a given time while coherently taking different attitudes towards them. The basis for this evaluation is, however, nothing but 'the Intuitive Criterion of Difference'; this criterion being thus presupposed by 'the Substitutivity Criterion', and supplementing the latter with the basis of evaluating whether two sentences express the same thought or not.

As regards de dicto interpretation of the sentence

(4) A believes that this steed is a horse,

suppose similarly as in the case of (3) that this interpretation is provided by A's not facing the object B is referring to while uttering (2), but A accepts (2) as true and utters it himself. The demonstrative is used in the same 'parasitic' way again.

Our estimation as to whether A's belief in the thought expressed by (1) conveys his belief in the thought expressed by (2), or vice versa, i.e., whether (3) and (4) are equivalent in the situation described, stems again from 'the Intuitive Criterion of Difference'. (We have seen above that the answer to the question whether (1) and (2) express the same thought in light of this criterion is affirmative).

'The Substitutivity Criterion' thus provides no ground for evaluation of whether two sentences express the same thought independent of the one provided by 'the Intuitive Criterion of Difference'.

The Principle of Parity of General Terms

On the basis of the judgement that (1) and (2) express the same thought the proof can be provided that both general terms in (1) and (2) are on a par as regards their semantical role, i.e., their role in singling out the referent of the demonstrative for the hearer, irrespective of whether either of them is directly attached to the demonstrative or not.

(i) (1) and (2) express the same thought, as argued.

(ii) If two sentences express the same thought, the corresponding referring expressions have the same sense and the same reference. (This is one of the basic principles of the Fregean theory).

Comment: The same syntactical structure of (1) and (2) enables us to determine the corresponding referring expressions in them. It is evident that the demonstrative 'this' is either a referring expression or a part of such an expression. Furthermore, according to the syntactical criterion, we won't take the referring expression in one of them to be simply 'this', while taking 'this horse' or 'this steed' to be the referring expression in the other. On the other hand, examples abound where this question is hard to settle, e.g., 'Jones is taller than Smith' and 'Smith is shorter than Jones' express the same thought but it is not clear what the corresponding referring expressions are.

(iii) The grammatical subjects of (1) and (2), i.e., 'this horse' and 'this steed' don't have the same sense, for their constituent general terms differ in sense. Namely, according to 'the Intuitive Criterion of Difference', it is possible to take different attitudes towards

(10) Bucephalus is a horse

and

(11) Bucephalus is a steed.

Given that (10) and (11) have the same subject expression, the difference in their senses stems from the difference in senses of their predicate expressions.[6]

(iv) As it is not possible, according to (ii), that in two sentences expressing the same thought the corresponding referring expressions differ in sense, it follows that in (1) and (2) the general terms in question are not parts of their subject expressions on a deeper analysis. The subject expression in (1) and (2) is, therefore, the simple demonstrative (type) 'this'.

(v) The bare linguistic meaning of the demonstrative 'this' does not constitute its complete sense on a given occasion. Other factors, some of them linguistic, some extra-linguistic, are also involved. Among the former, as regards (1) and (2), are the general terms 'horse' and 'steed'. In accordance with (ii), the sense of the demonstrative in question provided by (the senses of) these terms, as well by some extra-linguistic components,[7] must be the same in (1) and (2).

It is plausible to assume that the conjunction of both general terms constitutes this sense, not just one of them, for the following reasons. Suppose that somebody argues to the contrary, claiming for some reasons that this role is to be played by one general term, say, the general term which is a part of the grammatical subject in one of the sentences. The same term is, however, a grammatical predicate in the other sentence; the word 'horse' as a part of the grammatical subject in

(1) This horse is a steed

plays the role in question in

(2) This steed is a horse,

being its grammatical predicate. But, by the same token, it follows that the word 'steed' as a part of the grammatical subject in (2) and as a grammatical predicate in (1) plays, equally, the same role. This suggests that the conjunction of the two general terms, together with some extra-linguistic components constitutes the complete sense of the demonstrative 'this' in (1) and (2).

(vi) This implies the validity of 'the Principle of Parity of General Terms' as regards (1) and (2), i.e., both general terms in them are on a par regarding their semantic role in the hearer's singling out of the referent of the demonstrative in question.

* Corollary:

> *If the demonstrative 'this' in the situation described refers to the same object, regardless of whether one or the other general term is attached to it as a part of the grammatical subject, there is no need to treat either of them, in the logical analysis, otherwise than as a predicate. The predicative analysis is thus vindicated*.

Pairs of sentences having the same syntactical structure as (1) and (2), i.e., ones containing sentences differing from each other in the order of general terms. can be, prima facie, analysed analogously regardless of whether the general terms in question have the same extension or not.

Pairs of sentences like

(12) This is a horse and a steed

and

(13) This is a steed and a horse,

can be considered along the similar lines. On the other hand, the sentences containing attributives like

(14) This horse is a big steed

and

(15) This steed is a big horse,

don't seem to express the same thought, for there are standards according to which something can be a big steed but not a big horse, and vice versa.

Examples such as these show that not all pairs of sentences containing the same general terms, and the same demonstrative (type), express the same thought. In light of this, the conclusion of the above argument, i.e., (vi), which claims the validity of 'the Principle of Parity of General Terms', can be generalized only to some extent; in other words, 'the Principle of Parity of General Terms' has a limited application:

(vii) The general term attached to a demonstrative as a part of the grammatical subject, or the one playing the role of a grammatical predicate of a demonstrative sentence, need not have the semantical priority in question over the other general terms in the sentence.

Concluding Remarks

The account presented in this paper at least shows that in some demonstrative sentences all the general terms are semantically on a par as regards the singling out of the referent of the given demonstrative for the hearer. On the other hand, it at most shows that in every demonstrative sentence these terms have this feature, but due to..........................(the long list follows) reasons, in cases like.................., the shifting of the position of these terms influences the change of the thought expressed, so that the proofs analoguous to (i) - (vi) don't apply to these cases, for the analogues of (i) are in these cases false. As I find myself ignorant about all the factors that should be included into the foregoing list, I will accept the former, weaker, option. Remaining within its limits does not, however, make my account worthless or insignificant: even so limited, this account, I believe, points out one important way in which the referent of a demonstrative is being singled out.[8]

NOTES

1. For the sake of simplicity I will assume, contrary to Burge (/1/), that linguistic meaning is an ingredient of the Fregean sense.

2. This and related notions are vague, but I can't deal here with the question of their vagueness. I take this notion in its pre-theoretical sense. The same holds for the notions 'on the same occasion' and 'in the same situation'.

3. In accordance with the framework within which this paper proceeds, the corner quotes should be taken to denote the Fregean sense expressed by the symbols they enclose.

4. It should be noticed that according to original Frege's sense theory there is, prima facie, no room for belief de re, for every belief is completely conceptual. I accept the weaker sense theory as suggested by Evans (/2/3), which allows for de re attitudes towards the referent of a demonstrative. See also Burge's discussion (/1/) of what he calls the second function of Frege's sense.

   Given that my argument in the text rests on de dicto, not de re account, there is no need to discuss the weaker sense theory here .

5. According to the Kripkean theory, the difference between (5) and (6) is,

prima facie, accounted for by the distinction between a proper name ('nine') and a definite description ('Smith's favourite number').

6. In Geach's and Stoothoff's translation of 'Der Gedanke' the words which Frege mentions as having the same sense but different colouring, i.e., 'Pferd', 'Ross', 'Gaul', and 'Mahre', are translated as 'horse', 'nag', 'prad' and 'steed'. This last word (at least) differs from 'horse' not only in colouring but also in sense; namely, for two expressions to have the same sense it is necessary, though not sufficient, to have the same extension, and this is not the case with 'horse' and 'steed'. See /5a/ and /5b/.

7. If two expressions have the same sense it is understood that the relevant extra-linguistic components must be shared by them.

8. I want to express my gratitude to Barry Taylor for his helpful comments on the earlier drafts of this paper.

REFERENCES

1. Burge, Tyler. 'Belief De Re', Journal of Philosophy 74 (1977), 338-362.

2. Evans, Gareth. 'Understanding Demonstratives', Collected Papers Oxford, 1985, 291-321.

3. Evans, Gareth. Varieties of Reference, Oxford, 1982.

4. Frege, Gottlob. 'Sense and Reference', Copy & Gould, ed's., Contemporary Philosophical Logic, St. Martin's Press, New York, 1978, 65-83.

5a. Frege, Gottlob. 'Thoughts', Logical Investigations, Basil Blackwell, Oxford, 1977, 1-30.

5b. Frege, Gottlob.'Der Gedanke Eine Logische Untersuchung', Logische Untersuchungen, Gettingen, 1976, 30-53

6. Taylor, Barry. 'Truth-Theory for Indexical Languages', Platts, Mark, ed., Reference, Truth and Reality, Routledge and Kegan Paul, 1980, 182-198.

PHILOSOPHY AND PAIN RESEARCH

NIKOLA GRAHEK

The field of pain research, which has witnessed rapid development in recent years, comprises numerous psychological, physiological and clinical data gathered from the observation and experimental inquiries of various aspects of pain. As these data point to the complexity of the phenomenon of pain as well as to many of its perflexing features, they have to be taken into account by any acceptable general theory of the nature and function of pain. To put it in more precise terms, it is the set of the relevant psychological, physiological and clinical data which is considered by scientists to make a puzzling phenomenon and which is recognised by them to be the test of the descriptive, explanatory and predictive power of any broadly conceived theoretical model of pain. For instance, any adequate theory of pain should take cognizance of and be able to explain at least the following clinical data:

(1) the bizzare coupling of the amount of injury to the amount of pain.
(2) the triggering of pain by innocuous stimuli to normal tissue (tenderness, referred pain and trigeminal neuralgia).
(3) the sensation of pain in missing limbs (phantom pain).
(4) the failure of narcotics to influence some pains (brachial plexus avulsions).
(5) the success and failure of the various counter stimulation methods (acupuncture, transcutaneous stimulation).[1]

If data of this type play such an important role in scientific theorizing on pain phenomena, it was only to be expected that revisionary or more radical eliminative demands would be made with respect to those pain concepts which were found to be at odds with them. These demands actually came out of the conceptual disputes, during the past two decades, between the scientists working in various areas of pain research. In approximately the same period, revisionary or elininative demands concerning our pain concepts and discourse were also laid down by philosophers who were claiming that our sensations in general should be described and explained in physicalistic or at least in nonmentalistic terms. But upon closer examination, it may well turn out that their revisionary or eliminative demands, as well as their improvement and replacement recommendations, diverge considerably from, or even run counter to, those put forward by scientists.

A. Pavković (ed.), Contemporary Yugoslav Philosophy: The Analytic Approach, 85–96.

In order to prove this point, I will first examine those pain concepts which are almost unanimously thought by scientists to be in need of thorough revision or even outright displacement. Then I will consider the target of the revisionary or eliminative demands coming from philosophers of materialistic or antimentalistic turn of mind. The comparison of scientific and philosophical standpoints with regard to the subject matter of revision and elimination will have the following unexpected outcome: that exactly those pain concepts which are held by scientists to deserve major reconfiguration or complete abandonment are proposed by philosophers as the most appropriate to replace our ordinary discourse on pain or to improve our understanding of the real nature and function of pain. This outcome should be very surprising and disconcerting to philosophers because they claim that their revisionary or eliminative demands rely on the theoretical and experimental results of the scientific research on pain phenomena. Their general attitude towards scientific cognitive achievements is, as Rorty has pointed out, best captured by the slogan coined by Sellars: "science is the measure of all things."[2] But if the results of my comparative study prove to be valid, they will clearly show that the revisionary or eliminative demands of the philosophers, and especially their improvement and replacement recommendations, are not supported by scientific research into pain; that they were not put forward under the conceptual and empirical constraints arising from the field of pain research. So it seems that their proper source or motive is rather to be sought in the ontological, epistemological and semantical requirements of the general materialistic or anti-mentalistic doctrine which philosophers want to defend. How, then, one should explain the fact that they so much insist on their reliance on scientific or, to be more precise, neuroscientific research on pain phenomena? The right account is to be found in their claim that their revisionary or eliminative demands do not stem from the existing, "immature" neurosciences but will be supported in the future by the results of the advanced or mature neurosciences. But it may well happen that the idea of the advanced or mature neurosciences, which philosophers are invoking it here, is a purely normative one: namely, that only those scientific theories or conceptual frameworks which perfectly match the requirements of the strictly materialistic or antimentalistic philosophy will be recognised as advanced or mature. This seems to be confirmed by the fact that many philosophers appear to take no notice of the theoretical and empirical achievements of the "immature" scientific research on pain phenomena: research which has, during the past two decades, made major breakthrough in our knowledge and understanding of psychological, physiological and clinical aspects of pain. By ignoring these achievements and their conceptual implications philosophers were led to recommend the adoption of empirically inadequate, inherently confused or too simplistic concepts of pain. So the proper lesson to be learned is that science, as far as pain phenomena are concerned, has to be taken seriously even in its "immature" stage. Otherwise, there is a good chance that many empirical and conceptual mistakes will be committed while philosophers are waiting for the neurosciences to reach their nature or advanced stage; and the trouble is that the mature or advanced stage, postulated by philosophers, may in fact never be reached.

As far as the position of the scientists working in the field of pain research is concerned, they nowadays generally agree that there are three concepts or conceptions of pain which should be radically revised or literally banished from our discourse on people in pain:

1. the concept of pain by which it is postulated that there are specific pain receptors, pain fibers, pain pathways and, finally a specific pain center in the brain;

2. the conception of pain by which pain is considered to be invariably linked to injury or tissue damage, proportional to its intensity to the intensity of the stimulus, and having only a useful biological function;

3. the concept of pain which represents it as a primary, simple sensation.

I will in turn consider each one of these three concepts or conceptions of pain, presenting the main factual and theoretical reasons which have led scientists to demand their revision or elimination, and giving an outline of those concepts or models of pain which were suggested as modifications or replacements.

1. Though the concept of specific pain receptors, pain fibers, etc. has been used for a long time in medical and physiological textbooks, it is thought to be inadequate or mistaken for three reasons: first, because it presupposes that only one type or group of neurons can give rise to pain; second, because it implies that there is a direct, invariant relationship between the activity of a neural structure and some corresponding sensation; the third reason is that it postulates the existence of a fixed, straight-through channel projecting from the receptor via pain fibers and a pain pathway to a pain center in the brain.

It is presumably for these three reasons that the IASP Sub-committee on Taxonomy has explicitly given the following recommendation: "Avoid use of terms like pain receptor, pain pathway, etc."[3] Now,this is not to deny that there are nerve fibers which preferentially respond to noxious or potentially noxious stimulation and which play a special role in pain perception. Such fibers were identified within the range of high-threshold A-delta myelinated and C unmyelinated afferent fibers.[4] But, as the relevant clinical evidence shows, "severe pain can be triggered by slight tactile and weak thermal stimuli in various pathological state"[5] and "90% of patients with brachial root avulsions suffer severe pain in the absence of afferents let alone impulses."[6] On the other hand,pain can be absent in the cases of severe injury or tissue damage; furthermore, it has been experimentally demonstrated that the pain threshold does not coincide with the threshold for the firing of the fibers and that there is a mismatch between the duration of pain and the time course of the afferent barrage.[7] So to classify a cer- tain group of nerve fibers as pain fibers is to confuse the stimulation and electrochemical activity of the afferents with the sensory outcome or experience. To label them nociceptive fibers is, of the contrary, to keep clear the distinction between the notion of physiological specialisation and the notion of psychological specificity; for the term "nociceptive fibre" is a purely physiological term designating a fibre which preferentially responds to noxious or potentially noxious stimulation. As the same holds for other neural structures involved in pain perception, we should only use labels like "nociceptors", "nociceptive pathways", and so on. To put this point in more general terms: "Activity induced in the nociceptor and nociceptive pathways by a noxious stimulus is not pain, which is always a psychological state, even though we may well apprec-iate that pain most often has a proximate physical cause".[8] Once the equivalence 'nociceptor activity = pain' has been proved to be untenable, the model of the rigid, straight-through neural mechanism leading from the receptors to the centre in the brain will also have to be abandoned. Namely, the relevant physiological evidence unequivocally indicates that the nerve impulses of the nociceptive afferents are exposed to the influence of the descending inhibitory mechanisms, modulation or gating at all levels of the CNS and the influence of the activity of the other peripheral fibers: "The brain is not a passive receiver of sensory nerve impulses but a dynamic processing system that involves continuous modulation and gating of inputs so that the eventual perceptual experience is rarely the momentary ring of an alarm bell but more often a prolonged, complex experience that fluctuates in intensity, quality and spatial distribution."[9]

2. Nobody would be prepared to deny that pain is closely related to injury and that it has a biologically useful function. For pain is, indeed, standardly caused by injury and certainly serves as a kind of warning mechanism conditioning us to avoid injurious objects or situations. But to insist that pain is inextricably bound to injury disregards the relevant empirical counter evidence and blurs important distinctions between different types of pain. Furthermore, to stress only the beneficient function of pain ignores the cases where it no longer plays any biologically useful purpose. That injury, even a severe one, is not regularly followed by pain has been proved many times by clinical evidence: for instance, 37% of the patients who have arrived at an emergency clinic with variety of serious injuries "stated that they did not feel pain at the time of injury. The majority of these patients reported onset of pain within an hour of injury, although the delays were as long as 9h or more in some patients."[10]

The clinical data also suggests that there is mismatch between the amount of injury and amount of pain (the case of renal coliculi pain)[11] and that the site of injury does not coincide with the site of pain (the cases of referred or transferred pain). On the other hand, there are cases of pain where no injury is involved: for example, it is reported that "no apparent injury can be detected in 70% of people who suffer from chronic low back pain".[12] Besides, in many cases of chronic pain intensive, debilitating pain persists for months or even years after all possible healing has been completed; and this type of pain, in contrast to acute pain, "is no longer simply a symptom of injury or disease. It becomes a pain syndrome - a medical problem in its own right which requires urgent attention".[13]. What all this shows is that the link between pain and injury is a variable one: injury may occur without pain as well as pain without injury. So to claim that pain equals injury or to define pain as harmful stimulus "is to confuse... the physical event with the complex psychological process".[14] Not to make the distinction between acute and chronic pain just because the sensory outcome or experience can be described by the single word 'pain' is to think mistakenly "that in each of these different categories we are dealing with similar mechanisms and that observations and conclusion in one group are transferable to another".[15] The consideration of the cases of chronic or persistent pain is also important because it reveals that in some instances pain can be of no biological value. People who are suffering from the persistent pain gain nothing from it; on the contrary, "pain becomes debilitating and often produces severe depression... Patients are beset with the sense of helplessness, hopelessness and meaninglessness. The pain becomes evil - it is intolerable and serves no useful function".[16] This is why chronic pain deserves special clinical approach and understanding: it has to be taken as a separate medical entity, distinct from acute pain, and one must be aware that in many cases the treatment effective for acute pain will not be appropriate to chronic pain[17] Before I close my examination of the biological value of pain one more point is in order, and it has to do with its function as a reliable warning mechanism. Namely there are serious diseases or even those with terminal consequences where pain comes only when it is too late and presents an additional distress. So we should not exaggerate the role of pain as a warning mechanism reliably signalling that there is something wrong with the body. As Leriche has with bitterness remarked: "Defence reaction? Fortunate warning? But as a matter of fact, the majority of diseases, even the most serious, attack us without warning. When pain develops ... it is too late... The pain has only made more depressing and more sad a situation already long lost ...".[18]

3. With the development of sensory physiology and psycho-physics scientists mainly concentrated on the examination of the sensory aspect or dimension of pain; that is, pain was conceived as the primary, simple sensation distinguished by its unique modality or quality and varying only in intensity. Within this conceptual and research framework the measurement of pain was confined to its quantitative dimension: the laboratory measurements were concerned with the correlation between the intensity of pain and the intensity of stimulus, while the clinical measurements of different kinds of pain and the efficacy of various treatments were made only along the intensity scale. But to emphasize only the sensory dimension of pain is to neglect two other essential dimensions of pain; to regard it as a simple sensation having a single, unique quality is to ignore the qualitative variety of pain; finally, to measure exclusively the intensity of pain is to leave out of scientific consideration the more subjective features of pain. For pain is not only characterised by sensory factors but has a strong affective-motivational dimension and an important cognitive-evaluative aspect. Besides,

within the sensory dimension of pain there is a whole range of different qualitative features which distinguish various types of pain. So pain is rather to be considered as a complex experience comprising several dimensions and varying not only in intensity but also in quality,[19] The multi dimensional character of pain as well as its qualitative variety is mirrored in our language of pain, i.e. the set of verbal descriptors by which we describe the way we experience and react to pain. When these verbal descriptors were subjected to systematic analysis, it has been found out that they can be grouped along the sensory, affective and evaluative dimensions and that an important tool for the measurement of subjective components of pain has been devised.[20] The inherently affective character of pain distinguishes it from such perceptual experiences as seeing and hearing; but the fact that pain is susceptible to mental mediation - that it is influenced by past experience, cultural factors and the entire situation as conceived by the subject - is something that makes it close to them. This is the second reason why pain should not be conceived as a simple sensation but rather as a kind of perceptual experience.[21]

Now that the eliminative and revisionary demands of scientists, as well as their replacement and improvement recommendations, have been presented I will examine those that were put forward by philosophers and show wherein lie the main points of disagreement. The eliminative demands of the philosophers who advocate the general materialistic position will be first considered and assessed in the light of the replacement claims coming from the field of pain research. Then the revisionary suggestions proposed by philosophers of the antimentalistic turn of mind will be discussed and compared to those that were laid down by scientists under the constraints of the relevant psychological, physiological and clinical evidence.

1. The eliminative demands of the materialists are directed towards our pain discourse in general; that is, towards the very concept or language of pain which we commonly use to describe the internal states and explain the behaviour of the persons who are burned, cut or pricked, are yelling or writhing, removing quickly their hands from hot objects or making strong verbal complaints. According to eliminative materialists our ordinary pain discourse is to be replaced by the descriptions and explanations that would be set in the terminology of advanced neurosciences and would solely refer to the brain states or processes. If under this replacement conditions no descriptive, explanatory and predictive power would be lost, that would prove that we were strongly postulating the existence of some irreducible mental states or processes in order to describe and explain our internal states and behaviour; that there are actually no such states or processes as there are actually no such things as demons or witches.[22] This eliminative scenario is supported by the imaginary stories of the societies or linguistic communities whose members would, instead of using pain terminology, refer only to the firing of the C or A-delta fibers when describing and explaining the internal states and behaviour of somebody who has burned, cut or pricked his hand.[23] The lesson to be learned from these stories is that one day, when the

neurosciences reach their mature or advanced stage, we could start to behave like the members of the imaginary linguistic community, and that actually nothing would be lost. The more substantial argument in favour of the position of eliminative materialism is to be found in the claim that the ordinary concept of pain belongs to the categorial framework of the so-called folk psychology which will, once the neurosciences are completely developed, turn out to be confused, incorrect and poor in its descriptive, explanatory and predictive power. As such, it will be replaced by the superior conceptual framework of mature or advanced neurosciences.[24] So we should not be dogmatic and should give a chance to the new framework to develop if the categorial integrity, adequacy and power of the old framework is to be properly assessed. One might also expect that our internal discriminative powers will be enhanced, once the transactions and intricacies of our inner life are described in the language of the more penetrating conceptual framework of mature neurosciences; and for this to happen, we have to be patient and give time to the new language to develop.

If our ordinary concept and language of pain is considered to be mentalistic and if it represents pain as a subjective experience or state to be described in psychological terms, it seems that the scientists working in the field of pain research would find no major reason to discard it as confused or inadequate. For as I have already pointed out, the IASP Subcommittee on Taxonomy has explicitly remarked that we should view pain as a psychological state not to be confused with the activity of the relevant neural structure or noxious stimulation. Our ordinary concept of pain might be deficient, not sufficiently precise or completely coherent, but it certainly is not inherently confused or basically incorrect because it makes a clear distinction between the sensory outcome or experience and the stimulation and activity of some corresponding neural structure; between the physical cause or event and the psychological state or process.[25] If that is the case, it is almost certain that the scientists working in the field of pain research would disagree with the materialists on the question where the proper target for eliminative demands is to be found. They would also disagree with them on the issue of the descriptive and predictive power of our ordinary language of pain which is set within the categorial framework of the folk or, to use a less depreciatory term, common-sense psychology. As I have already remarked, when this language was subjected to systematic analysis it was shown that its pain descriptors could be classified into three groups along the sensory, affective and evaluative dimensions of pain. Each of these three groups was subdivided into different subgroups whose members were ranked along the intensity scale. When this systematically organised linguistic structure was put in the form of the questionnaire and subjected to experimental testing, it has turned out that it has a great descriptive, discriminative, predictive and diagnostic power. For instance, it has been demonstrated that major types of pain are characterised by distinct constellations of words; that our pain descriptors are sufficiently sensitive to detect the difference between the "organic" and "functional" low back pain; that various pain syndromes can be predicted and diagnosed on the basis of the words alone.[26] Now, this clearly shows how wrong we would be if we were to advocate the eliminative position towards our ordinary pain discourse. And this position is, indeed, intellectually unacceptable; for on the one hand, we are expected to be open-minded, benevolent and patient when the new categorial framework of mature neurosciences is concerned; on the other hand, we are supposed to regard the current categorial framework as poor, incorrect and confused even before it has been subjected to systematic analysis and its descriptive, discriminative and predictive power properly tested.

There is, as I have shown, no doubt that scientists would regard the eliminative demands of materialists to be completely misdirected; but it seems that they would be even more opposed to their replacement recommendations. In the imaginary stories which are supposed to support the eliminative scenario of materialists, the members of the scientifically advanced linguistic community will not use the concept of pain but will only talk about the activity of some corresponding neural structure. For instance, when a member of such community is burned his peers will soley refer to the firing of the C fibers in order to describe his internal state and explain his subsequent behaviour. But there is enough empirical evidence to prove that the firing of the C fibers (or of the A-delta fibers) is neither necessary nor sufficient for pain; that the pain threshold does not coincide with the threshold for the firing of the fibers, and that there is a mismatch between the onset and duration of pain and the onset and time course of the firing of the fibers. Let us suppose, for the sake of the argument, that somebody's hand is stimulated by the painful heat stimulus for 15 seconds. It might well happen, as the experiments have demonstrated, that the A-delta fibers will fire in the rapid high-frequency burst only for 5 seconds aned that the pain will be reported exactly when the firing has began to decrease.[27] So it follows that a factual mistake would have been made if the reference to pain had been replaced with the talk about the firing of the A-delta fibers. But this example obviously has wider implications, because it shows that the activity of the nerve fibers should not in general be confused with the sensory outcome or experience; that we have to make a clear distinction between the notion of the physiological specialisation and the notion of the psychological specificity. Besides, if we were to replace our pain discourse with the descriptions which referred solely to the activity of nerve fibers, we would postulate a too simplistic model of the neural mechanism subserving our pain perception. I have only recapitulated the main reasons which have led scientists to demand the elimination of the concept of specific pain receptors, pain fibers, and so on, as empirically inadequate, confused and too simplistic. These same reasons would lead them to reject as mistaken and misleading the replacement recommendations put forward by materialists; and the point is that they would be right if the relevant psychological, physiological and clinical evidence is taken into account.

2. Philosophers who argue for the antimentalistic interpretation of our sensation discourse are claiming that pain should be conceived as a form of sense perception. According to the perceptual view or model of pain, to feel

pain is to perceive or to be aware of the bodily disturbance or disorder.[28] If this model proved to be correct or adequate the door would be wide open to the objectification of pain. Namely, if the feeling of pain is to be understood only as the awareness of some objective state of affairs of our body, there would be no need to postulate some irreducible felt quality, sense-content or mental particular in order to account for its nature or status. From the philosophical point of view, that would mean that many intractable problems concerning the privacy and subjectivity of pain could be easily solved or resolved. Some philosophers who defend the perceptual model of pain are also asserting that pain should be exclusively considered as a warning mechanism having only useful biological function.[29] As far as the perceptual model of pain is concerned, it may at first sight appear that we have finally come to the point on which scientists and philosophers would agree. For haven't the scientists themselves claimed that pain should be considered as a kind of perceptual experience? But they have adopted this view for the reasons which are exactly opposite to those which have led philosophers to defend it. They have insisted on the similarity between pain and sense perception not because of its objective character but, on the contrary, because of its subjectivity. What they had in mind was the fact that pain, like seeing and hearing, is susceptible to mental mediation; that it is, like all perceptions individual "and modified by degrees of attention, emotional states and the conditioning influence of past experience".[30] On the other hand, scientists were cautious enough to point out that in other respects pain is quite different from sense perception; the difference being reflected in the fact that pain has a strong affective and motivational aspect.

This dimension of pain was for a long time neglected in favour of its sensory dimension; but to disregard this latter dimension of pain by giving it unity through the stimulus would mean to ignore the qualitative character of pain and its qualitative variety. Besides, the relevant clinical evidence has shown that the link between pain and injury or tissue damage is a highly variable one. Therefore, to define pain through the stimulus is to confuse the physical event and the complex psychological process as well as to ignore the difference between acute and chronic pain. The same would happen if we were to stress only the beneficient role of pain and insist on its function as a reliable warning mechanism. What all this shows is that the revisionary suggestions coming from philosophers considerably diverge from, or even run counter to, those put forward by scientists. While philosophers are tightening the link between pain and injury, scientists are loosening it; while the former are giving an objectivistic interpretation of pain, the latter are insisting on its subjectivity; while philosophers are trying to give unity to pain through the stimulus, scientists are pointing to its qualitative variety and casual diversity; finally, while philosophers are only stressing the beneficient role of pain, scientists are making us aware of the cases where pain has lost any biological function or value.

NOTES

1 P.D. Wall and C.J. Woolf, "What We Don't Know About Pain", Nature, 287 (1980), p.186.

2 R. Rorty, Philosophy and the Mirror of Nature, Princeton: Princeton University Press, 1979, p. 124.

3 IASP (International Association for the Study of Pain) Subcommittee on Taxonomy, "Pain Terms: A List With Definitions and Notes On Usage", Pain, 6 (1979), p. 251.

4 An overview of the types if receptors, fibres and central cells which have been identified as preferentially responding to noxious stimulation and playing a special role in pain perception was given by D.D. Price and R. Dubner. See D.D. Price and R. Dubner, "Neurons That Subserve the Sensory-Discriminative Aspects of Pain",Pain, 3 (1977 pp.307-338. The crucial step in the identification of such neural structures the introduction of the technique of intraneural microrecording and microstimulation.The thorough presentation and assessment of this technique and its results is to be found in P.D.Wall and S.B.McMahon "Microneuronography and its Relation to Perceived Sensation. A Critical Review",Pain, 21 (1985), pp. 209-229.

5 M. Zimmerman et al., "Recurrent Persistent Pain: Mechanisms and Models". In H.W. Kosterlitz and L.Y. Terenius (eds.), Pain and Society, Dahlem Konferenzen 1980, Weinheim: Verlag Chemie GmbH, 1980, p.369.

6 P.D. Wall and S.B. McMahon,"The Relationship of Perceived Pain to Afferent Nerve Impulses", Trends in Neurosciences, June (1986) p. 255.

7 Cf. P.D. Wall and S.B. McMahon,"The Relationship of Perceived Pain to Afferent Impulses", pp 254-255.

8 IASP Subcommittee on Taxonomy. op cit., p. 250.

9 R. Melzack,"Neuropsychological Basis of Pain Measurement". In L. Kruger and J.C. Liebeskind (eds.), Advances in Pain Research and Therapy, New York: Raven Press, 1984, p. 324.

10 R. Melzack, P.D. Wall and T.C. Ty, "Acute Pain in an Emergency Clinic: Latency of Onset and Descriptors Patterns Related to Different Injuries", Pain, 14 (1982), p. 33.

11 Cf. R. Melzack and P. D. Wall, The Challenge of Pain, Harmondswort: Penguin Books, 1982, pp. 20-21.

12 Ibid., p.15.

13 Ibid., p. 55.

14 Ibid., p. 70.

15 W. Noordenbos, "Prologue to Textbook of Pain". In P.D. Wall and R. Melzack (eds.), Textbook of Pain, Edinburgh: Churchill Livingstone,1984.

16 R. Melzack and P.D. Wall, The Challenge of Pain, p. 55.

17 Cf. R. Melzack and P.D. Wall, The Challenge of Pain, p. 55.

18 Quoted in R. Melzack and P. D. Wall, The Challenge of Pain, p. 24.

19 R.Melzack, "Neuropsychological Basis of Pain Measurement", pp. 327–329.

20 Cf. R. Melzack and W.S. Torgerson, "On the Language of Pain", Anesthesiology, 34 (1971), pp. 50–59.

21 Cf. R. Melzack, "Neuropsychological Basis of Pain Measurement", p. 324.

22 K. Wilkes, Physicalism, London: Routladge & Kegan Paul 1977, pp. 101–102; R. Rorty, Philosophy and the Mirror of Nature, p. 120.

23 Such science-fiction stories were constructed by K. Wilkes and R. Rorty. See K. Wilkes, op. cit., pp. 101–102 and R. Rorty, op. cit., pp. 70–77.

24 This way of arguing for the position of the eliminative materialism is given by P. Churchland; similar arguments were earlier presented by P. Feyerabend. Cf. P. Churchland, Matter and Consciousness, Cambridge Massachusetts: MIT press, 1984, pp. 43–45.; P. Feyerabend, "Materialism and the Mind-Body Problem", Review of Metaphysics, 65 (1963), pp. 49–66.

25 One could argue along these lines against Dennett's eliminativist contention that our ordinary concept of pain is inherently confused and relying on inconsistent intuitions. Cf. D. Dennett, "Why You Can't make a Computer that Feels Pain", Synthese, 38 (1978), pp. 415–456.

26 Cf. R. Melzack and P.D. Wall, The Challenge of Pain, pp. 66–69. The language of pain with its set of verbal descriptors was first subjected to systematic analysis and classification in 1971 by R. Melzack and W.S. Torgerson (op. cit., pp. 50–59). In 1975 the famous McGill Pain Questionnaire: was formulated by R. Melzack ("The McGill Pain Questionnaire: Major Properties and Scoring Methods", Pain, 1 (1975),

pp. 277–299); later a great number of experimental studies were carried out and the basic structure, reliability and validity of the MPQ has been proved. See C.R. Chapman, K.L. Casey, R. Dubner, K.M. Foley, R.H. Gracely and A.E. Reading, "Pain Measurement: an Overview", Pain, 22 (1985, pp. 21–22.27 Cf. P.D. Wall and S.B. McMahon, "The Relationship of Perceived Pain to Afferent Nerve Impulses", p. 255.

28 Cf. G. Pitcher, "Pain Perception", The Philosophical Review, 79 (1970), pp. 368–393; D.M. Armstrong, A Materialist Theory of the Mind, London: Routladge & Kegan Paul, 1968, pp. 306–320; N. Fleming, "The Objectivity of Pain", Mind, 85 (1976), pp. 522–541; K. Wilkes, op. cit., pp. 97–103.

29 Cf. K. Wilkes, op. cit., pp. 100–101.

30 K.W. Livingston, "What is Pain?", Scientific American 196 (1953), p. 60.

A DEFENCE OF RIGHTS–DUTIES CORRELATIVISM

MANE HAJDIN

1. INTRODUCTION

The thesis that I am going to defend in this paper is a thesis that, once upon a time, seemed uncontroversial to many moral philosophers but that is nowadays decidedly out of fashion. This is the thesis that moral claims made in terms of rights (right–ascriptions) are equivalent to moral claims made in terms of duties (duty–ascriptions)[1]. I shall refer to this thesis as 'rights–duties correlativism' or simply 'correlativism'[2]. I shall call the denial of this thesis anti–correlativism.

Throughout this paper I shall be presenting ascriptions of duties in the form:

$$\underline{A} \text{ has a duty to } \underline{P} \text{ that } \underline{A\Phi P}$$

where $\underline{A}$ is a place–holder for an expression referring to a moral agent, $\underline{P}$ for an expression referring to a moral patient[3] and Φ for a description of the relation that is morally required to obtain between the two (i.e. of the way in which the moral agent is supposed to behave towards the moral patient). When we are making a singular moral claim, $\underline{A}$ and $\underline{P}$ can be replaced by proper names of the relevant moral agent and the patient; at the more general level of moral rules and principles they are replaced by variables (bound by appropriate universal quantifiers) that range over the set of moral agents and the set of moral patients, respectively.

A moral patient is someone whose being affected by the behavior of some moral agent is directly relevant for the moral evaluation of that behavior. This is someone of whom it can be meaningfully said that he has been wronged by a violation of some relevant duty. In everyday situations, moral claims are often formulated in terms of duties without explicitly mentioning moral patients, e.g. 'You have a duty not to take that money' or 'Everyone has a duty to abstain from committing rape'. However when we ask someone who has members of the set of moral patients will have to emerge somewhere in his reply[4]. For example, a justification of the claim that I have a duty not to take certain money will presumably have to mention the owner of the money, while a justification of the claim about rape will have to mention how a rape affects the person raped.[5] Once we know whose being affected by the behavior

A. Pavković (ed.), Contemporary Yugoslav Philosophy: The Analytic Approach, 97–119.

in question (and in what way) is directly relevant for its moral evaluation we can reformulate the claim so as to make that explicit and this brings us to the 'canonical' form of duty-ascriptions that I have given above.

We say that the duty has been obeyed just in case the proposition that replaces AΦP is true, and that it has been violated just in case that proposition is false. Typically it will be impossible to replace by a single word of ordinary language, and we shall frequently encounter formulations of highly complex internal structure at the place of Φ.

The thesis of correlativism is that ascriptions of duties, when they are formulated in this way are equivalent to (and can be regarded as analyses of) ascriptions of rights that have the form:

P has a right (against A) that AΦP

The majority of anti-correlativists concede that such ascriptions of rights imply ascriptions of duties, but deny the converse because according to them the meaning of the word 'right' is such that the content of an ascription of a right is richer than the content of the ascription of the corresponding duty.

This paper is addressed only to anti-correlativists who belong to this majority. Consequently it takes for granted that we can move from right-ascriptions to duty-ascriptions and concentrates on showing that we can equally well move from duty-ascriptions to right-ascriptions (or, in other words, that replacing a right-ascription by a suitable duty-ascription does not involve any loss in meaning).

The paper does not attempt to argue directly with the minority of anti-correlativists who would deny that there is implication in either direction between right-ascriptions and duty-ascriptions.[6]. However, some of the points that I am going to make (especially in section 2) will have obvious consequences that will go against this view.

According to correlativism, as I have presented it here, being a moral patient for a particular duty is both necessary and sufficient for having the correlative right and it is not a feature of the concept of a moral patient that the duties that are owed to them are necessarily beneficial to them (although, as a matter of fact, they usually are). Correlativism therefore differs from Bentham's theory of rights (although both theories spring from similar theoretical motivation) in one respect: a correlativist, unlike Bentham, doesn't claim that respect for a right is necessarily beneficial to the holder of that right. In this way correlativism avoids some of the objections that were raised against the theory according to which being a right-holder is just being a beneficiary of a duty.[7]

Needless to say, if we choose to qualify our duty-ascriptions as *prima facie* or *sans phrase* ('all things considered') then we have to qualify the right-ascriptions in the same way; and we should expect a duty-ascription that is qualified in some such way to be equivalent only to a right-ascription that is qualified in the same way.

It should perhaps also be mentioned that both correlativism and various versions of anti-correlativism apply to only one of different senses that 'a right' has in moral contexts. In particular, we should note that there is another, perfectly legitimate, sense of the word such that saying that someone has a moral right in that sense to do something, is equivalent to saying that he does not have a duty not to do it. In order to avoid confusion, I shall follow the practice of using the expression 'liberty-right' when I need to invoke that notion (rather than simply 'right').

2. HOW TO TALK ABOUT RIGHTS

It may now be objected that my formulation of the issue between correlativism and anti-correlativism doesn't do justice to what various philosophers who hold these views have in mind because they are interested in providing a general account of right-ascriptions as we find them in actual moral discourse and my 'canonical' form of right ascriptions doesn't do justice to many of right-ascriptions that we encounter every day.

At first glance it really doesn't seem that the word 'a right' is, in virtue of its meaning, restricted to contexts of the form '*P* has a right (against *A*) that *AΦP*. Some uses of 'a right' may fit the model easily (e.g. 'Koviljka has a right that Janez pays her five thousand dinars') but many seem to be quite different. For example, we often speak about having rights to life, liberty, property, free speech, education, travel and so on.

I want to claim that each such expression, in so far as it has any definite meaning at all, *can be* replaced without loss by an expression of the form '*P* has a right (against *A*) that *AΦP*'. Moreover I want to claim that, for the purposes of a serious and fruitful philosophical discussion of rights, each of them *should* be so replaced.

In order to support these claims I am now putting forward the following principle: we don't understand the meaning of an ascription of a certain right to someone, unless we know what counts as violating (infringing) the right (and so what counts as respecting it). In other words, in order to know what a particular right is all about, we have to know what is the case when it is respected (or violated).[8]

This principle should be uncontroversial as it merely applies to right-ascriptions the reasoning that can be encountered in the well-known principles that one can't understand an assertion without knowing what is the case when it is confirmed or falsified, that a necessary condition for understanding an order is to know what constitutes obedience or disobedience to it, or that a promise can't be understood without knowing what counts as keeping or breaking it.

When a right is ascribed to someone by an expression of the form 'P has a right (against A) that AΦP' then it is very easy to see what its conditions of being respected are, and so what the whole ascription means. Obviously such a right is respected just in case that the relation that is designated by what we put at the place of Φ obtains between the appropriate individuals; otherwise it is violated. If someone says 'Koviljka has a right that Janez pays her five thousand dinars by Wednesday', then we understand immediately that a part of his meaning is that the right would be respected just in case that Janez does pay five thousand dinars to Koviljka by Wednesday, and that it would be violated if Janez does not pay five thousand dinars to Koviljka by Wednesday.

How do rights to life, education, etc. fare in the light of our principle about conditions of being respected? What has to be the case for my right to life to be respected? When is that right violated?

Could it be that someone's right to life is respected just in case he lives, and is violated otherwise? If he is about to die would that in and of itself constitute a violation of his right to life? Hardly. I don't know anyone who speaks correct English and who means that when he says that someone has a right to life. Deaths of incurable illnesses are not counted as violations of the right to life (at least in the absence of some special explanation), while clear-cut cases of murder usually are.

Similarly it is obvious that the condition of being respected for someone's right to travel cannot be that he actually travels. This is not what we mean by the 'right to travel'. Very many people have their right to travel respected without ever travelling anywhere.

It seems that we can speak about a violation of a right only when someone is somehow responsible for it. Saying that a right has been violated invites not only the question 'Whose right it is?' but also the question 'Who violated it?' An aspect of the conditions that make the right violated is that someone has violated it. When we say that a right has been violated we needn't be able to idenfity the violator immediately but we are implying that there is a violator, and that he is identifiable in principle.[9]

If violation of a right is always a violation of someone's right by someone else, then the violation itself can be described as failure of a certain relation to obtain between the two. And if this is what the condition of violation of the right is, then its condition of being respected consists in that same relation obtaining between the two individuals.

This means that, for any right, regardless of the linguistic form that we we might have initially chosen for asription of that right, the conditions of being respected always have the form of a certain relation obtaining between someone (bearer of the correlative duty) and the right-holder.

Now, suppose we take an ascription of a right to P which is not of the form 'P has a right (against A) that AΦP', and we analyze its meaning in terms of the conditions of being respected. By the above reasoning we shall come up with something that can be stated in the form 'that AΦP' as the condition of being respected for that right. We can then try to formulate an ascription of a right that would be different in form from the original one but give the same condition of being respected to the right it ascribes. Whatever other results this exercise might have, there will always be one formulation that will immediately suggest itself, and it will be of the form 'P has a right (against A) that AΦP'. This formulation is as much an ascription of a right as the original one, they both ascribe the right to the individual that is referred to by the expression that we put at the place of P and they both have the same condition of being respected ('that AΦP'). Therefore, they are for most ends and purposes interchangeable.

The preceding paragraph has provided us with a recipe for replacing all ascriptions of rights that are not of the form 'P has a right (against A) that AΦP' with ascriptions that have that form. What has been said in the preceding section can now be applied to all the ascriptions of rights that may result out of applications of this recipe.

The reader should note that I am not asking him to replace the formulations that he is used to with the ones that I am proposing, in order that my analysis may be applied to them. There are quite independent reasons for preferring my 'canonical' form of right-ascriptions.

Ascriptions of that form make it more obvious what their conditions of being respected are. Their meaning is therefore much clearer. Moreove, ascriptions of rights that don't have that form tend to be highly ambiguous[10] and these ambiguities have proven to be pernicious to moral argumentation.

Take, for example, the claim that we all have the 'right to life'. What does it mean? What are its conditions of being respected? To begin with, most people would agree that at least a part of these conditions is that others abstain from murdering us. But what else is included? Do the conditions include others helping those of us whose lives are in danger? Do they include providing us with everything we need to sustain life, or even someone's doing everything that is necessary to minimize the risks to our lives? Does the right to life entail the right to medical care (and if it does, what type and what amount of medical care)? Isn't the right to life already respected if the state takes some reasonable steps towards achieving some (or all) of the above conditions (e.g. by enacting appropriate laws and providing institutions that will enforce them)?

The above questions have no fixed answers. Everyone who uses the phrase 'right to life' has to answer them for himself and tell us what he means by the 'right to life'.

Trying to figure out what the 'right to life' really means is a pointless exercise. It can mean any or all of the above things and perhaps many more. It just happens to be highly ambiguous, and the same holds of all the other phrases that involve rights without following the form that I have advocated.

One might try to patch the things up by distinguishing three or four or seven different senses of such a phrase, but new ambiguities will keep popping up. Moreover a classification of senses of one such phrase would be of little general use as the ambiguities in the meaning of, say, 'right to travel' or 'right to education' will not follow the same pattern as the ambiguities in the 'right to life'. The only way to get rid of such ambiguities once for all is to put all our ascriptions of rights in the form 'P has a right (against A) that AΦP'.

Phrases like the 'right to life', that can be used to express very different claims about rights are pernicious, because they often lead us to forget that each of the different claims (different in the sense of having different conditions of being respected) that they can be used to express needs to be defended (or attacked) by different arguments.[11]

3. ALLEGED COUNTEREXAMPLES TO CORRELATIVISM

Having justified the specific form in which I will be presenting all claims involving rights, I can now go back to the dispute between correlativism and anti-correlativism.

Given that the anti-correlativist believes that the implication between rights and duties holds in one direction only, the most straightforward argument that he could produce for his position would be to come up with a counterexample to its converse. In other words he should come up with an example in which 'P has a right (against A) that AΦP' does not hold, although 'A has a duty to P that AΦP' holds.

In this section I shall discuss three such examples that have frequently been suggested.

3.1 Animals

The first one concerns our duties to animals. It has been alleged that we have some duties to animals, but that the animals don't have the corresponding rights.[12]

Both elements of that alleged counterexample are in fact far from obvious to me, indeed much less obvious than the thesis (correlativism) that the example is intended to refute. Careful ethical investigation seems to be necessary in order to ascertain whether we have any duties in which animals can be moral patients. Maybe we don't. Maybe all our seeming duties to animals are in fact duties to someone else, as was held by many respectable people, including Kant.[13] In that case animals not having rights is not a counterexample to correlativism, and it may well be that someone else who is the real moral patient of our duties concerning animals has the corresponding rights.[14]

Equally careful investigation is necessary to ascertain whether animals have rights: many contemporary philosophers have presented arguments to the effect that they do. Given that both animal rights and duties to animals are controversial, it seems quite possible that they will stand or fall together, which would support correlativism.

In fact it seems that the only people who firmly believe that animals don't have rights although we have duties to them are precisely those people who have already accepted a certain anti-correlativist conception of rights (which is such that it makes it impossible for animals to have rights). That's why the example of animals can't be of much use in converting anyone to anti-correlativism.

3.2 Charity

The second, much more powerful example is that of charity. Most people hold that there is a duty of charity although the potential recipients of charity don't have the right to receive it.

To fill in the details of the example, suppose that by charity we mean simply alms-giving, and that alms are typically given by individual donors directly to individual recipients (i.e. that there is no centralized system for distribution of charity). Suppose moreover that there is no problem with distinguishing those who are really in need of charity from those who would fraudulently present themselves as such.

In that case it seems reasonable to say that an individual beggar who approaches you asking for money really has no right to your money, although you might still have the duty to provide charity and your giving money to him would count towards satisfaction of your duty of charity.

In order to see whether this is really a counterexample to the rights-duties correlativism, we have to see what exactly is the content of your duty of charity. The duty of charity is such that we cacnnot ascertain whether you are satisfying or violating it just by looking at what you are doing at some specific occasion. The duty of charity does not require that you give money to some partiular beggar at a specific time. If I see you refusing to give anything to a particular beggar at a particular time, I am not entitled to conclude that you are neglecting your duty of charity. For all I know you might have given considerable amounts of money to several beggars the day before. Nor can I conclude that you are discharging your duty of charity simply on the basis of seeing you give something to a particular beggar on a particular occasion. Maybe that's the only case in your entire life of your doing such a thing. No particular instance of your giving alms constitutes the satisfaction of your duty of charity, although each such case counts towards satisfaction of that duty. It is only by observing your behavior over a longer period of time, and computing the total amount that you have given in alms over the period, that I can conclude whether you have neglected your duty of charity or not.

It is logically possible to discharge the duty of charity by giving a very small amount of money (perhaps less than a dinar) to every beggar you see, or even every beggar you know about. It is even logically possible to have a modified duty of charity that would require every donor to do precisely that. However that would be a highly unreasonable thing to do, as the costs, to both donors and recipients, of performing such numerous transactions, would far exceed the amount of alms involved. That's why the duty of charity has the form it has, namely the form that allows you to satisfy it by giving alms only to some, randomly chosen, beggars on some, randomly chosen, occasions. This enables donors to give a more reasonable amount of charity in each individual transaction (without affecting the total amount given over a period of time), and thereby reduces the total costs of performing the alms-giving transactions. Note that, if the selection of occasions on which alms are given is really random, and if the frequency of giving and the amounts given are properly adjusted, then each beggar ends up receiving approximately the same amount as he would under a system that would require each donor to give alms to every beggar.

The upshot of this discussion is that you are following your duty of charity if you are performing acts of alms-giving at randomly chosen occasions but with the right frequency. This is the content of the duty of charity. That a particular beggar who approaches you on a given occasion does not have the right that you give him anything on that occasion is, therefore, not a counterexample to the rights-duties correlativism as you don't have a duty that would correspond to it, namely the duty to give him something on that occasion.

Still, if correlativism holds, then we must be able to describe some other right that exists if and only if the duty of charity, as described above, exists. And I am going to claim that indeed there is such a right.

In order to discover what the right is, let's first ask ourselves who is the moral patient of the duty of charity. One possible answer is that it is a duty owed to the poor, taken together, as a group. If we conceive it in that way, then I don't see anything odd with saying that the poor as a group have the right to receive alms. On this account, no particular member of that group has such a right, but the group as a whole has the right to receive a certain portion of the income of each person who is reasonably well-off (the distribution of that amount on particular members of the group being random). However, the tenability of this line of argument is conditional upon it being possible for groups to be moral patients in the strict sense which seems questionable.

Fortunately, the duty of charity can be redescribed as a duty owed to particular beggars. What do you owe to a particular beggar who approaches you? What relation has to hold between you and him if you are not to violate your duty of charity? It can't be the relation of giving him a certain amount of money, as the duty of charity involves randomness. But you do have a duty to include him in the domain of over which you randomly distribute your donations. You don't owe him a donation, but you owe him a chance of a donation. Another way of putting it is to say that the relation that has to obtain beteen you an him is the relation of giving-a-certain-amount-of-money-with-a-certain-probability. And again, I don't see anything odd with saying that as much as you have a duty to stand in that relation to him, he has the right that you stand in that relation to him. The claim that a poor person has the right to a certain probability of receiving a charitable donation seems quite reasonable.

The rights that correspond to the duty of charity are on my account very similar to the rights of lottery-players. Buying a lottery ticket clearly gives you certain rights (both moral and legal). Your rights are not violated when you lose (this is a normal feature of a lottery) but they would be violated if the design of the machine used in drawing numbers made it impossible for your number to win. In other words, as a participant in a lottery you don't have the right to win the prize, but you do have the right to a certain chance of winning the prize.

3.3 Tax-Evasion

The third alleged counterexample to correlativism are the duties to abstain from 'free-riding', and the best example of 'free-riding' is probably tax-evasion. This example has been discussed by Hart [15] and Lyons [16], and I would like to suggest that they failed to find rights correlated with this duty because they were looking in the wrong direction. They assumed that the candidates for holders of rights correlated with the duty to pay taxes are the recipients of various tax-funded services, and then went on to point out that they either have no right to such services or that, when they do, the connection between the right and the duty turns out to be dubious. But it is quite possible to conceive the duty to pay taxes as a moral duty in which the moral patients are not the beneficiaries of the tax-funded services, but other tax-payers.

As an illustration of this point consider a country where there are taxes but no tax-funded services: an all-powerful monarch uses all the taxes to finance his immoral entertainment. I believe that in that country tax-evasion would still be morally wrong, assuming that it is not politically motivated, but that what is behind it is (as in most cases of tax evasion in actual societies) a simple wish of the individual tax-evader to save his money. This is because every act of tax-evasion diminishes the budget of the state and thereby contributes to creating the need that more taxes be collected in the next round of tax-collection. This means that the honest tax-payers are likely to end up paying more taxes than they would if there were no tax-evasion.

This reasoning applies to every society that has taxes and in which the government is not likely to be indifferent to the total amount of taxes that it succeeds in collecting. It is independent of uses to which the budget is put and of whether the government has <u>any</u> rationale for maintaining the budget on one level rather than the other.

Once, tax-evasion is viewed that way, it doesn't seem unreasonable to say that every tax payer has the <u>right</u> that everyone else pays his fair share of taxes (because that ensures that he won't be required to pay more than <u>his</u> fair share of taxes). This right can be viewed as the correlative of the duty to pay taxes (the duty to abstain from tax-evasion).

4. IS ANYTHING MISSING IN THE CORRELATIVIST ANALYSIS?

Alleged counterexamples to correlativism that I discussed in the preceding section were intended to appeal to the intuitions about the concept of rights that we have prior to accepting any particular analysis of it. I hope to have shown that the appeal was unsuccessful and that these examples don't compel us to reject correlativism. In this section I am going to look at some specific suggestions for an analysis of the meaning of 'a right' and see whether any of them presents a real threat to correlativism.

4.1 Limiting the Freedom of Others

In 1955 H.L.A. Hart published an article entitled 'Are There Any Natural Rights?'[17] and in the first part of the article he presented his anti-correlativist account of rights.

Crucial to his argument is the distinction between the duties that can be properly described as duties <u>to</u> someone (or obligations[18]) and duties that can't be so described. He seems [19] to be willing to concede to the correlativist that duties of the first kind (duties-to) imply rights, but he also believes that there are duties of the second kind.

From what I said in section 1, the reader can easily conclude that I believe that all our duties are duties-to. This is because the full justification of any claim that we have a duty to behave in a certain way will always involve reference to someone whose being affected by this behavior is directly relevant for its being a duty. Once we know how to identify that individual (the moral patient) we can describe the duty in question as a duty to him. I don't see anything that would prevent us from saying that it is a duty to him, and Hart's article doesn't contain anything that would convince me of the opposite.

Hart in fact doesn't provide any example of a duty that wouldn't be a duty-to. He mentions the duties concerning the treatment of animals and babies, but not in a way that would amount to such an example.

For the sake of argument we may grant him his assumption that duties to treat animals and babies in a certain way are not duties to animals or babies, but that still wouldn't entail that they are not duties-to at all. It may well be proper to describe them as duties to someone else, and Hart doesn't investigate that possibility at all.[20] The only thing that Hart was entitled to conclude from his treatment of the animals/babies example is that the way in which we formulate our claims about duties often fails to make it obvious to whom these duties are owed (who the moral patients are) and sometimes even misleads us about that. When a given claim about duties makes it difficult to figure out who is intended as the moral patient, then it will also be difficult to formulate the equivalent claim about rights. But although it can be difficult, I don't see any reason to believe that it will ever be impossible.

If we leave this difficulty in Hart's argument aside, and concentrate on the outcome of his analysis of rights, we shall see that he says that having them 'entails having a moral justification for limiting the freedom of another person and for determining how he should act'.[21] What he means is presumably that the holder of a right is morally justified in limiting the freedom of others in certain specific ways, namely the ways that are aimed at ensuring that the right is not violated. In other words, he seems to hold the view that 'P has a right (against A) that AΦP' is equivalent to

> A has a duty to P that AΦP and P is morally justified in limiting A's freedom so as to ensure that AΦP.

This analysis may seem plausible when we look at examples of some duties that are indeed accompanied by such justification. For instance, you have a duty to abstain from murdering me, and if you attempted to violate that duty of yours I would be justified in doing many things that are otherwise wrong (such as injuring or even killing you) in order to ensure that you comply with it. It is then tempting to say that both these elements constitute my right not to be murdered.

But if we look at some other examples, the analysis starts to lose its plausibility. If you borrowed five thousand dinars from me and promised to return them, then I have a right to receive five thousand dinars from you. But if you then threaten to violate that right there are very few things that I am morally justified in doing in order to ensure your compliance. I would be justified in reminding you of your promise, commenting on your unreliability, and after a while perhaps insisting that you pay. Grabbing a valuable piece of your property, capturing and then torturing you, abducting your children or threatening to do any of these things would all be much more efficient ways of exacting five thousand dinars from you, but I wouldn't be morally justified in doing anything like that.

So Hart obviously couldn't have meant that a particular right that I possess gives me moral justification to do anything that might prevent its violation. What I am morally justified in doing varies from case to case, and no amount of thinking about the *concept* of a right, nor about the contents of the proposition that replaces *AΦP* in the particular case will tell us what it is. In order to decide what are the morally appropriate reactions to possible or actual violations of our rights (or duties) we need to bring in moral principles, arguments and considerations that were not pertinent to deciding whether the right (or duty) in question really is a moral right (or duty). These are two totally different moral questions.

Someone might acknowledge that there are ways of limiting freedom of others that are justified in case of some rights that I possess and not justified in others, but still claim that there is a certain minimal reaction to an attempted or actual violation of my right that is justified in case of every right that I possess. I doubt that those who attempt this line of argument will ever be able to describe what that minimal reaction is.[22]

There is admittedly one trivial way in which *P*'s having a right that *AΦP* (and not having waived it) limits *A*'s freedom. It makes it impossible for *A* to behave morally while *AΦP*. But any duty that *A* has is sufficient to limit *A*'s freedom in that way. If that is all that is meant then the second conjunct of Hart's analysis (as I represented it) becomes redundant and the whole analysis collapses into correlativism.

My argument against Hart can now be summarized in the following way. He says that it is part of the meaning of 'a right' that the right holder is morally justified in limiting freedom of others. This can mean that every right justifies limiting freedom of others in (a) all the possible ways that could reduce the probability of a violation of the right, (b) certain specified ways, or (c) at least one way. Interpretation (a) is highly implausible, I don't see how (b) could be made sufficiently specific (while still remaining a part of the analysis of the concept of a right) and (c) is in danger of collapsing into correlativism.

The popularity of theories like Hart's seems to be due to the fact that the term 'right' in the sense of a moral right is very frequently used in arguments in favour of introducing the corresponding legal rights[23] and legal rights typically provide legal justification for certain forms of coercion. The frequency with which moral rights are mentioned in such contexts makes it understandable how it could appear plausible that the moral justification of coercion is somehow part of the meaning of 'moral right'.

4.2 Claiming

Another popular analysis of rights is Feinberg's[24] according to which 'P has a right (against A) that AΦP' is equivalent to:

> A has a duty to P that AΦP and P is in a position to make (valid) claims concerning AΦP.

A powerful criticism of Feinberg's account has been provided in William Nelson's article 'On the Alleged Importance of Moral Rights'[25] that frequently gets overlooked by admirers of Feinberg's analysis. Most of what I am going to say in this section can be regarded as a restatement or elaboration of Nelson's criticisms.

In analyzing the notion of rights in terms of claims, Feinberg purports to be using the word 'claim(ing)' in a special sense. The difference between the loose sense in which the word is often used (and in which it has been used elsewhere in this article) and the sense that Feinberg is interested in can be illustrated by comparing Peter's claiming in some casual conversation that a particular lost object that he has seen in the lost-and-found office is Paula's and should be returned to her with Paula's coming to the lost-and-found office and claiming the object back. Only the second occurrence of the word in the preceding sentence exemplifies the sense that is, according to Feinberg, relevant for analyzing the notion of a right.

To understand Feinberg's analysis it is important to notice that he conceives the activity of making claims (in the sense he is interested in) as an activity governed by constitutive rules. That someone has made a claim is for him an institutional fact. It can be understood only when viewed within the framework of the institution of claiming, just as the fact that the batsman is out can be understood only within the framework of the institution of cricket. Feinberg seems to think that there is a rule to the effect that the right-holders (and right-holders only) are in the position to make claims in morality and that this is the same type of rule as the rule that only an umpire can pronounce the batsman out.

Someone other than an umpire may attempt to go through the motions of pronouncing the batsman out, but according to the rules of the game of cricket, that simply doesn't count as pronouncing him out. According to Feinberg's analysis (as I have presented it), someone may go through the motions of making a moral claim (in the relevant sense of the word) but fail to make a valid claim either because:

1) there is no regulative moral rule imposing the appropriate duty on the appropriate individual; or

2) he is not in the right position to make a claim; i.e. he is not a right-holder in this particular case, either because someone else is a right-holder or because no rights are involved; or perhaps because

3) he violated some other constitutive rule of the institution of claiming.[26]

It seems obvious that Feinberg's vision of claiming (and so of rights) in morality is heavily influenced by the paradigm of legal claiming. In legal contexts there is indeed a clear difference between the two senses of claiming, that were illustrated by the example of the lost object, and claiming in the second sense is indeed a complex institution governed by all sorts of constitutive rules, and it may well be that the notion of legal rights should be analyzed in terms of that institution. Constitutive rules of legal claiming (in the strict sense) specify who will be recognized as a claimant for a certain type of cases, what are the deadlines for making claims, who will hear the claims, and even what size of paper is to be used for making claims. If your purported claim violates any of these rules then from the point of view of the law it is null and void, i.e. it is not a legal claim at all. If Janez has broken a contract he had made with Koviljka, and if Milorad is merely a busybody in no way related to Koviljka, then Koviljka can sue Janez (i.e. make a legal claim) for breach of contract, while Milorad can't. Nothing that Milorad can do <u>counts</u> as making a legal claim in that matter.

Feinberg's analysis assumes that in morality there is also a difference between the two senses of 'claiming' and that there is an institution of moral claiming in the strict sense which is roughly similar to the institution of legal claiming. Is there such an institution?

Suppose Janez is about to violate some <u>moral</u> right of Koviljka's. Milorad, the busybody, enters the scene and starts reminding Janez of his duties to Koviljka, pointing out the immorality of his intended conduct, and exhorting him to respect Koviljka's rights. Many things might be wrong with Milorad's conduct in this example, it might have been rude and even immoral, it might have violated many regulative rules, but we can't say that it violated any constitutive rules. We can't say that what he did doesn't <u>count</u>

as pointing out the immorality of Janez's intended conduct, or as exhorting Janez to respect Koviljka's rights. And it doesn't seem that Koviljka herself could do anything more than that. As has been remarked by Jan Narveson[27], it is not clear how making a claim that one's moral rights be respected could amount to anything more than pointing out the immorality of the conduct that would violate them.

In short, Feinberg hasn't provided us with any reasons to regard claiming in morality an institution similar to legal claiming. To be sure, moral claiming can be regarded as an institution in the way in which any form of speech at can be regarded as an institution, but it is not an institution with as stringent constitutive rules as are necessary for the success of Feinberg's analysis. In particular it doesn't have constitutive rules that would restrict the class of possible claimants in any interesting way. This means that the second conjunct of Feinberg's analysis of rights (as I have reconstructed it) will always hold, which means that the analysis will collapse into correlativism. It also means that in morality, unlike law, there is no substantial difference between the two senses of claiming.

I would now like to reinforce this argument by pointing out that not only is it a fact that the institution of moral claiming, as Feinberg conceives it, doesn't exist, but that it is also quite natural that it doesn't exist, as such an institution would have no point.

A rationale for having an institution can be either that its operations are fun in themselves, or that it enables us to achieve, in a more orderly fashion, something that is itself outside the institution. Examples of the former are various games, examples of the latter are legal institutions (including the institution of legal claiming). Constitutive rules of legal claiming determine what counts as a valid claim, just as constitutive rules of a game determine what counts as scoring a point. But making a valid legal claim is not merely like scoring a point; it has important further consequences. Only if I am in the position to make a valid legal claim will I be able to set in motion the system of legal coercion. The institution of legal claiming is normally used as means to that end. That's the point of having it.

What would be the point of having an institution of moral claiming as Feinberg conceives it? It is unlikely that such claiming would be great fun in itself. But in morality there is no system (analogous to the system of legal coercion) that one could set in motion by making moral claims (or fail to set in motion because one failed to satisfy all the constitutive rules of the institution). Therefore I don't see any further purpose that Feinberg's institution of moral claiming could serve. And if there is no such purpose then my ability or inability to make valid claims in it wouldn't make any difference. This pointlessness of moral claiming (when conceived in Feinberg's way) makes it quite unsurprising that there is in fact no such institution.

4.3 Waiving

It may seem to some people that the right-holder's power to waive his rights is an essential feature of the concept of a right and that it is left out in the correlativist account.[28] Two replies can be made to that suggestion.

First, it is not entirely obvious that every moral right involves such a power of waiver. In law there are rights that can't be waived and it is not clear that morals should be different in that respect. To be sure there may be good moral reasons for attaching such a power to every right, but I don't see why there should be anything conceptually wrong with a right that can't be waived.

Second, what we call the 'power to waive a right' can be redescribed without use of the word 'right'. To have a power to waive a certain right is to be in a situation to perform a speech act that changes the moral aspects of the situation in such a way that something which previously would have counted as a violation of a right doesn't count as a violation any more. Something that would have counted as immoral in the absence of that speech act is not immoral after it has been performed. Now, moral patients often have the power to change the moral aspects of the situation in exactly the same way, and this power of theirs is totally independent of whether they will also be described as holders of rights or not. A moral patient namely often has an option to perform the act of permitting the relevant moral agent to do to him something that would otherwise be wrong. A right-holder's waiving his right amounts to the same thing as a moral patient's permitting a moral agent to act contrary to what would otherwise be his duty. The idea that the rights can be waived is therefore perfectly compatible with the correlativist account of rights.

4.4 Protection of Liberties

Rights, if respected, usually make it easier for us to do certain things that we have a liberty-right to do. If I have a right that others abstain from forcibly preventing me from speaking publicly, and if others respect that right, then speaking publicly is likely to be an easier task for me than it would be otherwise. Moreover I have a liberty-right to speak publicly.

This might be viewed as suggesting that there is some conceptual connection between rights and liberty-rights. An analysis of rights that would attempt to incorporate this suggestion would have to say that 'P has a right (against A) that AΦP' is, in virtue of the meaning of 'a right' equivalent to:

A has a duty to P that AΦP and P has a
liberty-right that PYX and AΦP facilitates PYX

where X is a place-holder for a term for some moral patient(s) and Y for the description of the way in which there is a liberty-right to behave towards these patients.

Casual and imprecise way of talking about rights that was critized in section 2 promotes this line of analysis. In everyday moral discourse rights are frequently referred to in terms of the activities that are facilitated when their conditions of being respected are met (rather than in terms of these conditions themselves). For example my right that others abstain from forcibly preventing me from speaking publicly is often referred to as my 'right to speak publicly' and this promotes the idea that there is some sort of conceptual connection between that right and my liberty-right to speak publicly.

Although the way of talking about rights that I have advocated in section 2 has already greatly reduced the plausibility of this line of analysis, I would like to present two more arguments against it.

The first is that it can be argued convincingly that there are situations where we have a right, the respect for which facilitates our doing something that we don't have a liberty-right to do (something that it is wrong for us to do). For example it may be argued that I have a right that others abstain from forcibly preventing me from speaking publicly even in a situation where respect for that right facilitates my speaking publicly in a manner that promotes racism which is something that I don't have a liberty-right to do.[29]

One may reply to this argument by saying that although respect for my right can sometimes facilitate my doing wrong things it will always facilitate at least one thing that I have a liberty-right to do. This may be a sound reply to my first argument but it paves the way to my second argument.

Namely whenever someone has a duty to me I can find some trivial activity that I have a liberty-right to perform and that would be facilitated by respect for that duty. For example a great number of duties that others have to me in some way facilitate my scratching my ear, and scratching my ear is something that I almost always have a liberty-right to do. Some such trivial activity will make the last two conjuncts of the proposed analysis true in case of any duty. This trivializes the whole analysis, and makes it for most ends and purposes collapse into correlativism.

4.5 Rights as Grounds of Duties

It is sometimes suggested that there is an asymmetry between rights and duties because rights are grounds of duties, because they are somehow prior to duties.

One thing that this can mean is that they are somehow ontologically prior to duties. This is what could have been meant by some people who were talking about 'natural rights'. This naive idea of rights as something like an aura that surrounds human beings seems to be implicit in much of the popular, non-philosophical, thinking about rights. However, it is too obviously defective to be worth attacking.

Another way in which rights could be prior to duties is ordo cognoscendi i.e. it could be argued that we become acquainted with claims about rights first and that we then derive claims about duties from them. This is often true. For example, I found out that I have a right that others abstain from injuring me long before I found out that Jovan Miliradović has the duty to abstain from injuring me (and even long before I knew that Jovan Miliradović existed). If Jovan Miliradović asks me why should he abstain from injuring me I would quite naturally answer 'Because I have a right that others abstain from injuring me'. In that case it is indeed quite appropriate to say that my right is the ground of his duty and not the other way round.

However, such examples don't establish any general point about rights. In this example, the ascription of a right involved an element of universality that was absent from the claim about the duty, and that element of universality is responsible for the asymmetry rather than the fact one claim involves the word 'right' and the other the word 'duty'. The content of the right was that 'others' (i.e. everyone else) behaves to me in a certain way, while the content of the duty was that Jovan Miliradović (i.e. a specific individual) behaves to me in that way. I often find it convenient to express the moral norm requiring everyone to behave to me in a certain way by saying that I have a right to it but I could express it in terms of duties and nothing would be lost: the duty of everyone to behave so-and-so to me is prior to Jovan Miliradović's duty to behave so-and-so to me in exactly the same way in which my right that, everyone behaves so-and-so to me is prior to Jovan Miliradović's duty.

5. DISPENSABILITY OF 'RIGHTS'

If correlativism is correct then the word 'right's is dispensable: we could live without it. It could be eliminated from our moral language without any loss to its expressive power.

This doesn't mean that the correlativists claim that the notion of duty has to be regarded as prior to the notion of right. If correlativism is correct then the term 'duty' is as dispensable as the term 'right': talking about duties can always be replaced by talking about rights, without any loss.

Correlativists are likely to be accused of not 'taking rights seriously' but that accusation would be unjust: they can take rights very seriously, as seriously as they take morality itself. What they don't take too seriously is only the term 'rights' as they believe that this term is neither more nor less important to morality than 'duty' or 'ought' or 'good' or 'morally prohibited'.

Although having any one of these terms (without the others) would be sufficient for having a moral language as expressive as the present one, having all of them is still useful. Having the term 'rights' in addition to 'duty' in our moral language is useful in exactly the same way in which having the term 'moral prohibition' is: although none of them is logically indispensable, they enable us to say things that we want to say in a clearer, less cumbersome way. Moreover in some moral discussions we may want to focus our attention on the moral patient rather than the moral agent, and use of the word 'rights' (rather than 'duties') enables us to do that. Use of the word 'rights' also gives us the dubious benefit of being able to achieve rhetorical effects that couldn't be achieved otherwise.

In a language that contains both the word 'right' and the word 'duty' we can either take 'duty' as more basic and define 'right' in terms of it, or take 'right' as more basic and define 'duty' in terms of 'right'. It's a matter of choice and nothing hinges on which option we choose.[30]

Some moral positions might be easier to formulate in terms of rights and some in terms of duties. However, if correlativism is correct, then any moral position can be formulated in terms of rights and no moral position has to be so formulated. This has an important consequence that it is misleading to describe some disputes in contemporary normative ethics as disputes over 'right-based morality'.

If 'right-based morality' is to be understood as referring to a system of normative ethics in which principles involving 'rights' are taken as basic, then the label is totally uninformative, as any moral principles can be formulated so as to include the word 'rights'. If 'right-based morality' is to be understood as referring to a system of normative ethics formulated in the language that takes the word 'a right' as a primitive then the label is again uninformative as that language is equivalent to the language that takes 'duty' as primitive.

The only sense that we can make of the phrase 'right-based' morality that would be consistent with there being a dispute between its proponents and proponents of other normative-ethical views is that the phrase refers to a system of ethics that takes as its basic principles some principles that are easier to formulate in terms of 'rights' than in some other terms. But this is still very vague and that's why I think that the use of the label 'right-based morality' only confuses the issues to which it is applied. It is, by the way, interesting to note that many philosophers have engaged in the discussion over the 'right-based morality' without attempting to specify what exactly is supposed to be covered by the label.

NOTES

1. According to the often quoted formulation of this view that was given by Benn and Peters, '(r)ight and duty are different names for the same normative relation, according to the point of view from which it is regarded'. (Social Principles and the Democratic State, George Allen & Unwin, London, 1959, 89).

2. It should be noted that 'correlativity' and related terms are not used uniformly in the contemporary philosophical literature about the relation between rights and duties. For example, Lyons in 'The Correlativity of Rights and Duties' (Nous, 4 (1970), 45-55) uses them in a sense that is much wider than mine.

3. Although use of the term 'moral patient' for those to whom duties are (or can be) owed does not have as long tradition as use of 'moral agent', the term seems to be on its way to become a standard element of ethical vocabulary. Some of the authors using the term are :
   N. Fotion (Moral Situations, The Kent State University Press, Kent, Ohio, 1986),
   T. McPherson ('The Moral Patient'), Philosophy, 59 (1984), 171-183) and
   J. Narveson ('The How and Why of Universalizability' in N.T. Potter and M. Timmons (eds.) Morality and Universality, D. Reidel, Dordrecht, 1985, 3-44).

4. It might seem to some readers that my insistence that a justification of any moral claim formulated in terms of duties has to mention how the behavior in question affects someone is biased towards utilitarianism. However it is not: for a Kantian, a justification of a moral claim always has to mention how the behavior in question affects a member of the set of rational beings.

5. This is how the claim about rape would be justified today. Not so long ago, however, many people were inclined to say that everyone has a duty to abstain from committing rape because raping a woman is an affront to the dignity of her husband. For these people, he was the moral patient and not the woman. This example shows that very different moral attitudes can be hidden behind the same formulation such as 'Everyone has the duty to abstain from committing rape', and that we don't fully understand such a claim until it is made explicit who is intended as the moral patient in it.

6. An example of such anti-correlativism is the recent book by Allan R. White (Rights, Clarendon Press, Oxford, 1984).

7. For examples of these objections, see H.L.A. Hart, Bentham, British Academy, London, 1962, section IV; D. Lyons, 'Rights, Claimants, and Beneficiaries' in D. Lyons (ed.), Rights, Wadsworth, Belmont, 1979, 58-77; C. Wellman, A Theory of Rights, Rowman & Allanheld, Totowa, 1985, 25.

8. I use the word 'respected' here in a sense that makes it analytic that right is respected if and only if it is not violated.

9. The above consideration (we can summarize it as 'no violation without a violator') has the consequence that every right is a right against someone, in spite of McCloskey's repeated insistence on the opposite. McCloskey claimed e.g. that 'the right of the tennis club member to play on the club courts is a right to play, not a right against some vague group of potential or possible obstructors', and that the last person in the universe would still have rights H.J. McCloskey, 'Rights', The Philosophical Quarterly, 15 (1965), 118, 119). I haven't been able to discern any persuasive arguments for his position, but I believe that it is due to his neglect of the distinction between rights and liberty-rights.

10. This feature of talk about rights is sometimes called 'the dynamic aspect of rights' by philosophers who don't find it pernicious. (J. Raz 'On the Nature of Rights' Mind, 93 (1984), 200)

11. In spite of the proliferation of writings about rights in recent years the terminological problem that I discussed in this section has not received sufficient attention, apart from a remark on p. 38 of G.R. Grice's book The Grounds of Moral Judgment (Cambridge University Press, Cambridge, 1967).

    The way of talking about rights that I advocated in this section disposes of the paradox of 'a right to do wrong'. Saying 'It is morally wrong to defend your racist views publicly' and at the same time 'You have a moral right to defend your racist views publicly' may seem paradoxical; but the seeming paradox disappears when the latter is transformed into 'You have a moral right that no one forcibly prevents you from defending your racist views publicly'. Cf. J. Waldron 'A Right to do Wrong', Ethics, 92, 21-39 (1981).

12. E.g. by W.D. Ross in The Right and the Good, Clarendon Press, Oxford, 1930, 48-50.

13. See e.g. Lectures on Ethics, Hackett, Indianapolis, 1963, 239-241.

14. For example, if the duty to abstain from cruelty to animals is not owed to animals but to potential human spectators of such cruelties, then it seems quite reasonable to say that the potential human spectators have some right in the matter.

15. Bentham, 314.

16. 'Rights, Claimants, and Beneficiaries', 68–69.

17. The Philosophical Review, 64, 175–191.

18. See footnote 7 of his article.

19. At least this is my reading of his article. The article itself is not entirely explicit on this point. Cf. Frankena's remark on p. 213 of his article 'Natural and Inalienable Rights', published in the same issue of The Philosophical Review: 'I am not sure whether he /Hart/ denies that obligations entail rights or not'.

20. For example, if duty to abstain from cruelty to animals is not owed to animals then it may be that it is owed to potential human spectators of such cruelties.

21. 'Are There Any Natural Rights?', 183.

22. Hart's much later article 'Bentham on Legal Rights' contains something which might at first glance seem to be an answer to the question that troubles me here, namely the question as to what are the ways in which a right holder has control over the freedom of another. He says that the 'fullest measure of control comprises three distinguishable elements: (i) the right holder may waive or extinguish the duty or leave it in existence; (ii) after breach or threatened breach of a duty he may leave it 'unenforced' or may 'enforce' it by suing for compensation or, in certain cases, for an injunction or mandatory order to restrain the continued or further breach of duty; and (iii) he may waive or extinguish the obligation to pay compensation to which the breach gives rise' (p. 141 in D. Lyons (ed.) Rights). However one shouldn't forget that the article in which he says that is an article about legal rights. Although these remarks may be fine as parts of an analysis of legal rights, they are of little use for our understanding of moral rights which are the subject of his 1955 article. An attempt to include provisions like (ii) or (iii) in an analysis of moral right would fail for the same reasons for which Feinberg's analysis of moral rights (discussed in the next subsection) fails. Concerning provision (i) see subsection 4.3.

23. Hart himself notices this on p. 177.

24. 'The Nature and Value of Rights', The Journal of Value Inquiry, 4 (1970), 243–257.

25. Ratio, 18 (1976), 145–155.

26. Point 3 is in fact dubious as Feinberg doesn't make it clear what rules the institution of claiming in morality could have in addition to the rules for recognizing claimants.

27. See his commentary on Feinberg's article in The Journal of Value Inquiry, 4, 258–260.

28. See note 22, above. Carl Wellman's recent book A Theory of Rights also contains suggestions that some of the phrases involving the word 'right' that we use in ordinary moral discourse (e.g. 'right to social security' discussed on p. 181 of the book) should be analyzed in terms of someone's duties (as the 'defining core') and a power of waiver (as one of the associated elements). However this view of Wellman's is not opposed to correlativism, as he claims only that some particular phrases involving the word 'right' should be so analyzed and not that it is a feature of the meaning of the word 'right' itself.

29. This type of situation has recently been discussed in detail by Jeremy Waldron. See note 11, above.

30. Pace W.D. Lamont, who thinks that this is a significant question ('Rights' Proceedings of the Aristotelian Society, Supplementary Volume 24 (1950), 93–94)

In light of all this one is bound to be puzzled by some remarks in Christopher Arnold's article 'Analyses of Right' (in E.Kamenka and A.E. Tay (eds.) Human Rights, Edward Arnold, London, 1978, 74–86) as it seems to defend correlativism and at the same time contains a suggestion that we might need the concept of a right as a primitive.

SCIENTIFIC DISCOVERY : IS IT A LEGITIMATE SUBJECT FOR THE PHILOSOPHER OF SCIENCE ?

ALEKSANDAR JOKIĆ

Introduction

If we carefully examine claims about scientific discovery made in some of the most recent articles on this topic and compare them with that which has been said about scientific discovery in the classical works on the methodology of science, the need to distinguish between the following two questions becomes apparent: (1) Is there a logic of discovery? and (2) Should philosophers of science be at all interested in the subject of scientific discovery?

The following two questions must be distinguished: (1) Is there a logic of discovery? and (2) Should philosophers of science consider scientific discovery?

For my purpose in this paper it is of great importance that the distinction between (1) and (2) be made, since I will be concerned only with the latter. My emphasis on this distinction may be surprising to anyone uninformed about the problem of scientific discovery. Is it not obvious, one might say, that these questions are relevantly different; isn't it possible that even if there were no logic of discovery (whatever that means), the problem of scientific discovery could remain philosophically interesting? On the other hand, a reader acquainted with the history of the idea of the logic of discovery[1] would certainly be very surprised to see these questions separated[2]; and, I imagine, he would be particularly perplexed by the claim that this separation could be of any importance. The fact that such puzzlements can occur suggests that for a long time philosophers have had a one-sided approach to scientific discovery desiring only to produce a system of rules whose effortless application would result in new important discoveries. Precisely this circumstance, that the interest for scientific discovery has long been identified with the interest for developing a logic of discovery (in many papers explicitly represented as the only philosophical aspect of scientific discovery), explains the likely surprise of the reader acquainted with the appropriate literature, as well as the phenomenon that during the past century philosophers of science, on the whole, have considered scientific discovery only in order to banish it from philosophy into the more

A. Pavković (ed.), Contemporary Yugoslav Philosophy: The Analytic Approach, 121–149.

descriptive spheres of psychology, sociology and history of science. They did so on the basis of the argument that discovery is an irrational process, a "flash of insight", which philosophy cannot be expected to explain. these authors have identified rationality with logicality (to use Toulmin's term, 1972, p.84), and unable to develop the set of rules (or algorithm as it is usually called) whose mechanical application would produce scientific discoveries, they have removed discovery from the realm of philosophy.[3]

However, in order to make sense of the most recent discussions about discovery, it is more appropriate to begin with the question of whether philosophers of science should consider discovery, than with the question which necessarily narrows down our concern to the logic of scientific discovery. This seems even more appropriate if we are aware of the recent "shifts" in the meanings of some key terms. Here, first of all, the term "rationality" comes up which is no longer identified with "logicality" in the strict sense. Thus, it is not uncommon any more to come across the denial that every rational move must be an immediate instantiation of some principle or rule, which, in its turn, belongs to a consistent, hierarchized system of principles.

The essential difference between the traditional approach to scientific discovery and some recent considerations of this problem needs to be explained. The traditional approach began by conceiving logic as a system of rules for mechanical derivation of discoveries. As this system proved impossible, scientific discovery became philosophically insignificant. A number of recent writers have preferred to reflect on the second question concluding not only that discovery has some philosophically interesting aspects, but also that it is possible to talk about the logic of discovery in some sense. This latter approach has not been entirely unfruitful. On the contrary, it has generated valuable case-studies, distinctions to my mind problematic of various senses of the terms "discovery" and "logic"[4], and important analyses and models of reasoning processes which have been proposed as occurring in the so-called context of discovery. As a result, we have various arguments from the "friends of discovery" (to adopt Gary Gutting's label) that contend to demonstrate the legitimacy of the philosophic concern with discovery.

This revival of interest in discovery, however, has its opponents. According to L. Laudan (1980) this "newer program for the logic of discovery" (as he calls the recent methodological concern with discovery) "has yet to make clear what problems about science it is addressing"' The essence of Laudan's claim is that the older program, despite its failure, had a clear epistemic rationale, for logics of discovery were at once logics of justification, whereas the newer program has yet to show what is its philosophical rationale. Hence, Laudan's challenge comes down to the question: can a rationale be found? My main concern in this paper will be to propose one.

Now I believe that both the "friends of discovery" and Laudan only contribute to the current misformulation of the problem of discovery. Both positions involve major confusions regarding the meaning of "discovery". They all, mistakenly, conceive of the scientific enterprise as comprising several logically (and perhaps even temporally) distinct stages, one of which is discovery. Thus, a considerate and careful examination of the meaning of the word "discovery", and particularly its meaning in the context of science, is necessary. Professor Gutting was the first to suggest that an appreciation of the proper meaning of "discovery" may provide the response to Laudan's challenge. However, his entire argument consists of two wisely chosen examples, which are hardly sufficient to establish a rationale.[5] But, I believe that by further investigation in the same direction a rationale can be found: a thorough analysis of the concept of discovery will provide us with a response to Laudan's challenge by establishing that discovery is as philosophically interesting as science itself.

A remark here is in order. My arguments make the assumption, for which I will not argue in this paper, that conceptual analysis may sometimes provide us with some knowledge about the nature of the entity for which a given concept is meant to stand. The nature of the relation between concepts and the entities for which they are meant to stand is in itself a very difficult philosophical problem; hence it would be impossible to address it adequately here. In fact, however, the arguments of many analytic philosophers rely upon this assumption implicitly if not explicitly.

My inquiry falls into three sections: first, I present Laudan's challenge and argue for a specific interpretation of it; second, I examine several points found in the recent literature which could be used, though unsuccessfully, to respond to Laudan's challenge; and third, I propose my own solution to the problem.

Part I

Larry Laudan (1980) has presented all present philosophers of science, who advocate the thesis that the study of scientific discovery is methodologically important, with a serious challenge by requesting a precise specification of a <u>par excellence</u> philosophical problem (or problems) that the study of scientific discovery may help in solving or elucidating. The crucial motivation of the older program for the logic of discovery, according to Laudan, was the desire of the former philosophers of science, to properly formulate and solve the epistemological problem of how to justify scientific claims about the world. The logic of discovery was at that time envisaged as a solution to the problem of justification of scientific claims. Thus, the older friends of discovery, when writing about scientific discovery and its logic, have addressed an important epistemological problem: how are we to justify our beliefs, or more specifically, how are we to provide an epistemic warrant for accepting scientific theories? The newer concern with scientific discovery (from Hanson on), by contrast, Laudan claims, has yet to make clear just which philosophical problems about science it is addressing.

Two of Laudan's assumptions are already manifest. First, he believes that the analysis of the history of the idea of a logic of scientific discovery may decide what significance, if any, should be attributed to recent attempts to illuminate the philosophical problems about scientific discovery. Second, if the recent concern with scientific discovery is worth taking seriously, then, Laudan thinks, it must be because it might elucidate some important philosophical problems about the nature of science.

Before I go into details of Laudan's challenge, let us see in what sense Laudan uses the term "discovery". In accordance with the position developed in his first book (Progress and Its Problems, 1977) Laudan (1980) restates his view of scientific enterprise as consisting of three distinct stages: (1) the context of discovery, (2) the context of pursuit, and (3) the context of ultimate justification or acceptance. According to Laudan these three contexts mark the temporal, if not logical, history of a scientific hypothesis.[6] He explicitly binds the term "discovery" only with stage (1) (which is reflected in the way he selects the labels for the corresponding stages). Laudan construes discovery, as he himself puts it, "rather narrowly as concerned with 'the eureka moment'", i.e., the term "discovery" refers to the time when a new idea or conception first "hits" some scientist. Accordingly, he views the logic of discovery as a set of rules the mechanical application of which can generate new discoveries. In contrast to many philosophers of science, Laudan makes an effort to justify his way of using the term "discovery". Interestingly enough, his reason is not of a semantic nature, but rather heuristic. He believes that only if we construe discovery as he did, can any sense be given to the current debate about the existence of a logic of discovery; and this is because, according to Laudan's conviction, no one would deny that there certainly are general principles governing discovery, if the term is construed so "broadly" as to include stage (2).[7]

Laudan's scheme of the history of methodology should persuade us that discovery was important for the methodologists of science only when discovery itself represented the necessary form of justification. What this claim amounts to will soon become clear.

Laudan begins his historical survey with the statement that throughout the 17th and 18th centuries, both philosophers and scientists, believed in some logic of discovery. Bacon, Descartes, Boyle, Locke, Leibniz, and Newton all believed it was possible to formulate rules which would lead to the discovery of "useful" facts and theories about nature. A set of such rules, as we shall see, was considered to be the obvious, and most elegant, solution of some of the problems that worried philosophers of science at that time. However, by the last half of the 19th century, this undertaking was dead. It was unambiguously repudiated by such philosophers of science as Peirce, Jevons, Mach, and Duhem. A new task replaced the old one: to formulate the logic of post hoc justification of theories. But, how did this dramatic change take place? A look at the history of methodology will provide an answer to this question.

According to Laudan, the desire to develop a logic of discovery was based upon two different motives. On the one hand, there was the heuristic and pragmatic problem: how to accelerate the pace of scientific advances, how to increase the rate at which new discoveries were being made by articulating fruitful rules for invention and innovation. On the other hand, there was the epistemological problem: how to provide a sound warrant for our claims about the world. If it were possible to formulate a logic of discovery, it would solve both problems simultaneously. The logic of discovery, because infallible, would automatically guarantee that any theories produced by its use were epistemically well grounded. Thus, the logic of discovery, as conceived by most 17th-and 18th-century philosophers of science would function epistemologically as a logic of justification. Of course, this does not mean that these thinkers could not recognize the difference between discovery and justification, Laudan insists, rather, they were convinced that an appropriate, i.e., infallible, logic of discovery would automatically authenticate its products and that a separate logic of justification was therefore redundant and unnecessary. "They were pre-occupied", writes Laudan, "with developing logics of discovery, not because they were indifferent to the epistemological problem of justifying knowledge claims, but precisely because they took the justificational problem to be central."[8]

Why did our forebears have such a firm belief that it must be possible to formulate a logic of discovery? Laudan points out that this attitude is an unavoidable consequence of the conception of science as infallible knowledge for infallible knowledge is logically incompatible with any form of consequentialism. Consequentialism is the view that scientific theories or claims could be justified by comparing (a subset of) their consequences with observation. If an appropriately selected range of consequences proved to be true, this is taken to provide an epistemic justification for asserting the truth of the theory. Consequentialism was the prevalent doctrine among both ancient astronomers and ancient physicians. Eschewing the belief that theories could somehow be derived from observation, these thinkers stressed the necessity for the post hoc evaluation of a proposed theory in light of how well it stood up under the observational and experiential tests.

However, this view was quickly undermined by Plato and Aristotle, who conceived episteme as infallible knowledge. They saw a flaw in consequentialism: to argue from the truth of a consequence of a theory to the truth of the theory itself is a logical fallacy (the so-called fallacy of "affirming the consequent"). Hence, Plato and Aristotle argued that an infallibilist view of science was strictly incompatible with any form of consequentialism, i.e., with any form of post hoc empirical testing, since the latter was logically inconclusive. Therefore, if theories were to be demonstrably true, such demonstration could not come from any (nonexhaustive) survey of the truth of their consequences.

Thus, Laudan comes to the view that the majority of 17th- and 18th-century writers, who believed science was infallible knowledge, stressed the importance of the analysis of the ways in which theories are generated, precisely because consequentialism was generally acknowledged to be epistemically inconclusive. Namely, if one seeks infallible knowledge, and one grants the fallaciousness of affirming the consequent, then the only viable hope for a logic of justification will reside in the quest for truth-preserving logic of discovery.

However, there is general skepticism today about the possibility of a logic of discovery. This is so, according to Laudan, because most of the contemporary philosophers of science cannot conceive that there might be rules that would lead scientists from laboratory data to theories as complex as e.g., quantum theory, or the structure of DNA. But, if it is so easily recognizable that a logic of discovery is inconceivable, how was it possible that so many of the world's most brilliant minds had been obsessed by the desire to formulate such a logic? Laudan also has an answer to this question. The attitude regarding the possibility or impossibility of a logic of discovery depends crucially on our archetypes of significant science. If our archetypes involve theoretical entities and processes which are inferentially far removed from the data which they explain, it seems implausible that there might be rules to lead one from e.g., tracks on a photographic plate to claims about the fine structure of subatomic particles. On the other hand, if our archetypes were such lawlike statements as "All crows are black", or "All planets move in ellipses", then things might be quite different. If statements like these were viewed as the primary objects of scientific inquiry, the idea that there might be some rules for going from evidence to discovered theory would not seem so bizarre.

The main reason why inductive logics of discovery were ubiquitous among 18th-century philosophers (and scientists) lies in the fact that Bacon and Newton had persuaded most philosophers that speculative theories and unobservable entities were anathema. The turnover occurred in the 19th century, when empirical generalizations were viewed as mundane and uninteresting, and when theories were conceived chiefly as grandiose ontological frameworks, replete with unobservable entities. As a consequence, the quest for logics of discovery was abandoned, and their very existence denied.

From this historical survey it is clear, Laudan thinks, that the older program for a logic of discovery had a clear philosophical rationale: it was an attempt to solve the important philosophical problem of providing an epistemic warrant for accepting scientific theories. Laudan then makes the following crucial remark: "The newer program for the logic of discovery, by contrast, has yet to make clear what philosophical problems about science it is addressing."[9]

This apparently very clear statement of Laudan's has however, been misinterpreted in the current literature. Let us see, for instance, how in a very simple manner T. Nickles (1980b) rejects Laudan's challenge:

> Even if historical methodologists, from Bacon and Descartes to Herschel and Whewelle and after, had unanimously agreed that discovery was an appropriate methodological concern only insofar as discovery is necessary to justification, this agreement hardly would establish the claim and that philosophy of science in the fallibilistic, H-D era should confine itself to justification and exclude discovery (generation). For the claim itself is a normative claim, not a descriptive historical claim.[10]

A little further Nickles adds:

> ...while the appeal to history is thought-provoking and makes a prima facie case against the viability of a methodology of discovery, the historical evidence is not unambiguous and in any case cannot alone establish a methodological thesis.[11]

It seems that Nickles is accusing Laudan of inferring an "ought" from an "is" (or better to say from a "has been"). Nickles represents Laudan's alleged normative claim that in the H-D era philosophers of science should confine themselves to justification and exclude discovery, as a conclusion of some argument in which only historical claims occur as premises. Thus, Nickles reduces Laudan's challenge to an invitation to disclaim some methodological thesis. An almost identical interpretation can be found in the most recent attack on Laudan by R. McLoughlin (1982). In this article Laudan's alleged methodological thesis is called the "divorce thesis", and consists in the following:

> ...the "logics" or inference procedures involved in discovery and justification are different and independent. This thesis permits Laudan to argue that there is no rationale for the philosophical study of discovery--in contrast to justification, which is of central interest to epistemology.[12]

However, I believe that this interpretation does not have much textual support. Namely, it seems to me that Laudan is not committed to any of the normative theses attributed to him. He does not say ever that in the H-D era philosophy of science should confine itself to justification and exclude discovery. (All he is saying is that it is not clear why it should deal with discovery.) Nor does he explicitly endorse the "divorce thesis". He does not argue for any other claim on the basis of this thesis; if he accepts it at all, he sees it as a descriptive claim about the way recent authors write about discovery, rather than his own normative thesis, let alone a thesis for which he would have a further use. In order to understand Laudan's challenge properly let us repeat (in more detail) the basic assumptions that Laudan makes:

(a) An analysis of the evolution of ideas about the logic of discovery may put us in a position to see what significance, if any, should be attached to recent efforts to get clear about the philosophical problems of discovery.

(b) If any program for finding a logic of discovery is worth taking seriously, it must be because such a logic, once propounded, might illuminate or clarify some important philosophical problems about the nature of science.[13]

Here (b) is of crucial importance. Laudan does not support it by explicit arguments. He either takes it for granted that no one would challenge (b), or he perhaps believes that (a) provides (b) with sufficient plausibility. Whichever is the case, the important thing to realize is that (b) functions as a premise in Laudan's argument. The form of this argument is very simple:

Premise (1): Any program for the logic of discovery is worth taking seriously if, and only if, it helps illuminate some par excellence philosophical problem (regarding science).

Premise (2): The newer program for the logic of discovery (which, contrary to the old one, sees this logic as something very different from the logic of justification) has not yet made clear what philosophical problem about science it is addressing.

Conclusion: Therefore, at present, no significance can be attached to this program.

There are two major differences between my interpretation of Laudan's challenge on the one hand, and Nickles' and McLoughlin's on the other. First, they are committed to a different wording of (b): they would have to delete the words "some important philosophical problem about science," and write instead, "the problem of justification." Secondly, and more importantly, they would see this reinterpreted (b) as a conclusion, rather than as a premise, of some argument that they do not quite spell out, but nevertheless attribute to Laudan. As a consequence, the only way to answer Laudan's challenge, in their interpretation, is to show how (in the H–D era) the study of scientific discovery may indeed illuminate the problem of justification, while, according to my interpretation, connecting any par excellence philosophical problem with discovery would suffice.

Part II

If one wants to answer Laudan's challenge there are two viable approaches: (1) Accepting Laudan's assumptions, and adopting his way of using the term "discovery", one may attempt to show that there are clear philosophical problems the solution of which is more likely to be found if we include scientific discovery in our philosophical considerations; (2) On the other hand, one may attempt to prove that some of Laudan's assumptions (or his arguments based on them) are fallacious, or that his use of the term "discovery" is incorrect, and that philosophical problems about scientific discovery emerge as soon as these distortions are eliminated. In this section I shall present two arguments (found in recent literature) which imply, at least on first sight, that discovery is a bona fide topic for the philosophy of science. Both of these arguments, by contrast to the one offered in the last section, are characterized by approach (1). These arguments are: (a) the argument from understanding science; and (b) the argument from a new model of rationality.

Before examining these arguments I shall discuss another attempt by Nickles to answer Laudan's challenge. To my mind it is clearly an unsatisfactory answer. I mention it because it may turn out that both (a) and (b) suffer from the same sort of weakness. It is an argument of type (2). What, then, is Nickles' objection to Laudan? He questions Laudan's interpretation of history according to which methodology of generation is of little importance once one adopts a fallibilistic, H-D methodology. Starting from D. Shapear's claim that in developing science we have not only obtained new knowledge but that "we have learned to learn" also, [14] Nickles stresses that we may have legitimate interests in certain methodological topics which our predecessors did not have. Thus, Nickles claims that the fact that the interest in a logic of discovery coincides with the conception of scientific theories as infallible knowledge, and its abandonment with adoption of a fallibilistic methodology, does not permit Laudan to claim that discovery is philosophically insignificant. What this means, I suppose, is that since the historical evidence is not unambiguous, even the advocates of the fallibilistic conception of scientific theories may, at some point, become interested in the philosophical/methodological study of discovery. I think that what Nickles says is perfectly correct however it does not answer Laudan's challenge. His remarks establish, at most, that it is not necessary that in the H-D era scientific discovery should be philosophically uninteresting. But Laudan's challenge, as I interpret it, does not deny this; rather it claims that, as far as philosophy of science today is concerned, it is not clear what significance, if any, we should attach to the study of scientific discovery,

Now I turn to the presentation of the argument from understanding science. Occasionally we come across the thesis that the task of the philosophy of science is to offer a complete understanding of science, and since according to these thinkers discovery is a part of the scientific process philosophy of science ought to consider discovery (in the sense of theory generation). Thus, the Nobel Prize Laureate Ilya Prigogine writes:

> ...if there ought to be a philosophy of science, this philosophy has as its task to consider invention, it cannot, at pain of self-condemnation, surrender this question to other disciplines: psychology, aesthetics, sociology.[15]

Some philosophers express a similar thesis. Nickles says that establishing a complete, tested theory is only the last stage of a subtle process of skillful enquiry, and failure to consider the remaining aspects (i.e., discovery) "will result, and has resulted, in a skewed conception of scientific activity and flawed conception of rationality, learning and knowing -- both within science and without."[16]

A few years earlier F. Suppe (1974b), criticizing Reichenbach's distinction between the context of discovery and the context of justification, expressed the same thesis in a somewhat more precise manner. In opposition of Reichenbach, he claims that the epistemology of scientific knowledge must take into account the context of discovery as well as the context of justification if it is to arrive at a satisfactory account of the epistemic features of scientific knowledge. Suppe writes that scientists themselves conceive of science as an "ongoing social enterprise with common bonds of language, methodology, and so on." And further:

> Full epistemic understanding of scientific theories could only be had by seeing the dynamics of theory development, the acceptance and rejection of theories, the choosing of which experiments to perform, and so on. To understand a theory is to understand its use and development.[17]

Therefore, since science is an "ongoing social enterprise" which proceeds dynamically through time, in order to understand science we must come to grips with the dynamic process of theory development. And in order to investigate fully the dynamics of theory development, one has to examine the context of discovery.

Although reflections like these may appear as just empty phrases, some philosophers take them seriously. H. Siegel (1980) points to the deductive form of the argument from understanding science:

(1) Science is an ongoing, dynamic social process, and any adequate account of science must take this dynamicism into account.

(2) Therefore, the understanding of science requires attention to the dynamic dimension of science.

(3) Understanding of the dynamics of science can only be had through consideration of the context of discovery (as well as the context of justification).

(4) Therefore, the context of discovery is relevant to, and essential for, an epistemic understanding of science.[18]

One may be led to think, just on the basis of the presence of the term "epistemic" in (4), that we have achieved exactly what we were looking for: a rationale for the philosophic study of discovery. However, there are many problems with this argument. Siegel cites one: The phrase "understanding science" is ambiguous between the following two senses: (i) understanding the scientific process, or enterprise, in general; and (ii) understanding the status -- and especially the epistemological import -- of scientific theories. Siegel's point is that (4) does not follow from (2) and (3) unless "understanding" in (2) is taken as epistemic understanding -- that is, in the sense (ii). "But it is not clear," Siegel writes, "that this sense of 'understanding science' does in fact rely on the dynamic features of science. Establishing the epistemological import of a theory involves saying just what we have when we have a (well-supported) scientific theory. and doing that need not involve attention to the dynamic course of evolution any particular theory might have travelled."[19]

However there is a further complication, this time regarding the phrase "context of discovery." It is tempting to think that the term "context of discovery," in the scope of the argument, refers in fact to actual cases of scientific discovery. Accordingly the term "logic of discovery" cannot designate some as-yet-unknown set of rules the application of which would yield new discoveries, but rather a methodology of rational explanation or understanding of discoveries that have actually taken place in the historical development of science (historical discoveries). The argument from understanding science thus suggests the need to take historical discoveries as explananda. But we immediately face the following question: with regard to these discoveries, what things should we try to explain? There are many possibilities. We may try to explain the occurrence of the discovery. Or we might attempt to explain why it did not occur before. We may want to account for the intuition that the proper time for a discovery has come. Still other questions may seem important: Why does science sometimes manage to ignore insights of genius only to rediscover them years later? How should we account for 'premature' discoveries which fail? However important these questions may be, our main interest probably will be to explain the intellectual content of particular discoveries, that is, the course of reasoning that led a scientist (or a group of scientists) to make that discovery. And when we do this, we are mainly concerned with reconstruction of the arguments (which let to the discovery).

However, the following question immediately crops up: now that we have determined our explanadum what should our explanans be like? The most common response is "rational reconstruction." But, this phrase itself requires explanation. Most people speak of rational reconstruction in terms of some theory of rationality. Thus the argument from understanding science turned out to depend on the solution of the problem of rationality in science, which is really the topic of the argument from a new model of rationality. However, before I present this latter argument I shall point out that the suggestion of the argument from understanding science seems to be that there is a problem the solution of which cannot come from psychology (or other descriptive sciences): the task to express, precisely formulate, and elucidate the methodology of explaining historical discoveries.

The fact that a few historians of science have succeeded in making intelligible some of the most important discoveries strongly suggests that discovery may be much more logical, or at least a more rational procedure than most philosophers are prepared to admit, and gives us reason to hope that there are patterns of discovery. However, N. Koertge, fully agreeing with this attitude, emphasized that this sort of rational, historical understanding of discoveries does not imply that there is a logic of discovery,or that those patterns of discovery "would be rational in any standard sense of the term, although they may very well be cognitive i.e., reflect some of the typical ways in which we think."[20] But even where a discovery was made by means of some rational procedure, the problem of its historical explanation of historical discoveries, some would argue, remains a serious task for philosophers (although not only for philosophers). This task is serious because it can help the investigation of already established philosophical themes, e.g., the problem of rationality in science. My only comment, after considering this argument, would be that I am not quite persuaded that a philosopher is in any better position to perform this task than, e.g., a sociologist, historian or psychologist, nor that his contribution is essential or even necessary. At any rate, this is still a controversial issue. And we should not forget that the way to answer Laudan's challenge is to connect discovery with some par excellence philosophical problem about science.

Contemporary discussions of rationality in science suggest that the correct answer to the question of whether discovery is a philosophically significant subject depends on the model of rationality which is accepted. The old model of rationality implied that discovery was philosophically insignificant. However, many contemporary philosophers of science reject the old model, and attempt to formulate a more appropriate one. It is explicitly claimed by some of these philosophers that, according to a new model of rationality, the philosophical study of discovery may be of crucial importance.

The general theory of rationality which we inherited claims that every rational step must be a direct instantiation of some general principle or rule which, in turn, belongs to a consistent, hierarchized system of principles. A scientific explanation (of, e.g., the occurrence of some event) is, according to this model, simply a logical proof from hypotheses. Today, however, an ever increasing number of philosophers reject the traditional identification of rationality with "logicality," and yet still consider scientific enquiry to be, in some sense, a paradigmatic case of rational judgment.

If one looks at some of the most recent proposals for the formulation of a new model of rationality, one can recognize a whole spectrum of various conceptions in the discussion of rationality: on the one side of the spectrum there is the old conception of rationality, while on the other we find a new conception (defended by e.g., Wartofsky 1980, and McMullin, in Nickles 1980c) which claims that rationality is a matter of skill in making judgments (about such things as e.g., the preliminary plausibility of some hypothesis). This form of judgment involves the application of general maxims or rules to particular cases, but (contrary to the old model) not by deductive subsumption. According to Nickles' (1980b, p.34) understanding of Wartofsky it is a "form of rational judgment which is noninferential." (Wartofsky himself uses the term "craftsmanlike skill in judgment.")

The recent tendency to weaken the old model of rationality and show that rationality is more than "logicality" is a result, at least in part, of a new conception of what constitutes a problem situation. Philosophers have long supposed that the typical problem situation consists of a surprising fact which calls for explanation. The inescapable implication of this model of a problem situation is that hypothesis generation is a nonrational process, for there are no materials available to reason with except the data. (Of course, this is true only if we reject the old inductivist account of discovery, according to which the data are all scientists need to infer an hypothesis.) However, if a problem situation is understood so that it includes, in addition to experimental data, various constraints[21] (such as e.g., previous theoretical results, which function as consistency conditions, etc.) things may be very different. The idea is that "a particular constraint structure in a sense describes the problem situation being sought, albeit incompletely, indefinitely, or not in the desired form, and thereby determines the range of admissible answers."[22] In short, the problem itself more or less points to the way to its own solution. These constraints constitute a rich supply of premises and maxims (in addition to data) for reasoning towards a problem solution. (This makes possible even such cases as when the problem solution follows deductively from the constraints. But of course it does not always happen that way.)

It seems to me that precisely this modified model of a problem situation has made possible such a definition of rationality as, e.g., Toulmin's, in which rationality is identified with our ability to alter our belief systems adaptively in the face of changing circumstances. If such a definition of rationality were accepted then, by contrast to the old conception, it would be possible to talk about a rational generation of hypotheses. For rationality would be intellectual adaptability, and such adaptability does not implyu thinking and behaving must always be consistent with one's earlier set of intellectual presuppositions. And if Toulmin's view is correct, discovery becomes a matter of great importance for the investigation of rationality in science, for as Nickles noted, "it is precisely in creative problem solving that the intellectual flexibility or adaptability is most evident."[23]

However, the use of terms like "adaptation" in order to define rationality cannot be of much help. The notion of adaptability itself requires explanation -- adaptation is a problematic enough concept in biology, and its metaphorical application to theory f rationality cannot be expected to be any less problematic. Another problem with the argument from a new model of rationality is that it is still a matter of controversy exactly what rationality in science amounts to. There is no general agreement among philosophers of science concerning which model of rationality ought to be accepted. There are even those, like Ian Hacking, who question the appropriateness of the term "rational" to describe most scientific activity.[24]

So, both the argument from understanding science and the argument from a new model of rationality hardly establish more than does Nickles' argument: that it is conceivable, under certain conditions, for discovery to be philosophically interesting. This, as we have already seen, cannot be the answer to Laudan's challenge.

Part III

Both the philosophers who advocate the thesis about the methodological significance of scientific discovery and Laudan, who disputes this thesis, adopt the same general conception of science. According to this (now wide-spread) view, scientific inquiry consists of several logically and (according to some) temporally distinct stages or phases, one of which is discovery. I shall call this view "the stage picture of science." The fact that "friends of discovery" also adhere to this picture of science is the principal reason why they have been so unsuccessful in answering Laudan's challenge.

Anyone who, within his general conception of science, perceives scientific discovery as just a stage of the scientific enterprise is committed to the following two views: (1) discoveries take place at specific moments; and (2) within the scope of some general methodological scheme of the scientific enterprise it is possible to specify exactly when discoveries take place or have taken place. "When" as used in (2) needs clarification. When I speak of the possibility of specifying when discoveries take place or have taken place, I do not have in mind the possibility of predicting or foretelling the exact instant when a particular scientific discovery will occur - this we grant, is unfeasible - but of the possibility to establish within some general scheme of the scientific inquiry (such as, e.g., Laudan's) the exact position of scientific discovery in the sequence of various scientific activities and endeavors. Thus, according to Laudan, scientific discovery takes place before (a) the preliminary evaluation of hypotheses (context of pursuit) and (b) their final acceptance. We are entitled to say that discovery is a <u>phase</u> of the scientific enterprise only if both (1) and

(2) are true. Namely, on the basis of (1) alone we are perhaps in the position to maintain that discovery could be a phase of the scientific enterprise; however, (1) is not sufficient to warrant the claim that discovery is a phase, nor does it tell us which phase it would be.

(1) and (2), taken together, are acceptable only for those who, like Laudan, define the meaning of the term "discovery" as concerned with "the eureka moment", i.e., as referring to the time (or event) when a new idea or conception first "hits" a scientist. This way of understanding the meaning of "discovery" is precisely that which makes possible the abovementioned picture of science - according to which scientific discovery is some exactly specified and separate stage of the scientific enterprise. But, I believe, we must reject this picture of science. The appreciation of the proper meaning of the term "discovery" in ordinary language and particularly in the context of science shows the stage picture of science to be inadequate. In order to prove this I shall rely on the analysis of the concept of discovery, which will demonstrate that (2) is clearly false, and, consequently, that it is a mistake to view scientific discovery as a stage of the scientific enterprise. And once this has been shown, it is not difficult to answer Laudan's challenge.

In his classic The Concept of Mind, (1949) Gilbert Ryle noticed that "discovery" belongs to the class of words which are used to designate achievements or successes.[25] Some words of this class signify more or less sudden denouements; others signify more or less protracted proceedings, which can be seen from the following examples taken from Ryle:

> The thimble is found, the opponent checkmated, or the race won, at a specifiable instant; but the secret may be kept, the enemy held at bay, or the lead be retained, throughout a long span of time. The sort of success which consists in descrying the hawk differs in this way from the sort of success which consists inkeeping it in view.[26]

I take it everyone would think that "discovery" belongs to the former subset, i.e., "discovery" is used to designate more or less sudden denouements.

A characteristic feature of the verbs belonging to the class of "achievement words", according to Ryle is that in applying them we are asserting that some state of affairs obtains over and above that which consists in the performance of the subservient task activity. Ryle makes this obvious in the following example: for a runner to win, he must not only run, but run in such a way that his rivals reach the tape later than he. One important consequence of this fact is that "sucdess words", (and "discovery" is one), are not used to designate acts, exertions, operations or performances, but to point to the fact that certain acts, operations or performances, have had certain results.[27] This implies that we must be able to determine that a certain action or operation give the desired results, if we are to correctly use any of the "success words".

Thus, on the basis of his examples, Ryle teaches us that "discovery", as a "success word", always relates to a certain achievement, happening at a certain moment, as a result of performing the corresponding subservient task activity (or activities). If we continue the analysis of the concept of discovery we may arrive at some other interesting results.

In some cases of achievement it is quite clear that the result has been obtained. Such, for instance, are the cases in which certain constitutive rules unambiguously define the content of the achievement. Chess is such a case, where the task (or the desired outcome) is to checkmate the rival. And, checkmate is defined precisely by the rules of chess as a game. Let us now examine the situation wherein I checkmated my opponent on the 45th move. It is fully unnecessary in order to correctly apply the relevant "success word" to investigate whether or not my rival has been checkmated, while it is quite obvious that his king has no free field of movement. On the other hand, there are cases in which an agent produces a concrete result, but further research and verification are necessary in order to ascertain if, in that case, some of the "success words" are to be applied. Consider the following example: a lady loses her glove, searching for it, a gentleman finds a glove. We, however, can say that the gentleman has accomplished the task, and that he accomplished it at the moment he lifted the glove, only if it is known (or proved) that he lifted the glove which was lost, i.e., the lady's glove, which we know only if the lady identifies the glove as being her glove (or some similar process). Hence, there are cases in which some of the conditions necessary for a correct application of a word denoting achievement are realized later, or even considerably after, the moment which we single out as the moment when the achievement occurred. This is clear from the last example, because we certainly shall not say that the gentleman found the glove at the moment when the lady identified the glove as hers, though we consider her identification of the glove (or some relevantly similar process) as a necessary condition if we are to say that he is the discoverer of the lost glove. And that we are correct in considering this to be a necessary condition, is indicated by the fact that in an opposite case, if the lady said the glove was not the one she had lost a moment before -- we would not say that the gentleman was the discoverer.

The results of our analysis so far, suggest that the establishment of all the necessary conditions for applying a success word is not necessarily limited to the moment at which the performance of a subservient activity yields a certain result. And it seems to me, that in attributing a discovery to a scientist, it is usually possible to point at a whole series of conditions, all of which are necessary, if we are to speak of a discovery at all. However, it might happen that at the moment which we mark as the moment of discovery, not all of these conditions were satisfied.

A good illustration of this point is the usual attribution of the discovery of the planet Uranus to William Herschel.[28]

> On the night of March 13th, 1781, he registered that he caught sight of a shining object in the quartile near Zeta Tauri which he considered to be either a nebulous star or a comet. That is the moment which is generally taken as the moment of the discovery of the planet Uranus. However, we read in the books of the history of astronomy that between 1690 and Herschel's observation in 1781 the same object had been seen and recorded at least seventeen times by men who took it to be a star. Herschel differed from them only in the sense that he allowed the possibility that the object observed might be a comet. Two additional observations (on the 12th and 13th of March), confirmed that the object moved among the stars, which was taken as an indication that it was a comet. The news about the discovery rapidly spread throughout Europe. Mathematicians began to compute the new comet's orbit. Only everal months later, after all those attempts had failed to square with observation did Lexell suggest that the object seen by Herschel might be a planet. Thereafter, mathematicians began to use planetary orbit rather than that of a comet, and here new calculations showed congruence with prior observational results. In this example, we can distinguish, as the necessary conditions for the discovery, the following moments: (a) Herschel's interpretation of the observation (different from the previous one which treated the given object as a star), (b) the perception that the object, seen by Herschel, moved, which seemed to corroborate his reinterpretation of the observation, (c) the computations which tended to define the orbit of the comet, (d) Lexell's idea that the object might be a planet, (e) additional computations using a planet's rather than a comet's orbit, proved reconcilable with observation, and (f) the congruity of the computational results with observation was taken by the scientific community as sufficient reason to regard the object seen by Herschel as a new planet -- Uranus.

The word "discovery" can be applied correctly only in the case when all such (i.e., necessary) conditions are satisfied. Thus, if we were to apply the word "discovery", in the case of the discovery of the planet Uranus, (as in the case of the lost glove, in which it is necessary that the gentleman, after searching, lifts the glove, and that the lady identifies the glove as hers), it must be known that all the necessary conditions for the discovery are satisfied, regardless of the fact that these necessary conditions were occurring at different times, (for instance, (b) precedes (d), and (f) comes after (a)), and also regardless of the fact that these effects are the contributions of different people. Avoiding, for the moment, the question of whether the discovery of Uranus should be attributed to Herschel or Lexell, the important thing to realize is that none of the moments mentioned above, taken alone, would be sufficient to justify the application of the term "discovery".

The analysis of "discovery" made so far, apart from the result which I attributed to Ryle, has shown that in the case of many discoveries we may expect a whole series of conditions which have to be satisfied if we are to use the word "discovery" at all, and that the occurrence of all these conditions must not be limited to the moment which later on is marked as the moment of discovery.

Though the word "discovery" relates to an achievement being realized at a certain moment, the conditions necessary for the word "discovery" to be used correctly are not necessarily fulfilled at that moment, some may even be fulfilled later.[29] Actually, that may be the case with all important scientific discoveries. Particularly interesting and of considerable importance is the fact that the possibility of the correct use of the term "discovery" does not necessarily imply the possibility of locating the moment when the discovery took place. In fact, we might be in the position of knowing all the conditions which are necessary if we want to talk about a certain discovery; all of them, also, might be realized and known to be realized, and therefore, the word "discovery" correctly applied. However, we may still not be able to answer the question: when did the discovery take place?

Consider the discovery of X-rays. In 1895, the physicist Roentgen, during his investigation of cathode rays noticed a strange result. A barium platinocyanide screen far from his shielded apparatus glowed when the discharge was in process. He then proceeded to investigate the cause of this glow. He observed that the cause of the glow traveled in straight lines from the cathode ray tube, that the radiation cast shadows, that it could not be deflected by a magnet, and quite a few other things. Finally, Roentgen ascertained to his own satisfaction that the strange effect was not due to the cathode rays themselves, but to a new form of radiation and announced his discovery of X-rays. But, when were X-rays actually discovered? Let us now examine what T. Kuhn says in regard to this question. He argues that X-rays were certainly not discovered "at the first instant, when all that had been noted was a glowing screen. At least another investigator had seen that glow and, to his subsequent chargin, discovered nothing at all. Nor, ,,, can the moment of discovery be pushed back to a point during the last week of investigation. By that time Roentgen was exploring the properties of the new radiation he had already discovered."[30] Although we are not wrong in saying that X-rays have been discovered, we are still unable to locate the exact moment of the discovery. (Often, we are also unable to name the discoverer, unambiguously. I shall return to this problem at the end, where I discuss a possible objection to my analysis.)

Due to their structures, scientific discoveries are, as a rule, much more complicated undertakings that the search for the lost glove, and for this reason, we should not be surprised when we come across such difficulties. It may appear that the problems of pinpointing the exact moment of discovery and identifying the discoverer are completely due to the complexity of scientific discoveries themselves. But I shall suggest that these problems are also due to the fact that most scientific discoveries belong to a separate species of achievements. Therefore, we should regard the word "discovery" (in most instances) as being a specific "success word" belonging to a separate subset of "achievement words". The complexity of scientific discoveries is due to the fact that there is a whole series of different activities, all of which are of great importance for the final result. I take it that there are very few cases of scientific discovery similar to the case of finding the lost glove (in which we can easily agree about when and who achieved the corresponding result). The example of the discovery of X-rays highlights another feature which makes most scientific discoveries still more complicated, and/or rather specific achievements.

The argument by which Kuhn rejects the possibility of locating the moment of the discovery of X-rays may be questioned. In spite of the argument that Kuhn offers, it could seem quite natural to say that X-rays were discovered at the first instant when Roentgennoticed the strange effect. One could attempt to justify this by saying that the discovery of X-rays is reminiscent of the case of finding the lost glove. Similarity between these two cases, it may be held, comes from the fact that in the case of the lost glove, we consider the gentleman's lifting of a glove as the moment of the achievement, and everything that comes later, i.e., the lady's identification of the glove, only as a necessary condition if we are to correctly use the adequate achievement word. In the case of the discovery of X-rays we could, in the same manner, take the moment when Roentgen noticed, for the first time, the strange effect, i.e., the glowing screen, as a moment of discovery, and everything that happened later as necessary conditions, if we were to say that X-rays had been discovered.

There is, however, a great and relevant difference between the finding of the lost glove and the discovery of the X-rays. The gentleman searching for the lost glove knew what he was searching for, and also all the conditions which were to be fulfilled in order for him to be given credit for finding the lost glove were known in advance. In the case of the X-rays, however, the object of discovery had not been defined at all, so it was not known in advance which conditions had to be fulfilled before it could be considered a discovery. We found out what those conditions were only ex post facto. Contrary to the gentleman, Roentgen did not conduct his experiment searching for something known, and obviously the necessary conditions for the discovery of X-rays could not be known in advance, simply because X-rays were not the object of the research. In my opinion, this circumstance itself complicates our efforts to define the moment when the discovery took place, and prevent us from linking it with the moment when Roentgen initially noticed the strange effect, or, for that matter, with any other moment in the discovery process. On the other hand, when we are searching for something that is well known and defined, the moment of our first acquaintance with it may be taken as the instant when the corresponding achievement occurs. But, it is very likely that only a few, if any, important scientific discoveries could be numbered among such cases. In the case of most scientific discoveries, a long period of time is necessary to pass (and perhaps whole new theories developed and accepted in the interim) before we are in the position to use the word "discovery".

On the basis of the analyzed examples, we can differentiate at least three types of cases in which we use some of the "achievement words": (a) cases in which it is immediately quite clear that the result has been achieved (e.g., the case of checkmate), (b) cases in which, at the moment of achievement, it is not quite clear that the result has been achieved, as some of the necessary conditions (if we are to use the adequate word at all) are realized afterwards, although all these conditions are known as necessary in advance, which makes possible the identification of the exact moment of the achievement (e.g., the case of the lost glove), and (c) cases when we know of an achievement, but cannot define the moment when it occurred, because only ex post facto do we know which conditions were, in fact, necessary if we were to be in the position to use the adequate achievement word (e.g., the example of the discovery of X-rays). Consequently, many scientific discoveries should be classified as (c)-cases.

In light of this analysis, the contemporary efforts to create some kind of schematic conception of scientific enterprise in which the moment of the discovery would be fixed in advance, seems hopeless. In those schematic conceptions, science is viewed as a two- or three-stage process, discovery being one of the stages. The fallacy of identifying discovery with the phase of the hypothesis generation (Laudan's context of discovery), or the phase of preliminary evaluation (Laudan's context of pursuit) has been made `repeatedly. However, it seems to me that the above analysis has shown that all scientific activities should be considered as different aspects of the discovery process itself. Accordingly, when philosophers of science consider a certain discovery they would not necessarily restrict themselves to only one stage in the scientific enterprise, or only one moment in the life of some scientist, as implied by e.g., Laudan's "narrow sense of 'discovery'", construed as concerned with "the eureka moment".[31] But contrary to Laudan's view, when concerned with scientific discovery, our interest must be much wider to include considerations of the whole series of operations and undertakings, such as experimental observation, new theoretical conceptions, mathematical calculations, reinterpretation of the established results, and so on. The justification of such an attitude may become plain if one realizes how strange it is to apply the word "discovery" at the levels of hypothesis generation, and preliminary evaluation. Gary Gutting made this point obvious using the following examples:

Being the first to think of a new idea is not sufficient for making a discovery. Nor is it necessary. Scientists have discovered that the Earth revolves around the Sun, but this discovery came hundreds of years after Aristarchus of Samos, who is usually regarded as the first to think of this idea. The idea of defining "discovery" as the preliminary evaluation (as, e.g., plausible) of a hypothesis is equally incorrect. A physics graduate student may be advised by his thesis director to try to derive a result that the director sees to be plausible, but if he successfully carries out the task, it is the student, not the director who has discovered the result.[32]

Thus, my objection to the prevailing view of science as consisting of three logically (and according to Laudan temporally) distinct stages is that the analysis of the meaning of the word "discovery" demonstrates itself, the impossibility of separability of those three stages as logically distinct phases of the scientific endeavor. And, if one has to use the usual terminology, however misleading it may be, I would certainly agree with Gutting that, properly understood,the problem of discovery includes the problem of justification. Of course, to say this does not imply that the assertion "X has been discovered" means that X has been accepted or justified. This would be similar to saying that the gentleman found the glove at the moment when the lady identified it as hers, which is obviously a mistake. However, the acceptance of X is a necessary condition which must be satisfied if we are to talk about the discovery of X at all Similarly, the lady's identification of the lost glove is necessary if we should talk about the gentleman as the discoverer. Therefore, a scientific discovery is not just one phase in the scientific enterprise, but its outcome.

Some might find it a puzzling implication of this conceptual analysis that only in some exceptional cases is it entirely clear who made a certain discovery and when. Even a cursory look at an encyclopedia seems to show that everything is in the best order, all discoveries are confidently attributed to various scientists, everything seems indisputable. However, the results of the analysis presented in this paper contradict what we may call the everyday practice of the unambiguous location and ascription of scientific discoveries. But, any kind of confusion disappears as soon as we realize that, in our everyday practice, we are led by certain practical aims in locating and ascribing discoveries to a particular scientist. I shall try to make this plain by using an example.

The Nobel Prize Committee is informed of an important discovery, and is ready to ascribe the adequate recognition to certain individuals, or a group of scientists. Since discovery is the outcome of the scientific enterprise as a whole, each discovery will be a result of many different and varied scientific efforts--experimental observations, production of new and more powerful instruments, new theoretical conceptions, mathematical calculations, reinterpretation of observation, critical evaluation of established results, etc. It could happen that the competent Nobel Committee, for a given case, finds that certain of these elements are the most important for the discovery. In that case, it would seem quite naturalto proclaim as discoverer the person who contributed that particular element to the whole discovery process. In most cases, the greatest importance is ascribed to the work done earlier in the discovery process, and for that reason we (or better--historians of science) often talk about these scientists as the discoverers, because they were the first to conceive some hypothesis and/or realize its plausibility. This being so, it is easy to commit the mistake of identifying those early,and often crucial contributions, with the discovery itself. But, regardless of its usefulness for assigning recognition to scientists (or similar practical purposes, e.g., writing of encyclopedia articles) this identification of specific stages in the scientific enterprise with the discovery process itself is disastrous to a proper philosophical understanding of scientific knowledge. Continuing such a practice would be to ignore results achieved in the conceptual analysis of "discovery".

We also have to consider the following question: Can a role for the idea of an approximate time (if not moment) of discovery be preserved in the idea of scientists coming, for any reason, to agree that a discovery has been made? Or more specifically: How are we to preserve, and should we, our intuition that discoveries are such events that happen at a specific moment, and yet avoid contradicting the results of the conceptual analysis of "discovery" which suggests that scientific (as opposed to ordinary) discoveries do not happen at a specific moment, but are diffused over a longer period of time? Namely, I have a strong intuition that the question of when exactly was, e.g., the planet Uranus discovered, is not a conceptually defective question. Nevertheless, I believe that this is reconcilable with the results of my analysis, if we apply to the temporal location of discoveries something similar to what I said about the ascription of various discoveries to particular scientists, and then make all this a part of the conceptual analysis of "discovery". According to such analysis it would be perfectly appropriate to ask when did a discovery exactly occur, but this question, as well as its answer, would be tied to particular institutions, e.g., quizzes, encyclopedia articles, Nobel Prizes, etc. The temporal location of discoveries would thus be (just like the ascription of discoveries to particular scientists) a semi-institutional fact. In other words, I think that it can be claimed that the moment of the discovery of Uranus is precisely that moment which has been proclaimed to be the moment of the discovery of Uranus. Such institutional proclamations are determined by historical facts. Writers of the history of science could have proclaimed Lexell's suggestion from the second half of 1781 as the moment of the discovery of Uranus, but they didn't: they decided instead that Herschel's observations in March be considered the moment of discovery.

It seems to me that the analysis of scientific discovery as something which is partially connected with some institutions provides a good compromise, for it preserves both the linguistic intuition according to which discoveries have a precise temporal fixation, and my thesis that in the process of scientific investigations nothing of great value happens at any specific instant.

However, a different solution may be even more appropriate: to reject our linguistic intuition that the question "When exactly was X discovered?" always makes sense, then give an explanation of how does this intuition generate, and why science departs from the ordinary usage of the word "discovery". Only an outline of this solution can be presented here. The fact that "discovery" is used to refer to a type of success which is realized at a specific moment could be attributed to the fact that, in most everyday contexts, discovery is made by sense-perception, so that sense-perception discovery has become theparadigm case in terms of which we tend to think of discovery and to use the word. Such "discovery" has been associated with finding "brute facts", which are there to be found, by simply looking. To follow this suggestion, then treating scientific discovery in the Rylean (= ordinary) way would have a deeper motivation: a conception of science as being based on the sense-perception of brute facts.

Prior to concluding this paper, I think it is in order to consider in more detail the reasons which led me to discuss the discovery of the planet Uranus and the discovery of X-rays, in particular. By means of these examples the soundness of a certain philosophical thesis was supposed to be tested; and the final purpose of that thesis was to provide a satisfactory answer to Laudan's challenge. In the course of philosophical argumentation the best test of the soundness of some philosophical thesis is certainly its confrontation with examples that, at first sight, least favour it. In order to see whether my selection of examples from the history of science satisfies this condition, let us recall that the purpose of the arguments from the last section of the paper was to show that the stage picture of science is inadequate. As we have already seen anyone who views scientific discovery as a stage of the scientific enterprise is making the following two commitments: (1) discoveries take place at specific moments; and (2) within the scope of some general methodological scheme of the scientific enterprise it is possible to specify exactly when discoveries take place or have taken place. It is precisely (2) which was the target of my criticism in which the selected examples of scientific discoveries had to play the crucial role. Examination of those examples was supposed to show that (2) is false and, consequently, that the stage picture of science is untenable.

Scientific endeavours which result in the discovery of new entities or new facts are, at first sight, very similar to everyday cases of discoveries made by sense-perception. We tend to think of sense-perception discoveries as having the following two characteristics: (1') they take place at specific moments; and (2')it is possible to know in advance – before undertaking the project itself – what conditions must be satisfied in order to obtain the result (cf., the lost glove example). The fact that there is a prima facie similarity between scientific discoveries of, e.g., a new planet or a new form of radiation and cases of everyday discoveries made by sense-perception indicates that the chosen examples of scientific discoveries are suitable for the test of the form described above. (Namely, we can see those examples as least favouring the claim that (2) is false.) It may seem natural to think, because of the similarity in question, that the discovery of Uranus and the discovery of X-rays are also characterized by (1′) and (2′); and, thus, since (1′) is the same as (1) and (2′) can be regarded as a variant of (2), that the stage picture of science is true. However, it turned out that even these cases of scientific discoveries - which because of their prima facie similarity with everyday sense-perception discoveries should be unproblematic when it comes to determining the exact moment when they occurred - are such that they go against the thesis that within the scope of some general methodological scheme of the scientific enterprise it is possible to specify exactly when discoveries take place or have taken place.

In my opinion, the following are the three main reasons why the mistake of identifying discovery with some stage in the scientific enterprise is made so often: (i) the way in which historians of science talk about discoveries is, although natural, misleading, (ii) it is not difficult to notice that "discovery", as a term, is used to refer to a type of success which is realized at a specific moment, and (iii) it was not realized that, although the word "discovery" relates to a type of achievement occurring at a certain moment, the fulfillment of all of the necessary conditions which result in the ability to apply the word "discovery" is essential (while the process of satisfying these conditions may have extended over a long period of time).

In conclusion, I would like to state my answer to Laudan. The results of the examination of the meaning of "discovery" clearly suggest that discovery is not only a legitimate subject for the philosopher of science, but that it is among the most important. For scientific discovery (being the outcome of the entire scientific enterprise, rather than just a phase) is as philosophically interesting as science itself.

BIBLIOGRAPHY

Achinstein, Peter: 1970, "Inference to Scientific Laws", in R. Steuwer (1970).

Achinstein, Peter: 1980, "Discovery and Rule-Books", in T. Nickles (1980a).

Blackwell, Richard: 1980, "In Defense of the Context of Discovery", Revue Internationale de Philosophie no. 131-132, pp. 90-109.

Burian, Richard: 1980, "Why Philosophers Should Not Despair of Understanding Scientific Discovery", in T. Nickles (1980a).

Campbell, Donald: 1974, "Unjustified Variation and Selective Retention in Scientific Discovery", in F.S. Ayala & T. Dobzhanski (eds.), Studies in the Philosophy of Biology, MacMillan, London, pp. 139-161.

Curd, Martin: 1980, "The Logic of Discovery: An Analysis of Three Approaches", in T. Nickles (1980a).

Gruber, Howard: 1981, "On the Relation Between 'Aha Experiences' and the Construction of Ideas", History of Science 19, pp. 41-59.

Gutting, Gary: 1973, "A Defense of the Logic of Discovery", Philosophical Forum 4, pp. 384-405.

Gutting, Gary: 1980a, "The Logic of Invention", in T. Nickles (1980a).

Gutting, Gary: 1980b, "Science as Discovery", Revue Internationale de Philosophie no. 131-132, pp. 26-49.

Hanson, Norwood Russell: 1958a, Patterns of Discovery, Cambridge University Press, Cambridge.

Hanson, Norwood Russell: 1958b, "The Logic of Discovery", Journal of Philosophy 55, pp. 1073-1089.

Hanson, Norwood Russell: 1960, "Is There a Logic of Discovery", in H. Feigel & G. Maxwell (eds.), Current Issues in the Philosophy of Science, Holt, Rinehart, Winston, New York, pp. 20-42.

Jokic, Aleksandar: 1986, "Naucno otkrice i filozofija nauke", Filozofske Studije.

Kiesel, Theodore: 1980, "Ars Inveniendi: A Classical Source for Contemporary Philosophy of Science", Revue Internationale de Philosophie no. 131-132, pp. 130-155.

Kordig, Carl: 1978, "Discovery and Justification", Philosophy of Science 45, pp. 110–117.

Kuhn, Thomas: 1977, "The Historical Structure of Discovery", in T. Kuhn, The Essential Tension, University of Chicago Press, Chicago.

Laudan, Larry: 1977, Progress and Its Problems, University of California Press, Berkeley.

Laudan, Larry: "Why Was the Logic of Discovery Abandoned"? in T. Nickles (1980a).

McLoughlin, Robert: 1982, "Invention and Induction, Loudan, Simon and the Logic of Discovery", Philosophy of Science 49, pp. 198–211.

Monk, Robert: 1976, "The Logic of Discovery", Philosophical Archives 3, pp.1–50.

Monk, Robert: 1980, "Productive Reasoning and the Structure of Scientific Research", in T. Nickles (1980a).

Nickles, Thomas: 1908a, (ed.) Scientific Discovery, Logic and Rationality, D. Reidel Publishing Company, Dordreht, Holland.

Nickles, Thomas: 1980b, "Introductory Essay: Scientific Discovery and the Future of Philosophy of Science", in T. Nickles (1980a).

Nickles, Thomas: 1980c, Scientific Discovery: Case Studies, D. Reidel Publishing Company, Dorhreht, Holland.

Prigogine, Ilya & Stengers, Isabelle: 1980, "Le probleme de l'invention et la philosophie des scionces", Revue Internationale de Philosophie no.131–132, pp. 1–25.

Reichenbach, Hans: 1938, Experience and Prediction, University of Chicago Press, Chicago.

Ruse, Michael: 1980, "Ought Philosophers of Science Consider Scientific Discovery", in T. Nickles (1980c).

Ryle, Gilbert: 1949, The Concept of Mind, Hutchison, London.

Schaffner, Kenneth: 1974, "Logic of Discovery and Justification in Regulatory Genetics", Studies in History and Philosophy of Science 4, pp. 349–385.

Shapere, Dudley: 1980, "The Structure of Scientific Change", in T. Nickles (1980a).

Simon, H.A.: 1973, "Does Scientific Discovery Have a Logic?" Philosophy of Science, 40, pp. 471–480.

Stuewer, Roger: 1970, (ed.) Historical and Philosophical Perspectives of Science, Minnesota Studies in the Philosophy of Science, vol. 5, University of Minnesota Press, Minneapolis.

NOTES

1. But not with the most recent literature on this issue.

2. That these two questions are not always distinguished even nowadays is best shown by a quote from M. Curd (1980, p. 201):

> Is there anything about the discovery of scientific theories that is of legitimate concern to the philosopher of science? The answer to this question is stillcontroversial. Popper, Hempel, and Braithwaite, for example, have categorically denied that anything like a logic of discovery is possible. On the other hand, the notion has been vigorously defended byC.S. Peirce and,more recently by Hanson, Achinstein, Simon, Shaffner and others.

   It is evident that Curd identifies that which represent the "legitimate concern to the philosopher of science" with a logic of discovery.

3. The view that despite the fact that philosophy of science attempts to explain the nature of scientific progress, it should not deal with discovery is called by E. Pietruska-Madej (1985) "the nihilistic attitude towards scientific discovery".

4. Cf., for instance, M. Curd (1980), T.Nickles (1980b), and G. Gutting (1980a).

5. Cf., Gutting (1980b), p. 31.

6. For a detailed discussion of this problem, see Laudan (1977), pp. 108–114.

7. Cf., Laudan (1980), pp. 174–175.

8. Op. cit. p. 176.

9. Op. cit. p. 182.

10. T. Nickles (1980b), pp. 5–6.

11. Op. Cit. p. 7.

12. R. McLoughlin (1982), p. 198.

13. Laudan (1980, p. 173) explicitely makes three assumptions.

14. Shapere (1980), and in the discussion with Laudan (Nickles, 1980b, p. 104).

15. Prigogine & I. Stengers (1980), p. 1. Translation from French is mine.

16. Nickles (1980b), p. 7.

17. F. Suppe (1974b), p. 125.

18. I have slightly altered the argument Siegel (1980, p. 306) gives,for my reasons to consider it are very different. Siegel discussed this argument in order to argue in favour of the original Reichenbach's distinction.

19. H. Siegel (1980), pp. 307-308.

20. Cf., N. Kertge, panel discussion (Nickles 1980c, p. 47).

21. This term is ever more present in the contemporary discussions of the generation of scientific theories (especially case studies), although, to my knowledge, it has not yet been precisely defined. For a somewhat detailed discussion of the role constraints play in the search for problem solutions see, for instance, T. Nickles (1980b), pp. 33-38, R. Monk (1980), pp. 343, 347.

22. T. Nickles (1980b), p. 37.

23. Op. cit. p. 40.

24. Cf., I. Hacking (Nickles 1980c, p. 56).

25. Ryle labels those words "achievement words", "success words", or "got it words".

26. Ryle (1949), p. 149.

27. Op. cit. p. 151.

28. The example is cited from T. Kuhn (1977), pp. 181-182.

29. The idea that for some terms the conditions required for their correct application may obtain after the time to which these terms refer, is not novel. I first learned of this insight at the conference: Current Issues in the Philosophy of Mind: Functualism anf Explanation, held in Dubrovnik (March 1983), when M. Slote read his paper "Moral Luck". Discussing the ship of Theseus problem he proposed that whether the term "---is identical with" applies to some objects a and b at t_1 may depend on what happens at a latter time t_2. D. Davidson treatment of some mental predicates also makes use of this insight.

30. T. Kuhn (1977), p. 172.

31. Philosophers who see discoveries as sudden moments of insight, dramatic reorganizations of ideas, i.e., as "Eureka experiences" seem to think of the phenomena as though (1) they are very rare in the life of the thinking person, (2) they are very rapid, like a millisecond flash, having no inner structure, and (3) they are singularities, marking a rapture with the past. Now,it is an empirical question whether this conception of "Eureka experience" is correct. Some recent studies provide evidence that this view is wrong on all three points. Cf., for instance, H.E. Gruber (1981).

32. G. Gutting (1980b).

HEGEL AND LOGIC

SVETLANA KNJAZEV-ADAMOVIĆ

A lot has been written concerning Hegel's critical attitude to the so-called formal logic[1].

One should note at the very beginning that Hegel rarely uses the expressions 'formal logic' and other derived from this one; most often he speaks of the 'old' or 'previous' logic or of 'common' logic or the logic of 'understanding'; later we shall see why he uses these expressions and in particular to the 'laws of thought' on which the traditional logic is based (the law of identity, of contradiction, of the excluded middle and of sufficient reason); in particular Marxists have written a lot about this. Perhaps one could exhaust the whole topic of this essay by simply listing this enormous literature. If, in spite of this, I intend to go back to this seemingly unfashionable topic, this is not only because I would like to offer yet another reading of Hegel's texts; true, due to their extraordinary complex structure and non-standard terminology, his texts always invite a diversity of readings. My intention is much more specific; let me now explain it.

A great number of Marxists have in one way or another a negative attitude to contemporary logic. Of course, the time has passed - and let it not return - when the so-called formal logic, both traditional and contemporary, was considered as a useless or even harmful game, as empty formalism - and so was not offered as a subject at the universities in the USSR. Yet it is still widely and vigorously believed that logic is needed by the 'positive', 'uncritical' science, and that any attempt to appeal to logic in philosophical debate or in the 'critical theory' shows only a lack of understanding: logic cannot be applied in this way. The Marxists who implicitly or explicitly believe this, rarely try to offer an original argument in its support; thus they proceed in their work as if on this topic Hegel said all there is to be said. It is this attitude of contemporary Marxists that makes it both useful and interesting to go back to Hegel's original argument.

I hope that even a superficial analysis - the only kind which is feasible within a short essay - can show how Hegel's arguments against that which he calls the logic of understanding are of unequal strength. Some of them are purely doctrinaire arguments because they directly or indirectly follow from the main theses of Hegel's absolute idealism. One could well ask how Marxists

A. Pavković (ed.), Contemporary Yugoslav Philosophy: The Analytic Approach, 151–173.

could accept these arguments given that they have mercilessly criticised Hegel's absolute idealism. Hegel's other arguments may be considered valid only if directed against the views which are found in the traditional logic textbooks but could not be used – at least not without significant modifications – against contemporary logic. Some of his arguments might have appeared to be well-grounded in his own time simply because the logic of the time had no developed apparatus within which one could show their invalidity. To accept these arguments in our time may seem to exhibit an elementary illiteracy in logic. Some of his arguments only appear to be directed against formal logic until one realises that Hegel attributes a quite peculiar meaning to the terms which are of crucial importance for these arguments. Finally, among Hegel's arguments one finds some which are acceptable and which have been accepted by logicians of later times – although the logicians came to these arguments not through Hegel but by their own efforts.

We shall examine Hegel's arguments by going from more general to more particular ones: first, we shall consider Hegel's arguments concerning the traditional logic as a logic of understanding and as a formal logic; then we shall consider his arguments against the so-called laws of thought of the traditional logic and at the end we shall deal with his arguments against the treatment of concepts based on the understanding alone.

The Traditional Logic as the Logic of Understanding

Philosophy, in Hegel's opinion, is the science which aims at a conceptual and so logically necessary reconstruction of that which Christianity presents in the form of images – of the doctrine of God as the only truth and of the doctrine concerning the nature and human spirit in their mutual relation and in their relation to this unique and absolute truth.[2] Since logic is the first philosophical science – the 'science of the Idea in itself and for itself'[3] – he has to construct a conceptual apparatus which would be as rich as the main task of philosophy demands. According to Hegel, logic in its form contains the following three aspects or moments of everything that is logically real: the Abstract or that of understanding, the Dialectical or that of negative reason and, finally, the Speculative or that of positive reason.[4]

With the help of the moment of understanding, one grasps the determinedness of the finite, with the help of the dialectic one grasps that the finite determinations by themselves transform themselves into their opposites; and with the help of the speculative moment one grasps the unity of these opposite determinations so that their oppositeness is resolved in an affirmative way.[5] But as Hegel advocates the unity of logic and metaphysics, these logical moments must not be viewed only as something subjective: on the contrary, the understanding, dialectical and speculative moments are found in everything real and objective. The moment of understanding is the very existence of things as finite items; the dialectical moment is their being subject to disappearance, destruction and change into something which is again finite; and the speculative moment is the presence of something infinite (of God, spirit, truth) in that which is finite. Only the last – the speculative moment – is the rational one but at the same time mystical.[6] For this triad of logical moments or aspects the same holds as for all of Hegel's triads: the moment of understanding as the thesis and the dialectical as the anti-thesis are not true if taken separately – only the speculative as their synthesis contains their truth. But the converse also holds: the synthesis would not be true without its thesis and anti-thesis.

What could Hegel in this context mean by saying that the 'old logic' – that is, logic before him – is the logic of understanding? Hegel often repeats that his speculative logic contains the 'old logic';[7] Hegel also accepts the forms which the 'old logic' discovered – the concept, judgment, inference, the classification of judgment and of syllogism etc. – but he interprets them not only as forms of human thought but as moments of the very Idea, as forms present in objects. In view of this treatment of the 'old logic' I think that Hegel regards it as the logic of understanding because it interprets the forms which it discovered in the manner of understanding. If so, Hegel's demand that the logic of understanding be 'transcended' may be understood as a demand that a new, speculative sense should be given to the forms discovered by the old logic. To put it in the terminology closer to our contemporary one: his criticism does not concern logic alone but a philosophy of logic – which, he believes has to be transcended; his criticism thus concerns a series of philosophical issues concerning the contents and meaning of logical forms.

But Hegel did not always follow this line of thought. One finds passages in his texts in which the old logic seems to be the logic of understanding because it describes the aspects of thought which belong to the understanding.[8] Even so, one could not claim that understanding and 'its logic' are good enough for science but useless to philosophy because Hegel explicitly says that understanding is necessary but not sufficient in philosophy too.[9]

But if Hegel is to be interpreted in this way, it would follow that in addition to the logic of understanding there is or there ought to be a logic of the dialectical and a logic of the speculative moment of the Idea. This, however, obviously contradicts both Hegel's conception of the sides or moments of everything that is logically real,[10] and that which he in fact does; for he does not write – nor pretends to be writing – one or two allegedly missing logics but only one Logic.

Whichever of the two ways of interpreting Hegel one takes, the Marxists clearly should not appeal to this argument of Hegel in their critique of the so-called formal logic. For this argument essentially depends on Hegel's main assumption that there is a subject of the speculative moment of thought, i.e., that God exists (whatever other term we may want to use to refer to God).

If one rejects this subject of speculative thought which is essential for Hegel, then either there is no need for the speculative moment of thought [11] or one would have to construct a new speculative infinite object which corresponds to the speculative moment.[12]

The Traditional Logic as Formal Logic

As I mentioned above, Hegel seldom uses the term 'formal' in connection with logic; he does so only when he writes about – in his opinion inadequate – interpretations of logical forms. The understanding holds that the form and content are contrary one to another in that the content is something 'essential and self-subsisting' while the form is 'inessential and not self-subsisting'; contrary to these views, Hegel insists on their unity – that neither there is any content which in a way peculiar to itself would not be subject to a form nor are there any forms which would carry with themselves a certain content. As an example of this he gives us art and claims that 'the true art works are just those in which the form and content are completely identical one to another'.[13]

But Hegel also mentions the two assumptions required for a correct understanding of this unity of form and content: first, by a form one should not understand only something external and indifferent to the content, for example, whether the copy of a book which contains a work of art is 'bound in leather or paper'.[14] Second, by the content one should not understand that which could be apprehended by the senses alone, because there is no such content either in philosophy or in logic.

In view of this, the main context in which Hegel could talk of logic as formal is the context of his critique of Kant's view according to which the forms are empty and they have to be filled by a content whih they themselves do not possess and which comes to them from outside.[15] But since such talk of empty forms assumes that the forms are understood as something subjective, Hegel's explicit criticism of the traditional logic as a formal one is primarily directed to the subjective aspect of it. When thinking is understood in its 'common subjective sense as one of the spiritual activities or powers among many others',[16] (and not as the basis of everything existing, as Hegel understands it), then logic is regarded as a science of this subjective activity. Aristotle did this first and this is how this is commonly done; having discovered the forms of thought, the 'common logic' considers them only as forms of subjective thought and not as forms of the very reality and of that which is most real, i.e., of the very Idea. Since the old Aristotelian logic discovered the forms of thought but, in Hegel's opinion, interpreted them wrongly, it could not have brought to the self-consciousness its true speculative content. Due to that, it was possible for Kant to regard them as 'empty' and to hold that their content – a sense-content foreign to them - has to come from outside. According to Hegel, this explains why the common logic is formal.

As we have seen, Hegel's criticism of the old logic as formal is based on the main assumption of his system - on the assumption of the identity of the subjective and objective and of the reason as the foundation of the world; Hegel's own doctrine of the relation of form to the content is not a part of his argument here. On the contrary, in view of his own doctrine of form, Hegel could not have criticised any science on the ground that its subject of investigation is a logical form.[17]

As in the previous criticism of the 'old logic' as that of understanding, Hegel's criticism of the traditional logic as formal does not concern the very forms of the traditional logic and purports neither to point out the incorrectness of these forms nor to restrict their scope of application; his criticism is directed against certain philosophical interpretations of those forms and so it concerns not the logic but rather the philosophical understanding of its subject-matter.

And as in the previous case, could the Marxists - who criticise Hegel's main assumption of the mind as the foundation of the world as a form of idealism - accept his thesis of the identity of the subjective and objective? For this thesis stems directly from his main assumption. And if they do not accept it, could they accept Hegel's criticism of the traditional logic as formal, in the same sense of the term 'formal' which Hegel employs? And if they accept the thesis but interpret it in a non-Hegelian fashion, they should show what consequences their interpretation of the thesis of the identity of the subjective and the objective has for a philosophical interpretation of logical forms.

Hegel's Critique of the Laws of Thought of the Traditional Logic

Hegel's critique of formal logical laws certainly rests on the assumptions which are not, strictly speaking, logical but rather metaphysical and ontological: his critique rests on a few fundamental theses of his system. Hegel's aim was to construct a consistent concept of the Absolute such that would not only enable him to explain the movement but would force, so to speak, the Absolute to self-movement. He wanted to have at his disposal a dynamic concept of the Absolute so as to avoid the static picture characteristic of all the idealist systems until then; in all those systems, starting from Plato's and ending in Schelling's the movement would turn out to be only an appearance and the truth would be found in an immobile and self-identical Something which one could call God or the Idea or the Absolute and so forth.

So Hegel took upon himself to 'provide' the Absolute with laws which would make the self-movement necessary; these are the well-known dialectical laws of the unity and struggle of opposites, and of the negation of negation and so on. Does a critique of formal logical laws necessarily follow from this? Yes, but only under the following two fundamental assumptions: first, that the formal logical laws are interpreted not as psychological or logical laws of the thought itself but as the laws of everything that exists; that is, one is to attribute to them a metaphysical and ontological significance.[18] Second, that the general laws of thought and the general laws of everything existing have to coincide, i.e., have to be identical in themselves; that is, the correct thought only reflects the existing laws which may have its origins in the nature of the realm of ideas.

The usual 'metaphysical' (in Hegel's and Engels' sense of word) static picture of the world results only from these two assumptions: if the law of identity is one of the fundamental laws of everything existing and if the law is interpreted in an absolute way, then what exists – the existing – is conceived as immobile so that it does not have the principle of movement in itself but, at best, outside itself.

A close reading of Hegel's texts would, I think, show that he had in mind primarily these very formulations of the formal logical laws in which one attributes to them the metaphysical and ontological significance – in spite of giving them the name 'the laws of thought'. In Hegel's own words:

> Thus the essential category of identity is enunciated in the proposition: everything is identical with itself, A=A. Or negatively: A cannot at the same time be A and not A... Only, since they are enunciated as universal laws of thought, they still require a subject to their relation, and this subject is: everything, or an A which equally means each and every existent.[19]

Of course, Hegel had good reasons for his criticism of the so called 'laws of thought' in this metaphysical and ontological version in which they are valid for 'each and every existent': almost the whole of the traditional logic interprets these 'laws' precisely in this fashion. True, in Aristotle one could find various other formulations of these laws but even he formulates and grounds these laws mainly in the above metaphysical and ontological fashion. A particularly strong emphasis on the metaphysical and ontological aspect of the formal logical laws could be found in Wolff and his followers – in the tradition almost immediately antecedent to Hegel; this is neither historically nor psychologically insignificant for understanding Hegel's own approach.

For the mentalistic and logicistic interpretations which do not demand metaphysical and ontological grounding of the formal logical laws were put forward after Hegel's death (although Kant has to some extent anticipated approaches such as these); this holds in particular of the operationalism and other contemporary approaches which deny that there is any metaphysical and ontological aspect to these laws. This may lead us to doubt that Hegel's arguments directed against the interpretation of the formal logical laws characteristic of the traditional logic are valid against a logic which interprets and grounds its laws in a quite different way.

Let us now consider Hegel's arguments. One could devide them into two groups: the first concerns all formal logical laws, i.e., the view which takes them all together and accepts them as a group; the second concerns particular laws of this kind. The first general objection – which concern all the formal logical laws – is directed against their formulation as propositions. In order to explain the meaning of these objections one should see what Hegel means by the term 'proposition'. In his Science of Logic Hegel writes:

> For the difference between proposition and judgment is mainly that in the former the content constitutes the relation itself or is a specific relation. The judgment, on the contrary, transfers the content to the predicate as a universal determinateness which is for itself and is distinct from its relation, the simple copula.[20]

In the Encyclopaedia of Philosophical Sciences Hegel formulates the difference in another way:

> A judgment is, however, distinguished from a proposition. The latter contains a statement about the subject, which does not stand to it in any universal relationship, but expresses some single action, or some state or the like...Again it is absurd to say that such statements as 'I slept well last night' or 'Present arms!' may be turned into the form of a judgment,[21]

As we can see the definitions offered in the Encyclopaedia are wider: a twofold division of all sentences into judgments and propositions is offered. In judgments such predicates are used which are general in their relation to the subject, and propositions are all other sentences even those which assert nothing. But we find among propositions even those sentences which assert something – whether they are relations (and these alone are mentioned in the Science of Logic), states or actions etc.

It is interesting to note that Hegel not only grasps the difference between sentences in which generality is being asserted (i.e. judgments of the S–P structure) and sentences of a relational structure but he also recognises the logical peculiarity of the latter. Of course, the question as to whether the latter are reducible to the former he formulates in an unclear way: his example shows only that the sentences which do not contain assertions could not be reduced to judgments. But the text of the Science of Logic seems to point out to the irreducibility of relational propositions to judgments in principle.[22]

If one takes into account that Hegel's texts originate in 1812 (the Science of Logic or 'The Great Logic') or 1817 (Encyclopaedie der philosophischen Wiessenschaften or the 'Little Logic') and that the debates concerning the logical peculiarities of relational propositions started only in the eighteen fifties, one has to grant to Hegel an uncanny ability to anticipate some of these debates.

Let us, however, not digress. As we have seen, Hegel claims that formal logical laws are formulated as propositions; this means that they are relational structures, assertions concerning relations, and not concerning generality which would be attributed to the subject of a sentence. This claim in the Science of Logic is followed by two objections. First, that the form of propositions is superfluous because identity, differences, etc. are determinations of reflexion and 'in that they are determinatnesses which are in themselves relations, to that extent they already contain within themselves the propositional form.'[23] The second objection is that these relations (identity, difference, contrariety, contradiction) are being wrongly put back into the sphere of being because one takes "A" to be the subject of the propositions which express these relations and "A" means 'each and every being'; but these relations in fact (i.e., within Hegel's system) belong to the sphere of essence.

In his Encyclopaedie der philosophischen Wissenschaften (the 'Little Logic') there is no mention of these objections. It seems that Hegel did not consider them important; perhaps he realised that they have no great significance because they are formal rather than essential in nature. The first is fairly unclear in any case and the second acquires sense only within Hegel's system and so it is not generally valid.

Hegel's next objection is a principled one and concerns the whole conception within which the four laws and the so-called law of difference are affirmed:

> The several propositions which are set up as absolute laws of thought, are, therefore, more closely considered, opposed to one another, they contradict one another and mutually sublate themselves. If everything is identical with itself, then it is not different, not opposed, has no ground. Or, if it is assumed that no two things are the same, that is, everything is different from everything else, then A is not equal to A, nor is A opposed to A, and so on. The assumption of these propositions rules out the assumption of any others.[24]

As we see, Hegel here appeals to the authority of the law of contradiction because his objection reduces to the claim that these laws contradict one another. Another assumption of this argument is the ambiguity of meaning of the term 'all'. For if 'all' means the whole, the totality of all 'things' - of all possible existing entities - and if this totality, the whole of all existing, is identical with oneself, then it is not different - there is nothing to be different from since it comprises everything; it is not contrary to anything either because again there is nothing to be contrary to and it has no basis because there is nothing outside of it. In this case, the difference and everything else (the contrariness, basis, etc.) can exist only in it, in the totality and thereby can exist within the identity and not as a difference external to it. But if we take another meaning of 'all', this part of Hegel's argument is valid. This is the meaning of 'all' as 'every particular thing' or 'every individual entity'; This is the sense in which Hegel in fact takes it in the clause which starts with 'and if one takes it that there are no two identical things...' If we take it in this sense, then the proposition of identity has another meaning too: that every A is identical with A; or in the symbolic notation:

$$(x)(A(x) \leftrightarrow A(x))$$

Hegel thinks that one can show the falsity of this proposition if one starts from the proposition that there are no two identical things in the world; for, according to him, A on the left of the identity relation is one 'thing' and A on the right is another 'thing', and so they cannot be identical. Consequently, Hegel could have maintained that the law of identity in this case could not be expressed at all.

The argument, however, rests on two assumptions which contemporary logic regards as serious confusions: first, Hegel fails to distinguish the level of signs from the level of things signified, and, second, he fails to heed the type-token distinction. As for the first, one should note that signs or symbols are used to talk about objects but one cannot talk about the signs or symbols themselves using the same signs or symbols. What the law of identity asserts is not that the two signs (let us for the moment assume that these are two signs and not two occurrences of the same one) on both sides of the identity sign are identical to each other; but with their help one asserts that the object signified by one sign is identical to the object signified by the other. Correspondingly, when we claim that the triangle is a flat geometrical figure with three sides, we are not thereby saying that the word 'triangle' is identical to the words 'geometrical figure...' but are asserting that the ideal object signified by the word 'triangle' is identical with the ideal object signified by the words 'flat geometrical figure....".

As for the second assumption, it partly depends on a conception of identity (absolute identity or an abstraction of identity) of which we shall say more below. For the moment, against the assumption one can point out that if one treats every occurrence of a symbol as an independent symbol, as a separate thing on its own, one thereby precludes any communication. For instance, on this view, one could not say that an alphabet has so many letters, because there are infinitely many individual letters - as many as they were ever used and will be used from the beginning of that particular language and its alphabet until the language and alphabet cease to exist. On this view, every particular use of a letter (in writing or in speech) is a letter.

In connection with this objection we shall discuss yet another possible interpretation. If we accept the law of identity in the form of 'every A=A', could we perhaps infer from this the proposition: 'The whole (totality) is identical with itself', the proposition which Hegel has shown to be untenable. For one could reason as follows: if everything is identical with itself, then the totality is also identical with itself because it is also a 'thing'. Against this the contemporary logic has a powerful objection;: the theory of types prohibits any talk of totalities in the same way and the same sense as one talks of the elements of the totalities. In other words, the theory of types in the name of logical precision prohibits a too 'universal' language. If every 'thing' within a totality – within the whole of everything existing – is identical to itself, one is not allowed to infer that the totality is identical to itself; a statement concerning the identity of individual items does not and must not have the same sense as a statement of identity concerning a set, class or totality of these items.

Having advanced these general objections against all formal logical laws – i.e., against the thought construction within which they are contained – Hegel proceeds to state objections against some of these particular laws. Contrary to what one might expect, his criticism focuses not on the law of contradiction (that A cannot simultaneously be A); he considers the law of contradiction just a negative form of the law of identity. In the 'Great Logic' and in the 'Little Logic' he objects to the law of identity on the ground that it does not correspond to its form; in the 'Great Logic' his argument is longer but more vague while in the 'Little Logic' the argument is presented in a laconic but clear form: 'The propositional form itself contradicts it: for a proposition always promises a distinction between subject and predicate; while the present does not fulfil what its form requires'[25]

There are two ways in which one could interpret this objection: the first a psychologyzing way refers to our habit of finding the subject and the predicate in one proposition different; and thus it is contrary to this habit to find one term on both of these places. The second interprets the objection as logical, based on the failure to distinguish the function a term has in a proposition from the term itself. To be a predicate and to be a subject are indeed two different functions but this does not imply that one and the same term cannot perform both of them – and so, in the proposition expressing the law of identity one and the same term functions both as a subject and as a predicate. To put it in another way: the use of two variables neither requires nor implies the prohibition that both variables take the same value from their common domain. One would introduce such a prohibition only in multisortal logic: if each of the two variables belongs to a different 'sort', i.e. if one would take its value from one domain and the other from a different domain, then one would rule out the possibility that two such variables take the same value. But the law of identity is not expressed with the help of such a logic; and, moreover, it is hard to see how one could express it in such a way.

Hegel's next objection aims to show that the identity separated from the difference is an empty, abstract identity based on the understanding and that the true dialectical identity includes the difference in itself. In criticizing the form of propositions which the laws of formal logic take, Hegel, as we have seen, maintained that the identity, difference, etc., are not determinations of the being but essences. Contrary to this view of his, his arguments in favour of the thesis of the unity of identity and difference find their support in experience. This appeal to the experience he justifies by reference to the fact that the adherents of formal logic do the same: they, Hegel says in jest, appeal to the never experienced experience that every mind agrees with these laws. In opposition to them, Hegel appeals to the 'genuine' experience which as yet has not been altered by reasoning:

> (F)or the fact is that experience contains identity in unity with difference and is the immediate refutation of the assertion that abstract identity as such is something true, for the exact opposite, namely, identity only in union with difference, occurs in every experience.[26]

There are many other places both in the 'Great Logic' and in the 'Little' one in which Hegel argues in a similar way in order to prove that in the experience the identity is always given in the unity with the difference.

Now one should note that following the traditional logic Hegel distinguishes equality from identity; identity is the relation which an entity has only to itself while two entities may stand in the relation of equality. In consequence, Hegel's arguments should be interpreted as arguments concerning the occurrence of identity together with the difference in experience; Hegel's views on equality and inequality we shall consider later. Thus Hegel holds that the identity of a thing to itself always occurs in experience as the identity tied to difference; and this appears prima facie untenable.

Here Hegel seems to have had in mind that every concrete entity is complex and so identity does not occur as a simple identity of an abstract entity with itself, but as an identity of a complex whole of different properties, sides, aspects, relations, etc... I say 'seems to have had in mind' because Hegel nowhere says this explicitly; I infer this from his claim that absolute identity is simple. It is in this sense that the identity occurs in a concrete experience as an identity which includes difference in itself. If one interprets Hegel's argument in this way, it would be both valid and tenable.

But this argument advanced on the level of experience, i.e., on the level of empirically given particular objects was for Hegel certainly unimportant. For his whole critique of formal logical laws is presented not in the section on being but at the beginning of the section on essence; and Hegel also criticises any attempt to 'take back' the determination of identity, difference, etc., into the realm of being; and so these arguments which concern the realm of being and not of essence must have been quite unimportant for him. But it is interesting to note that Hegel's argument at this level is quite tenable: one can agree with him that in experience identity always occurs united with difference.

But since in Hegel's work identity, difference, etc., are treated as relations in the realm of essence, it is more appropriate to represent the question of the correctness of formal logical laws as the question 'Is every ideal object identical with itself?' than as 'Is every empirical given individual object identical to itself?' This way of raising the question is of crucial importance only for a narrow circle of philosophers whose point of view one may regard as extreme realism; only these philosophers claim that numerically unique ideal objects exist or subsist or have some other mode of existence separate from the individual empirically given objects which exist in the usual sense of the word. These ideal objects are absolutely identical to themselves and Hegel's criticism of this view is both apposite and successful. The assumption of the absolute identity is not, or at least need not be, a part of any other conception of ideal objects whether it be a moderate realist doctrine of universals (ideal objects, essences) in things or an extreme nominalist doctrine denying their existence; all these conceptions of ideal objects can accept a relative concept of identity. All such views can accept Hegel's argument concerning the complexity and they can view identity as the identity of different sides, properties, relations, etc. Moderate realists may in particular accept the thesis that one and the same universal 'exhibits itself' in various concrete cases in a somewhat different manner. Nominalists accept or may accept the thesis that the same words are used in a partly identical and a partly different way and that in consequence identity occurs as an identity of difference.

But Hegel apparently did not intend to prove only that all concrete exhibitions of identity, whether in the realm of being or in that of essence, at the same time include an exhibition of differences too. Within his system the thesis of the unity of identity and difference has two further possible points: to prove the unity of identity and difference in the framework of the relation of being and essence (namely, to prove that being and essence are both mutually identical and different); and to prove that the very concept of identity includes in itself the concept of difference and vice versa. As Hegel's arguments in support of these two points are interdependent, we shall expound his arguments in continuo and then by analysing it indicate the separate points. And since in Hegel's critique of formal logical laws, arguments of this kind are found in many places I shall discuss only a few of them so as to show how contemporary logic could deal with them.

At one point, for example, he writes:

> In other words, identity is the reflection-into-self, that is identity only as internal repulsion, and is this repulsion as reflection-into-self, repulsion which immediately takes itself back into itself. Thus it is identity as difference that is identical with itself. But difference is only identical with itself in so far as it is not identity but absolute non-identity. But non-identity is absolute in so far as it contains nothing of its other but only itself, that is, in so far as it is absolute identity with itself.[27]

In this text conclusions are drawn in a way which obviously breaks the restrictions imposed by the theory of types. For when we speak of the identity and difference between the objects or entities of any kind, the names of these entities are logical constants or variables of the zero order (the names of individuals); and so the difference and identity of those entities are first order relations (relations among individuals); but the relations between the very concepts of difference and identity are second order relations. In other words, the relations among individuals - which Hegel mentions in the first place - must not be identified with relations among the concepts of these relations. Hegel in fact does so and the apparent persuasive force of his arguments rests on this confusion. Thus he saysthat 'the difference is identical with itself...' etc. This difference is a relation of the first order because it is a relation among individuals (whatever they are - individual objects, ideal objects or whatever else); it is also non-identity which is in this context again a first order relation. But the identity of the concept of difference with itself (i.e., with the concept of difference) is a second order relation. The same occurs in the next sentence where Hegel tries to demonstrate that non-identity is in fact identity with itself: non-identity is here a first order relation (the non-identity of individuals); the identity of the concept of non-identity with itself (i.e., with the concept of non-identity) is a second order relation. In any case, the very phrase 'non-identity...contains only itself' (the author's emphasis) reminds us of the paradoxes such as Russell's paradox, which the theory of types was intended to eliminate.

There is now only the first part of the quoted paragraph left in which one cannot detect any infringement of the theory of types; in this part, Hegel tries to demonstrate that identity is difference. If one takes into account the claims Hegel makes elsewhere concerning reflexion, the relation of being and essence and concerning essence as an identity, one can take this part to be concerning the relation of being to essence; according to this interpretation, Hegel here maintains that being and essence are and are not identical - that is, that the relation of being and essence contains both difference and identity. One can easily accept such a claim; but one can then offer a close analysis which would show in what sense, in which aspect and to what extent being and essence are identical and what sense, aspect and extent they are different; and by such an analysis one would be able to avoid the claim that identity is difference.

Let us have a look at another similar place; there Hegel writes:

> It is thus the empty identity (i.e., the concept of identity which is employed in the law of identity, S.K.) that is rigidly adhered to by those who take it, as such, to be something true and are given to saying that identity is not difference, but that identity and difference are different. They do not see that in this very assertion they are themselves saying that <u>identity is different</u>; so they are saying that <u>identity is different</u> from difference; since this must at the same time be admitted to be the nature of identity, their assertion implies that identity, not externally, but in its own self, in its very nature, is this, to be different.[28]

Here we can recognise our old acquaintance - the infringement of the theory of types. Once again the identity and difference are first order relation and the difference between the concepts of identity and of difference is a second order relation which does not have the same meaning or sense as a first order relation. And even if there were no such infringement, it would be easy to show that the argument is not valid; its form is:

$$\underline{x}R\underline{y} \rightarrow \underline{x}R\underline{x}$$

and it is easy to show that this is not a valid form of inference. If R is any relation, then from $\underline{x}$ is the father of $\underline{y}$ it does not follow that $\underline{x}$ is the father of $\underline{x}$. The argument - which due to Hegel's way of talking could be interpreted in various ways - may be interpreted in the following way too:

$$\underline{x}R\underline{y} \rightarrow \underline{x}R$$

But either the left hand expression or the right hand one is here senseless. For the 'difference' in question is either a relation (a two-place predicate) or a property (an one-place predicate); if it is a relation, the left side has sense but the right does not and if it is a property then the converse is true. One and the same word ('difference') cannot in the same context be both a property and a relation. There are a number of similar places in both the 'Great Logic' and in the 'Little Logic'; since they exemplify the above two types of confusion, I shall not quote them here.

Whether or not Hegel's arguments in support of the claim that the concept of identity contains in itself the concept of difference and conversely, are valid, they point out to the following problem: How to define identity and difference without circularity? That is, how to get definitions in which the identity would not be defined as the absence of difference, as non-difference, and the difference would not be defined as the absence of identity, i.e., non-identity. Were identity and difference only correlative concepts (such as the concepts of father and son, of husband and wife and of master and slave) then an occurrence of one in the definition of the other - even if it could not be avoided - would not lead to a circular definition; for in such a case both concepts would have been defined by the relation which holds between them and so, provided that the relation is not defined by reference to either of them, there would be no logical fallacy of circular definition. But identity and difference are not simple correlative concepts but contradictory co-ordinated ones: one of them is a negation of the other (such as white and non-white). If so, the rules of logic require that one of them be defined independently, and that the other can be defined negatively, as the absence or negation of the one which has already been independently defined. And in contemporary logic identity is defined independently, by using implication or equivalence; and in this way one avoids circularity.

Hegel's critique of the law of excluded middle contains two interesting arguments. The first runs as follows: according to the usual interpretation of the laws of excluded middle 'it means nothing more than that, of all predicates, either this particular predicate or its non-being belongs to a thing. The opposite means here merely the lack (of a predicate), or rather indeterminateness; and the proposition is so trivial that it is not worth the trouble of saying it.[29]

This objection to the traditional view of the law is well grounded; for the traditional logic neither distinguishes grammatical from logical (or semantic) meaningfulness nor explicates the difference between absolutely contradictory and contradictory co-ordinated concepts; and so in the framework of the traditional logic, one of the pair of propositions - 'the spirit is sweet', 'the spirit is non-sweet' - would have to be true. Moreover, Hegel could have shown that this traditional view leads to the absurd conclusion that to one subject belongs one and only one predicate.

But in contemporary logic such paradoxical conclusions and trivial applications of the law of excluded middle are avoided by the means we have already mentioned: for every predicate one determines the domain of its definability - to use the usual phrase - or meaningfulness; or, alternatively, one can say that one determines its 'universe of discourse'. So the predicate '...is even' is defined or meaningful with regard to positive and negative integers (in other words, this is its 'universe of discourse'); the term is not defined or meaningful not only with regard to plants or something similar but it is not defined for, say, irrational numbers. In this way, in the set of grammatically correct sentences (such as the sentence 'This plant is even') one finds a set of logically (semantically) meaningful and a set of logically (semantically) meaningless sentences. The sentence 'The spirit is sweet and non-sweet' would be logically (semantically) meaningless, so that the question of truth of any of its components would not arise.

The distinction between the absolutely contradictory and co-ordinated contradictory concepts and the restriction of the application of the law of excluded middle only to the latter, serves the same purpose. For in case of the co-ordinated contradictory concepts, that negative concept is not 'only the absence or indefiniteness' but the absence of one predicate within the framework of first generic concept; for example, 'non-white' does not mean any other predicate but the absence of white colour, that is the presence of some other colour (and not of any other quality). Therefore the question of definition and meaningfulness arises first in relation to the generic concept; so, if the application of the generic concept is meaningful in a given context (for example, if an object can have a colour at all), then the question of which of the two co-ordinated contradictory concepts can be truly applied is also meaningful (e.g. the question of whether a coloured object is white or non-white is meaningful).

Russell's theory of descriptions (or any theory equivalent to it in its logical 'strength') is yet another example of the restriction of validity of the law of excluded middle in contemporary logic; in this case, the law is inapplicable to the judgments the subject of which is a so-called 'empty concept' (the concept which denotes nothing or, in other words, the concept which designates null or empty class). According to the traditional logic, one of the pair 'The present king of France is bald' and 'The present king of France is non-bald' must be true; but within the framework of contemporary logic it could be shown that both are false - and that corresponds to our intuitions in this case.

Therefore, this objection to the law of excluded middle by Hegel is a justified objection against the traditional view but not justified if one takes into account contemporary logical theories.

Hegel's second objection to the law of excluded middle runs as follows:

> The law of excluded middle 'implies that there is nothing that is neither A nor not-A, that there is not a third that is indifferent to the opposition. But, in fact, the third that is indifferent to the opposition is given in the law itself, namely A itself is present in it. This A is neither +A nor-A and is equally well +A as -A. The something that was supposed to be either -A or not A is therefore related to both +A and not-A; and again, in being related to A, it is supposed not to be related to not-A, nor to A, if it is related to not-A. The something itself, therefore, is the third which was supposed to be excluded'.[31]

This is a masterpiece of sophistry. First, Hegel substitutes the logical operation of negation with a mathematical one, and then he equates with +A; this gives rise to the illusion that the argument is a convincing one. But in the very statement of the law of excluded middle there is no mention of any +A but only of A; so one cannot infer that this A is something third because it is only one side in the contradiction of A with not-A. But when one is discussing examples in which quantities (and not concepts) -A and +A are mentioned, then the introduction of A alone without any sign, as something third, results from a confusion of contexts; this is a confusion of vector with scalar quantities or natural with positive numbers.

This completes our statement and analysis of the part of Hegel's critique of the so-called formal logical laws of thought. Let us now see what conclusion can we draw from it.

Hegel's critique concerned primarily the application of these laws to the whole of reality; the interpretation against which it is directed thus either explicitly proclaims them to be the laws of everything existing or at least finds metaphysical and ontological foundations for them. In this sense Hegel's critique does not concern the views which do not infer logical laws from the laws of reality in one way or another; and modern views on logic are of this, second, type. This does not mean, of course, that these views are perfect and not subject to any criticism; this only means that they cannot be successfully criticised by arguments of the Hegelian kind. Given that no metaphysical and ontological foundations are offered for Laws of Logic by these views, they are obliged to find some other logical foundations, i.e., they are under obligation to point out to the source from which, according to them, the laws of logic derive their authority.

But suppose that the laws of identity, of contradiction and of the excluded middle, conceived as metaphysical and ontological laws, do in a sense, result in a static conception of the world. One can then ask whether the same is the case mutatis mutandis when they are conceived only as laws of thought. Does the acceptance of such laws require or lead to a static conception of thought as a non-evolutionary structure given once and for all? As far as I can see, an objection of this kind may be removed once one grasps the huge difference between the Absolute and thought. For the Absolute has to contain the principle of movement in itself and its movement has to be self-movement; this is required if one is to be able to explain movement at all. But thought does not have to contain the principles of its 'movement' within itself; its 'movement' (the progress of knowledge, the rejection of old and creation of new theories) is not 'self-movement' but it is mediated by something outside thought; that something we can call 'reality', 'experience', 'praxis' or anything we like. And the formal logical laws are, after all, not laws of the 'movement of thought' but only norms for structuring of the results of thought.

As for Hegel's objections to particular formal and logical laws, his critique of the law and of the concept of identity - to which Hegel devoted a lot of effort,time and space - surely deserves most attention. At first this crtique appears so radical that one easily gets the impression - widely shared among commentators - that it concerns not some of the formulations of the law given at the time but the very possibility of formulating such a law and of employing such a concept. But a closer look at his critique reveals that it is directed only against some conceptions of identity, viz., the conceptions in which identity is an absolute concept and in which it is not acknowledged that identity is only an abstraction which in all the concrete cases of its realisation or occurrence is accompanied by some difference. But contemporary logic rarely uses the absolute concept of identity. Although this assertion may appear too bold, it could be proven easily. One should first note the fact that in contemporary logic the law of identity does not occur in the way it does in the traditional logic. Of course, in the contemporary logic one can find some analogues of the traditional law of identity; in fact, one can find quite a number of them – one in the propositional logic, second, in the first order predicate calculus, third, in the second order predicate calculus, fourth, in the modal logics, etc. Apart from those, the metalogical or semantic requirement that the terms employed retain the same meaning throughout may be viewed as yet another analogue to the traditional law of identity.

But is one employing or has one to employ the absolute concept of identity in any of these cases? Let us consider each in turn. The metalogical or semantic requirement that there be the same meaning of terms (or that one should avoid the ambiguity) means only that these terms in one context should be used in a way which avoids the fallacy traditionally known as quaternio terminorum. This requirement does not rule out the possibility that the meaning of terms should change in the course of time or depending on the context of their use nor does it demand the absolute samness of meaning.[32] Of course, in modern logic we can find more one-sided and exclusive views concerning the principle of the identity of meaning; one such view is the concept of an 'ideal language' developed in the works of early Wittgenstein, Russell and the logical positivists; in relation to this view, Hegel's critique has not lost all of its relevance. But only the demand for an absolute identity of meaning would be the analogue to the traditional concept of identity.

The second analogue of the law of identity is the following theorem in the propositional logic: $p \rightarrow p$ (every propositional implies itself). This theorem as well as its corresponding theorems in the predicate calculus of the form

$$Fn\ (x_1 \ldots . x_n) \rightarrow Fn\ (x_1 \ldots . x_n)$$

are in fact often called the law of identity. But all these theorems do not directly employ the concept of identity or of equality and mean roughly the same as the above requirement of the sameness of meaning or the requirement of consistency.

Another group of analogues to the law of identity are the axioms which introduce the concept of equality into the predicate calculus. For the first order predicate calculus (which allows only the quantification over individual variables) the following are the schemes of these axioms:

$$a=b \Rightarrow F(a)\Rightarrow F(b)$$ [36]

Equality is introduced into the second order predicate calculus (which allows for the quantification over predicate variables) by a similar axiom-scheme:

$$a=b \equiv \forall F(F(a)\Rightarrow F(b))$$

And if one is stating the theorem as a definition the equivalence sign is replaced by the sign indicating equality by definition.

Now it is commonly said that both of these formulas express in the language of modern logic Leibniz's principium indiscernibilium. And the relation so defined has the formal properties of reflexivity, symmetry and transitivity; and it licenses the employment of the rule of substitution of identicals. Should we, on the basis of that, conclude that we are here dealing with an absolute concept of identity?

Let us digress a little into history. In 1892 Frege considered the question of whether the phrase a=a means the same as a=b on the assumption that a and b refer to the same object; he concludes that a=a is a tautology which does not expend our knowledge whereas a=b is a synthetic proposition because the terms a and b do not have the same sense in spite of referring to the same object; each of the terms represent the same object in a different way, from a different aspect. The identity of a and b, according with this view, includes in itself both the identity (the same object) and difference (the different ways or aspects of comprehending the objects). The last sentence does not state the famous logician's own thought but it could be validly inferred from his thought.

In our contemporary logic one usually speaks of identity or equality 'to a certain degree' or 'in relation to a set of properties' and the like; this way of speaking emphasizes the relative character of the contemporary concept of identity or equality (for contemporary logic does not distinguish the two). As an example, let us quote a contemporary interpretation of the concept:

> The concept of identity in philosophy and logic has often been exhausted by the traditional formula of the law of identity x=x; but this formula expresses only the reflexivity of the relation of identity (and of the relations of that type) and this is insufficient for definition of the concept (predicate) of identity. Every non-trivial use of this concept is based on the principle of identification of indistinguishables.[33] i.e., it presupposes the identification of those objects which cannot be distinguished within the framework of some abstraction of identification (within the interval of abstraction of identification); this is typically the case with applied logical and mathematical calculi. Because of this, the law of identity as it is applied has lost its former fundamental meaning of 'the principle which expresses the unchangeability of the existing' which was attributed to it with no justification. This was replaced by various definitions of the predicate of identity (equality) which have nothing in common with the 'metaphysical foundations' of formal logic.[34]

One could object to my quotation from the Soviet work on logic because Soviet logicians are Marxists and their interpretation may thus presuppose the acceptance of Hegel's critique and the reformulation of the law in the light of it. But in fact, their interpretation and their quite successful work in contemporary logic only show that Hegel's critique concerns not the very foundations of the so called formal logic - and the contemporary symbolic logic is certainly formal in that usual sense of the word - but rather the conceptions which employ the absolute concept of identity. The development of contemporary logic does not necessarily require such an absolute concept; contemporary logic has been developing quite well while using the relative concept of the abstraction of identity; that is, it has been developing under the assumption that the same objects are and can be distinguished outside the framework of the system within which we have been comparing them in respect of a set of relevant properties and within which we have asserted their identity or equality. Of course, contemporary logic - in particular modal logic - has at its disposal means powerful enough to define the concept of absolute identity but usually no such concept is defined. This is because the concept of relative identity (equality) is quite sufficient for its application and development.

From this, it does not follow, of course, that the very concept of identity contains in itself the concept of difference nor that the very concept of identity contains in itself the concept of non-identity. In any case, a similar relation holds between other pairs of contrary concepts which in ordinary life, in reality we do not encounter in their 'pure form'. For example, a real man is neither absolutely free nor absolutely bound by necessity; but this does not mean that freedom contains in itself necessity (not even in its known form, as Hegel believed) but only that freedom is the absence of necessity.

And although Hegel sometimes attempted to prove the contradictoriness of concepts used in the traditional logic by force and sophistry, one should not be unjust to him and identify him with the lazy dialecticians for whom periodical swinging between the contrary assertions seems to be the pinacle of wisdom; this kind of 'dialectic' Hegel considered as a form of sophistry, maintaining that 'one cannot remain with a contradiction because it annuls itself, and in effect transcends itself'.[35]

Translated from Serbo-Croat by
Aleksandar Pavković

NOTES

1. One should note at the very beginning that Hegel rarely uses the expressions 'formal logic' and other derived from this one; most often he speaks of the 'old' or 'previous' logic or of 'common' logic or the logic of 'understanding'; later we shall see why he uses these expressions.

2. Hegel's Logic, Oxford 1873, #1.

3. Hegel's Logic translated by William Wallace, Oxford, 1873 (Abbreviated henceforth as 'HL'), #18

4. HL #79 and notes on this paragraph.

5. HL, #81 and 83.

6. Appendix to #82 of HL.

7. On this the most explicit is section 9 of HL in which Hegel says: 'Speculative Logic contains all previous logic and metaphysics: it preserves the same forms of thought, the same laws and objects (my emphasis) while at the same time remodelling and expanding them with wider categories.' See his note to #82.

8. For example, the appendices to #20, #28 and to #80 of HL.

9. 'That Philosophy can never get on without the help of understanding hardly calls for special remark after what has been said. Its foremost requirement is that that every thought be grasped in its full precision and nothing allowed to remain vague and obscure'. Appendix to #80 of HL, p. 146 of the Wallace translation. See also Appendix to #36.

10. 'These three sides...may all be put under the first stage, that of understanding, and so kept isolated from each other; but this would give an inadequate conception of them.' , #79.

11. This is what the Diamat, the orthodox Soviet Marxists have done. In rejecting the speculative which reconciles the understanding and the dialectical, they have remained on the level of contradistinction of understanding and dialectics. And since they opted for the dialectics alone, this forced them to treat the understanding and the formal logic in a purely negative way.

12. The candidates for these objects are History, The Sense of History and the like found in the 'Western Marxism' (Lukacs, Bloch).

13. See #133 of HL as well as a note and appendix to this section.

14. True, Hegel calls this 'something external' a form too, but only an 'External Form'; he also offers a criterion of distinguishing the 'External Form' from the real one: 'There is thus a doubling of form. At one time it is reflected into itself; and then is identical with the content. At another time it is not reflected into itself, and then is the external existence, which does not at all affect the content.' HL, #133. No doubt, it is not easy to understand this criterion.

15. HL, #43 and its appendix in which Hegel ridicules Kant's view that forms in themselves are empty.

16. 16 HL, #20; see also #24 and appendices to this section.

17. 'If the logical forms of the notion were really dead and inert receptacles of conceptions and thoughts, careless of what they contained, knowledge about them would be an idle curiosity which the truth might dispense with. On the contrary, they really are, as forms of the notion, the vital spirit of the actual world.' HL, #162.

18. I shall use the philosophical terms in the meaning which Hegel attributes to them only when quoting or paraphrasing Hegel's texts; this, I hope, will prevent various ambiguities which arise when one forgets that Hegel attributes to most of the terms quite peculiar meaning.

19. Wissenschaft der Logic Erster Teil, Die Objective Logik. pp. 505 and 506–507 of the Samtliche Werke, 1965, pp. 409 and 410 of Hegel's Science of Logic, translated by A.V. Miller, London, Allen & Unwin, 1069. All page references are to this translation.

20. Ibid. p. 410 of the translation.

21. HL, #167.

22. Here the terms 'proposition' and 'judgment' have been used in Hegel's sense; in a more modern terminology one would first divide sentences into declarative and non-declarative (imperative, questions, etc.). The first, which could be called propositions, could be further classified according to their structure; and the terms 'proposition' and 'judgment' would be used synonmously.

23. The Science of Logic p. 410.

24. Ibid. p. 411.

25. The Encyclopaedie der philosophischen Wissenschaften (HL) #115 (tr. W. Wallace).

26. The Science of Logic, p. 415.

27. Ibid. p. 413.

28. Ibid. p. 413.

29. Ibid. p. 238.

30. Ibid.

31. Ibid. pp. 438–439.

32. See for example S. Zuszczewa Romanova 'Wieloznacznosc a jazyk nauki' in wartalnik filozoficzny 1948; and M. Kokoszynska 'Nazwy nieostre' in Institut filozofii i sociologiji. Sprawozdania z prac naukowych wydzialu nauk spolecznych, zesc. 3–4 1959).

33. That is, on Leibniz's principium identitas indiscernibilium (the author's note).

34. Philosophical Encyclopaedia (in Russian), Volume V, Moscow, 1970, p. 237.

35. HL #119 Appendix.39. HL #119 Appendix.

TEMPORAL MODALITIES AND MODAL TENSE OPERATORS

ALEKSANDAR KRON

1. INTRODUCTION

Let A be a proposition and let us consider the following phrases:
(1) *It is always necessary that* A;
(2) *It is always possible that* A;
(3) *It is sometimes necessary that* A;
(4) *It is sometimes possible that* A.
Our first aim is to construct a model theory for (1)-(4).
Let us now consider

(1') *It is necessary that always* A;
(2') *It is possible that always* A;
(3') *It is necessary that sometimes* A;
(4') *It is possible that sometimes* A.
Our second aim is to construct a model theory for (1')-(4').

Let us write $a\Box$, $a\Diamond$, $s\Box$, and $s\Diamond$ for temporal modalities *always necessary, always possible, sometimes necessary,* and *sometimes possible.* Also, let us write $\Box a$, $\Diamond a$, $\Box s$, and $\Diamond s$ for modal tense operators *necessary that always, possible that always, necessary that sometimes,* and *possible that sometimes*, respectively. How temporal modalities and modal tense operators are related? In particular, is (i) equivalent to (i'), $1 \leq i \leq 4$? Our third aim is to answer this question.

2. THE LANGUAGES

Let $\mathcal{L}_1$, $\mathcal{L}_2$, and $\mathcal{L}_3$ be the languages of temporal modalities, modal tense operators, and of both of them, respectively, and let $\mathcal{L} \in \{\mathcal{L}_1, \mathcal{L}_2, \mathcal{L}_3\}$. $\mathcal{L}$ is defined on the language of the classical propositional calculus **PC**. Let $A, B, C, \ldots$ range over the set of formulas of **PC**. The primitive symbols of $\mathcal{L}$ are: $\Rightarrow, \wedge, \vee, \neg$ (the connectives) and parentheses. Moreover, in $\mathcal{L}_1$ we have $a\Box$ and $s\Box$, in $\mathcal{L}_2$ we have $\Box a$ and $\Diamond a$, and in $\mathcal{L}_3$ we have all of them.

The sets $\mathcal{F}_1at$, $\mathcal{F}_2at$, and $\mathcal{F}_3at$ of atomic formulas of $\mathcal{L}_1$, $\mathcal{L}_2$, and $\mathcal{L}_3$, respectively, are the smallest sets such that if A is a formula of **PC**, then (i) $a\Box A$, $s\Box A \in \mathcal{F}_1at$, (ii) $\Box aA$, $\Diamond aA \in \mathcal{F}_2at$, and (iii) $\mathcal{F}_3at = \mathcal{F}_1at \cup \mathcal{F}_2at$. Let $\mathcal{F}at \in \{\mathcal{F}_1at, \mathcal{F}_2at, \mathcal{F}_3at\}$. The sets of formulas of $\mathcal{L}_1$, $\mathcal{L}_2$, and $\mathcal{L}_3$ are denoted by $\mathcal{F}_1$, $\mathcal{F}_2$, and $\mathcal{F}_3$, respectively. Let $\mathcal{F} \in \{\mathcal{F}_1, \mathcal{F}_2, \mathcal{F}_3\}$; then $\mathcal{F}$ is the smallest set such that (iv) $\mathcal{F}at \subset \mathcal{F}$ and (v) if $U, V \in \mathcal{F}$, then $(U \Rightarrow V)$, $(U \wedge V)$, $(U \vee V)$, $\neg U \in \mathcal{F}$. We assume that

A. Pavković (ed.), Contemporary Yugoslav Philosophy: The Analytic Approach, 175–183.

$L, M, N, \ldots$; $P, Q, R, \ldots$; $X, Y, Z, \ldots$ range over $\mathcal{F}_1, \mathcal{F}_2$, and $\mathcal{F}_3$, respectively. Also, we assume that $\Rightarrow$, $\wedge$, and $\Longleftrightarrow$ are definable in terms of $\vee$ and $\neg$, as in **PC**. Furthermore, let us define:

$$s \diamond A \Longleftrightarrow_{Df} \neg a \Box \neg A$$

$$a \diamond A \Longleftrightarrow_{Df} \neg s \Box \neg A$$

$$\diamond s A \Longleftrightarrow_{Df} \neg \Box a \neg A$$

$$\Box s A \Longleftrightarrow_{Df} \neg \diamond a \neg A.$$

3. MODELS FOR $\mathcal{L}_1$

By T we denote a non-empty set of moments (intervals) of time, by W a non-empty set of possible worlds, and by $\mathcal{P}(W)$ the power set of W. Let f be a function from T to $\mathcal{P}(W)$ such that for any $m \in T$ $f(m) \neq \emptyset$. If A is a variable, then either $w \models A$ or $w \not\models A$; furthermore, $w \models \neg A$ iff $w \not\models A$ and $w \not\models A \vee B$ iff $w \not\models A$ and $w \not\models B$, for any $w \in W$.

The quadruple $\mathcal{M}_f = < T, W, f, \models >$ is called a *temporal modal structure (tms)*. For a tms $\mathcal{M}_f$ and a formula of $\mathcal{F}_1$ let us define:

$$\mathcal{M}_f \models a\Box A \text{ iff } \forall m \in T \, \forall w \in f(m) w \models A;$$

$$\mathcal{M}_f \models s\Box A \text{ iff } \exists m \in T \, \forall w \in f(m) w \models A;$$

$$\mathcal{M}_f \models \neg L \text{ iff } \mathcal{M}_f \not\models L \text{ (not } \mathcal{M}_f \models L);$$

$$\mathcal{M}_f \models L \vee M \text{ iff either } \mathcal{M}_f \models L \text{ or } \mathcal{M}_f \models M.$$

What is the intuitive meaning of a tms? The set T can be regarded as a period of time. For any $m \in T$ there is exactly one state of affairs w that takes place at m. However, thinking of $m \in T$, we consider all possible states of affairs (possible worlds) that can be actualized at m. Of course, what states of affairs are possible at m is an empirical question. Since we take into account no empirical restriction, we allow any non-empty set $f(m)$ to be the set of states of affairs possible at m. Therefore, we consider any function $f : T \rightarrow \mathcal{P}(W) - \{\emptyset\}$.

By a *frame* $\mathcal{M}$of a tms we understand $< T, W, \models >$. Let Γ be the set of all functions $f : T \rightarrow \mathcal{P}(W) - \{\emptyset\}$; we say that L is *valid in* $\mathcal{M}$ (in symbols: $\mathcal{M} \models L$) iff for any $f \in \Gamma \mathcal{M}_f \models L$. If L is valid in all frames, then we say that L is *valid* and we write $\models L$.

4. THE SYSTEM TM

In this section we axiomatize the set of valid formulas of $\mathcal{F}_1$. Let us define the set T_1 of theorems of **TM**, where **TM** is an axiomatic system in the language $\mathcal{L}_1$. First, all instances of any tautology B of **PC**, obtained by substitutions of formulas of $\mathcal{F}_1$ for propositional variables in B are axioms. Second, we accept further axioms given by the following schemata:

$$TM_1 \quad s\Box A \Rightarrow s \diamond A$$

$$TM_2 \quad a \diamond A \Rightarrow s \diamond A$$

$$TM_3 \quad a\Box(A \Rightarrow B) \Rightarrow .a\Box A \Rightarrow a\Box B$$

$$TM_4 \quad a\Box(A \Rightarrow B) \Rightarrow .s\Box A \Rightarrow s\Box B.$$

The set T_1 is the smallest set containing all axioms and closed under modus ponens (MP) and the following rule:

$$N_1 \qquad \text{if } A \text{ is a } \mathbf{PC}\text{-tautology, then } a\Box A \in T_1.$$

We omit the proof of

THEOREM 4.1 *If* $L \in T_1$, *then* $\models L$.

Before we state and prove the completeness theorem for **TM**, let us note that in **TM** we have the standard deduction theorem. This is due to the fact that any $L \in \mathcal{F}_1$ is modalized (i.e. every occurrence of a propositional variable in L is in the scope of either $a\Box$ or $s\Box$), and that no $L \in \mathcal{F}_1$ contains iterated modalities ($a\Box$ or $s\Box$ cannot occur in L in the scope of either $a\Box$ or $s\Box$). The reader should be familiar with the concepts of deducibility ($\vdash$), consistency of a set of formulas, and maximal consistent (mc) sets of formulas. We recall that any consistent set of formulas of **PC** can be extended to a mc set. This applies to consistent sets of formulas of many other propositional systems including **TM**.

THEOREM 4.2 *If a set* $H \subset \mathcal{F}_1$ *is consistent, then there is a tms* $\mathcal{M}_f$ *such that* $\mathcal{M}_f \models L$ *for any* $L \in H$.

PROOF . We shall construct a tms with the desired property. Let T' be the set of non-negative integers and let W' be the set of all mc sets of formulas of **PC**. Suppose that H is given and let $G \subset \mathcal{F}_1$ be a mc set containing H. Since G is a mc set, either $s \diamond A \in G$ or $\neg s \diamond A \in G$. If $\neg s \diamond A \in G$, then using TM_2 and the properties of G, we infer that $\neg a\diamond \in G$. Hence, $s\Box\neg A \in G$ and by $TM_1 s \diamond \neg A \in G$. Therefore, for any formula A of **PC** either $s \diamond A \in G$ or $s \diamond \neg A \in G$. Since there are countably many formulas of **PC**, there are countably many formulas of the form $s \diamond A$ in G.

Let us enumerate all members of G of the form $s \diamond A$ or $s\Box A$, and let E be such an enumeration. For any $m \in T'$ we define a family I_m of sets of formulas of **PC** according to the following recipe:

(1) if L_m in E is $s \diamond A$, then the members of I_m are

(1.1) the set $w_1 = \{A\} \cup \{C : a\Box C \in G\}$;

(1.2) for any $a \diamond B \in G$, the set $w_2 = \{B\} \cup \{C : a\Box C \in G\}$.

(2) If L_m in E is $s\Box A$, then the members of I_m are

(2.1) for any $a \diamond B \in G$, the set $w_3 = \{A, B\} \cup \{C : a\Box C \in G\}$.

Now we prove that every member of I_m is consistent.

(1.1) If w_1 is inconsistent, there are $C_1, \ldots, C_n \in w_1$ such that $C_1, \ldots, C_n \vdash \neg A$. Using the properties of **PC**, N_1, TM_3, and the definition of $s \diamond A$, we prove $G \vdash \neg s \diamond A$, and thus G is inconsistent.

(1.2) If w_2 is inconsistent, there are $C_1, \ldots, C_n, B \in w_2$ such that $C_1, \ldots, C_n \vdash \neg B$. As above, we prove $G \vdash \neg s \diamond B$. Hence, by TM_2 and the properties of G, we prove $G \models \neg a \diamond B$, and thus G is inconsistent.

(2.1) If w_3 is inconsistent, there are $C_1, \ldots, C_n, B \in w_3$ such that $C_1, \ldots, C_n, A \vdash \neg B$. Using properties of **PC**, N_1, and TM_3, we prove $G \vdash a\Box(A \Rightarrow \neg B)$. Now, by TM_4, $G \vdash s\Box A \Rightarrow s\Box\neg B$, and hence, by using MP and the definition of $a\diamond A$, $G \vdash \neg\diamond B$; therefore, G is inconsistent.

This proves that every member of I_m is consistent.

For any $w \in I_m$ there is a mc set containing w. For any $m \in T'$, let $J_m \subseteq W'$ be the smallest set such that for any $w \in I_m$ there is exactly one mc set in J_m containing w. By the axiom of choice, for any $m \in T'$ there is such a J_m, $J_m \neq \emptyset$. It is obvious that we have defined a function $f' : T' \to \mathcal{P}(W')$ such that $f'(m) = J_m$.

For any $m \in T'$, any $w \in J_m$, and any propositional variable p we define $\models'$:

$$w \models' p \text{ iff } p \in w.$$

It is easy to prove that for any formula A we have $A \in w$ iff $w \models' A$. Hence, $\mathcal{M}_{f'} = < T', W', f', \models' >$ is a tms. By the construction of $\mathcal{M}_{f'}$, using standard methods, we can prove that $L \in G$ iff $\mathcal{M}_{f'} \models' L$, and therefore, for any $L \in H$, $\mathcal{M}_{f'} \models' L$. This completes the proof of the theorem.

As a corollary we have

THEOREM 4.3 *If* $\models L$, *then* $L \in T_1$.

5. MODELS FOR $\mathcal{L}_2$

Let T and W be as before, and let Δ be the set of functions $g : T \to W$; then $\mathcal{M}_\Delta =< T, W, \Delta, \models >$ is a *modal tense structure* (mts). For a mts $\mathcal{M}_\Delta$ and any formula of $\mathcal{F}_2$ we define:

$$\mathcal{M}_\Delta \models \Box a A \text{ iff } \forall g \in \Delta \forall m \in T \; g(m) \models A$$

$$\mathcal{M}_\Delta \models \diamond a A \text{ iff } \exists g \in \Delta \forall m \; g(m) \models A$$

$$\mathcal{M}_\Delta \models \neg P \text{ iff } \mathcal{M}_\Delta \not\models P$$

$$\mathcal{M}_\Delta \models P \vee Q \text{ iff } \mathcal{M}_\Delta \models P \text{ or } \mathcal{M}_\Delta \models Q.$$

The intuitive interpretation of a mts is as follows: the set T is a period of time and W is a set of possible states of affairs that can take place in the period T. Again, at each $m \in T$ exactly one state of affairs is actualized. A function $g \in \Delta$ gives us a sequence of states of affairs. In fact, the range of g is a set of states of affairs, and if T is ordered by the binary relation "not later than", then the range of g is a sequence which can be called a *course of events*. Hence, Δ can be viewed as a set of courses of events possible in T. Again, what courses of events are possible in T is an empirical question.

$\mathcal{M}_\Delta \models \Box a A$ means that in any course of events g and at any moment of it A is true in the state of affairs that takes place at that moment.

$\mathcal{M}_\Delta \models \diamond a A$ means that there is a possible course of events in T such that at any moment m of it A is true in the state of affairs that takes place at m.

The meaning of $\mathcal{M}_\Delta \models \neg P$ and $\mathcal{M}_\Delta \models P \vee Q$ is obvious.

In the sequel, since we take into account no empirical restriction, we allow any course of events to be possible in T. Therefore, we take Δ to be the set of all functions $g : T \to W$. In this case we write $\mathcal{M}$ instead of $\mathcal{M}_\Delta$ and we say that P *is valid in* $\mathcal{M}=< T, W, \models>$ iff $\mathcal{M} \models P$. P *is valid* ($\models P$) iff P is valid in all frames.

6. THE SYSTEM MT

In this section we axiomatize the set of valid formulas of $\mathcal{F}_2$. **MT** is the corresponding axiomatic system in the language $\mathcal{L}_2$. Let us define the set T_2 of theorems of **MT**. First, all instances of any tautology B of **PC** obtained by substitution of formulas of $\mathcal{F}_2$ for propositional variables in B are axioms of **MT**. Second, we accept further axioms given by the following schemata:

$$MT_1 \qquad \Box sA \Rightarrow \Diamond sA$$

$$MT_2 \qquad \Diamond aA \Rightarrow \Diamond sA$$

$$MT_3 \qquad \Box a(A \Rightarrow B) \Rightarrow .\Box aA \Rightarrow \Box aB$$

$$MT_4 \qquad \Box a(A \Rightarrow B) \Rightarrow .\Box sA \Rightarrow \Box sB.$$

The set T_2 is the smallest set containing all axioms and closed under MP and the following rule:

$$N_2 \qquad \text{if} A \text{ is a } \mathbf{PC}\text{-tautology, then } \Box aA \in T_2.$$

THEOREM 6.1 *If* $P \in T_2$, *then* $\models P$.

THEOREM 6.2 *If a set* $H \subset \mathcal{F}_2$ *is consistent, then there is a mts* $\mathcal{M}_\Delta$ *such that* $\mathcal{M}_\Delta \models P$ *for any* $P \in H$.

PROOF . We shall construct a mts with the desired property. Let T' and W' be as in a tms. Suppose that H is given and let $G \subset \mathcal{F}_2$ be a mc set containing H. We can show that that there are countably many members of the form $\Diamond sA$ in G. Let us enumerate all members of G of the form $\Diamond sA$ or $\Diamond aA$, and let E be such an enumeration. Also, let us enumerate all members of G of the form $\Box sA$, and let E' be such an enumeration. For any $m \in T'$ we define a family I_m of sets of formulas of **PC** according to the following rules:

(1) if P_m in E is of the form $\Diamond sA$, then the members of I_m are

(1.1) the set $w'_= \{A\} \cup \{C : \Box aC \in G\}$;

(1.2) for any $\Box sB_n$ in E' the set $w'_n = \{B\} \cup \{C : \Box aC \in G\}$;

(1.3) the set $w'_0 = \{C : \Box aC \in G\}$;

(2) if P_m in E is of the form $\Diamond aA$, then the members of I_m are

(2.1) for any $\Box sB_n$ in E' the set $w'_{m,n} = \{B\} \cup \{C : \Box aC \in G\}$;

(2.2) the set w'_m.

Now we prove that every member of I_m is consistent.

(1.1) If w'_m is inconsistent, then for some $C_1, \ldots, C_k \in w'_m$, $C_1, \ldots, C_k \vdash \neg A$. Using the properties of **PC**, N_2, MT_3, and the definition of $\Diamond sA$, we prove $G \vdash \neg \Diamond sA$ and thus G is inconsistent.

(1.2) If w'_n is inconsistent, then for some $C_1, \ldots, C_k \in w'_n$, $C_1, \ldots, C_k \vdash \neg B$. Using **PC**, N_2, MT_3, and the definition of $\Diamond sA$, we prove $G \vdash \neg \Diamond sB$. Hence, by MT_1 and the properties of G, $G \vdash \neg \Box sB$, and G is inconsistent.

(1.3) It is obvious that w'_0 is consistent.

(2.1) If $w'_{m,n}$ is inconsistent, then for some $C_1, \ldots, C_k \in w'_{m,n}$, $C_1, \ldots, C_k, B \vdash \neg A$. Using **PC**, N_2 , and MT_3 we prove $G \vdash \Box a(B \Rightarrow \neg A)$. By MT_4, $G \vdash \Box sB \Rightarrow \Box s\neg A$, $G \vdash \Box s \neg A$, and by the definition of $\Box sA$, $G \vdash \neg \Diamond aA$. Hence, G is inconsistent.

(2.2) As in (1.1) of this proof, we obtain $G \vdash \neg \Diamond sA$, and by MT_2, $G \vdash \neg \Diamond aA$. Thus G is inconsistent.

For any $w' \in I_m$ there is a mc set containing w'. For any $m \in T'$, let $J_m \subseteq W'$ be a set of mc sets of formulas of **PC** such that for any $w' \in I_m$ there is exactly one mc set in J_m containing w'; such a member of J_m we denote by w', too. Now for each $m \in T'$ we define a function $g^m : T' \to J_m$ as follows.

(1) If P_m in E is $\Diamond sA$, then $g^m(l) = w'_0$ for any $1 \leq l < m$;

(1.1) $g^m(m) = w'_m$;

(1.2) if E' is infinite, then $g^m(m+n) = w'_n$ for any $n \in T'$;

(1.3) if E' is finite and Q_n is the last member of E', then $g^m(m+l) = w'_l$ for all $1 \leq l \leq n$, and $g^m(m+l) = w'_0$ for $l > n$;

(2) if P_m in E is $\Diamond aA$, then $g^m(l) = w'_m$ for all $1 \leq l \leq m$;

(2.1) if E' is infinite, then $g^m(m+n) = w'_{m,n}$ for all $n \in T'$;

(2.2) if E' is finite and Q_n is the last member of E', then $g^m(m+l) = w'_{m,l}$ for all $1 \leq l \leq n$, and $g^m(m+l) = w'_n$ for $l > n$.

Let $\Delta' = \{g^m : m \in T'\}$; it is obvious that each $g' \in \Delta'$ is a function from T' to W'. If we define $\models'$ as in a tms, then

$$\mathcal{M}_{\Delta'} = < T', W', \Delta', \models' >$$

is a mts. It is easy to show that $P \in G$ iff $\mathcal{M}_{\Delta'} \models' P$. Hence, for any $P \in H$, $\mathcal{M}_{\Delta'} \models' P$.

As a corollary we have

THEOREM 6.3 *If* $\models P$, then $P \in T_2$.

There is an obvious connection between **TM** and **MT**. Let $h : \mathcal{F}_1 \to \mathcal{F}_2$ be a function such that $h(a\Box A) = \Box aA$, $h(s\Box A) = \Box sA$, $h(\neg L) = \neg h(L)$, and $h(L \vee M) = h(L) \vee h(M)$. It is trivial that h is an isomorphism and that $L \in T_1$ iff $h(L) \in T_2$.

7. MODELS FOR $\mathcal{L}_3$

Let T, W, f, Δ , and $\models$ be as before; then $\mathcal{M}_{f,\Delta} =< T, W, f, \Delta, \models>$ is a *combined structure* (cs). Obviously, $\mathcal{M}_f =< T, W, f, \models>$ is a tms and $\mathcal{M}_\Delta =< T, W, \Delta, \models>$ is a mts. For a cs $\mathcal{M}_{f,\Delta}$ and any $X \in \mathcal{F}_3$ we define

$$\text{if } X \in \mathcal{F}_1 at, \text{ then } \mathcal{M}_{f,\Delta} \models X \text{ iff } \mathcal{M}_f \models X$$

$$if X \in \mathcal{F}_2 at, \text{ then } \mathcal{M}_{f,\Delta} \models X \text{ iff } \mathcal{M}_\Delta \models X$$

$$\mathcal{M}_{f,\Delta} \models \neg X \text{ iff } \mathcal{M}_{f,\Delta} \not\models X$$

$$\mathcal{M}_{f,\Delta} \models X \vee Y \text{ iff either } \mathcal{M}_{f,\Delta} \models X \text{ or } \mathcal{M}_{f,\Delta} \models Y.$$

Let Γ be as in Section 3, and let $\overline{\Delta}$ denote the set of all functions $g : T \rightarrow W$. X is Γ *-valid in the frame* $\mathcal{M}=< T, W, \models>$ iff $\mathcal{M}_{f,\overline{\Delta}} \models X$ for any $f \in \Gamma$. X is *valid* iff X is valid in any frame M.

8. THE SYSTEM TMT

Now we axiomatize the set of valid formulas of $\mathcal{F}_3$. Let **TMT** be the corresponding system.The axioms and rules of **TMT** are those of both **TM** and **MT**. Let T_3 be the set of theorems of **TMT**. Obviously, we have

THEOREM 8.1 *If* $X \in T_3$, *then* $\models X$.

THEOREM 8.2 *For any consistent* $H \subset \mathcal{F}_3$ *there is a cs* $\mathcal{M}_{f,\Delta}$ *such that* $\mathcal{M}_{f,\Delta} \models X$, *for any* $X \in H$.

PROOF . Let G be a mc set containing H. We already know that there are a frame $\mathcal{M} =< T, W, \models>$, a function $f : T \rightarrow \mathcal{P}(W) - \{\emptyset\}$, and a set Δ of some functions $T \rightarrow W$ such that $\mathcal{M}_f \models X$ for any $X \in G \cap \mathcal{F}_1$ and $\mathcal{M}_\Delta \models X$ for any $X \in G \cap \mathcal{F}_2$. By an inductive argument we show that for any $X \in \mathcal{F}_3$, $\mathcal{M}_{f,\Delta} \models X$ iff $X \in G$.

THEOREM 8.3 *If* $\models X$, then $X \in T_3$.

9. SPECIAL MODELS FOR $\mathcal{L}_3$

In a cs $\mathcal{M}_{f,\overline{\Delta}}$, $\overline{\Delta}$ was the set of all functions $g : T \rightarrow W$. However, we may choose to consider other sets of functions instead. Let

$\Delta_f = \{g : T \rightarrow W | \forall m \in T \; g(m) \in f(m)\}$.

THEOREM 9.1 *For any* $f : T \rightarrow \mathcal{P}(W) - \{\emptyset\}$, $\Delta_f \neq \emptyset$.

PROOF . The theorem is just a formulation of the axiom of choice.

THEOREM 9.2 *For any non-empty set* Δ *of functions* $g : T \rightarrow W$ *there is a function* $f : T \rightarrow \mathcal{P}(W) - \{\emptyset\}$ *such that* $\Delta = \Delta_f$.

PROOF . Given Δ , let us define $f : T \rightarrow \mathcal{P}(W) - \{\emptyset\}$ by $f(m) = \{g(m) : g \in \Delta\}$. It is clear that $\Delta = \Delta_f$ for f just defined. Moreover, given Δ , the function f is unique.

THEOREM 9.3 $\forall m \in T \, \forall w \in W \exists g \in \Delta_f \; g(m) = w$.

PROOF . By 9.1, there is a $g \in \Delta_f$; given $m \in T$ and $w \in W$ let us define g' as follows: $g'(m') = g(m')$ for all $m' \neq m$ and $g'(m) = w$. Obviously, $g' \in \Delta_f$.

THEOREM 9.4 *For any scs,*

(1) $\forall m \in T \forall w \in f(m)\ w \models A$ iff $\forall g \in \Delta_f \forall m \in T\ g(m) \models A$

(2) $\exists m \in T \forall w \in f(m)\ w \models A$ iff $\forall g \in \Delta_f \exists m \in T\ g(m) \models A$.

PROOF . The proof of (1) is omitted.

Suppose that (a) $\forall g \in \Delta_f \exists m \in T\ g(m) \models A$ and (b) $\forall m \in T \exists w \in f(m)\ w \not\models A$. By the axiom of choice and (b), there is a $g_0 \in \Delta_f$ such that

$$\forall m \in T g_0(m) \not\models A.$$

Obviously, this contradicts (a). The remaining "half" of the proof of (2) is trivial.

10. THE SYSTEM **ST**

We axiomatize the subset of $\mathcal{F}_3$ of formulas valid with respect to scs. The axioms and rules are those of **MT**; additional axiom-schemes are:

$$ST_1 \qquad a\Box A \Rightarrow \Box a A$$

$$ST_2 \qquad \Box a A \Rightarrow a\Box A$$

$$ST_3 \qquad s\Box A \Rightarrow \Box s A$$

$$ST_4 \qquad \Box s A \Rightarrow s\Box A.$$

Let T_4 be the set of theorems of **ST**.

THEOREM 10.1 *If* $X \in T_4$, *then* $\models X$.

PROOF . Omitted.

THEOREM 10.2 *If a set* $H \subset \mathcal{F}_3$ *is consistent (with respect to* **ST**), *then there is a scs* $\mathcal{M}_{f,\Delta_f}$ *such that* $\mathcal{M}_{f,\Delta_f} \models X$ for any $X \in H$.

PROOF . Let $G \subset T_3$ be a mc set, $H \subseteq G$; by 4.2, there is a tms $\mathcal{M}_f$ such that for any $X \in G \cap \mathcal{F}_1$, $\mathcal{M}_f \models X$ iff $X \in G$. By 9.1, Δ_f is non-empty and hence there is a scs M_{f,Δ_f}. Let us show that $\mathcal{M}_{\Delta_f} \models X$ iff $X \in G \cap \mathcal{F}\_2$.

Let $X = \Box a A$; by $ST_1 - ST_2$, $a\Box A \Rightarrow \Box a A, \Box a A \Rightarrow a\Box A \in G$. Hence, $\Box a A \in G$ iff $a\Box A \in G$ iff $\mathcal{M}_f \models a\Box A$ iff $\mathcal{M}_{\Delta_f} \models \Box a A$.

Let $X = \Diamond a A$; by $ST_3 - ST_4$, $\Diamond a A \Rightarrow a \Diamond A, a \Diamond A \Rightarrow \Diamond a A \in G$. Hence, $\Diamond a A \in G$ iff $a \Diamond A \in G$ iff $\mathcal{M}_f \models a \Diamond A$ iff $\mathcal{M}_{\Delta_f} \models \Diamond a A$.

By an easy inductive argument one can prove that $M_{f,\Delta_f} \models X$ iff $X \in G$.

As a corollary we have

THEOREM 10.3 *If* $\models X$ *(with respect to scs), then* $X \in T_4$.

11. THE ANSWER TO THE THIRD QUESTION

The answer to the third question depends on the definition of validity. It is clear that in **ST** (i) and (i') are equivalent, $i \in \{1, \ldots, 4\}$, for $a\Box A \Longleftrightarrow \Box a A$, $a \Diamond A \Longleftrightarrow \Diamond a A$, $s\Box A \Longleftrightarrow \Box s A$, and $s \Diamond A \Longleftrightarrow \Diamond s A$ are members of T_4. Moreover, this equivalence is equivalent to the axiom of choice, as it is seen from 9.4 and 10.3.

In **TMT** (i) and (i') are non-equivalent, as will be shown in the next few theorems.

THEOREM 11.1 (a) *There is a* $\mathcal{M}_{f,\Delta}$ *such that* $\mathcal{M}_{f,\Delta} \not\models a\Box p \Rightarrow \Box ap$, (b) *there is a* $\mathcal{M}_{f,\Delta}$ *such that* $\mathcal{M}_{f,\Delta} \not\models \Box ap \Rightarrow a\Box p$, (c) *there is a* $\mathcal{M}_{f,\Delta}$ *such that* $\mathcal{M}_{f,\Delta} \not\models s\Box p \Rightarrow \Box sp$, and (d) *there is a* $\mathcal{M}_{f,\Delta}$ *such that* $\mathcal{M}_{f,\Delta} \not\models \Box sp \Rightarrow s\Box p$.

PROOF . For any variable p there are mc sets w_1 and w_2 such that $p \in w_1$ and $\neg p \in w_2$. Let $T = \{1\}$, let W be the set of all mc sets of formulas of **PC**, let $f(1) = \{w_1\}$, $g(1) = w_2$, and $\Delta = \{g\}$. Obviously, $\mathcal{M}_f \models a\Box p$ and $\mathcal{M}_\Delta \not\models \Box ap$. This proves (a) and (c).

To prove (b) and (d), let T, W, w_1, and w_2 be as before, let $f(1) = \{w_2\}$, and $g(1) = w_1$.

Nevertheless, in **TMT** we can prove the following weak equivalences.

THEOREM 11.5 *For any frame* $\mathcal{M}$, $a\Box A$ *is* Γ - *valid in* $\mathcal{M}$ *iff* $\Box aA$ *is valid in* $\mathcal{M}$ *(i.e.* $\mathcal{M}_{\overline{\Delta}} \models \Box aA$).

PROOF . Omitted.

THEOREM 11.6 *From any frame* $\mathcal{M}$, $s\Box A$ *is* $\Gamma -$ *valid in* $\mathcal{M}$ *iff* $\Box sA$ *is valid in* $\mathcal{M}$

PROOF . If we assume that $s\Box A$ is Γ-valid in $\mathcal{M}$ and that $\Box sA$ is not, then it is easy to derive a contradiction. Let us assume that $\Box sA$ is valid in $\mathcal{M}$ and that $s\Box A$ is not Γ-valid; then

(a) $\forall g \in \overline{\Delta} \exists m \in T\ g(m) \models A,$

(b) $\exists f \in \Gamma \forall m \in T \exists w \in f(m)\ w \not\models A.$

Let $f_0 \in \Gamma$ be such that

c) $\forall m \in T \exists w \in f_0(m)\ w \not\models A.$

By the axiom of choice,

(d) $\exists g \in \overline{\Delta} \forall m \in T \forall ! w \in f_0(m)\ (g(m) = w \wedge w \not\models A).$

It is now easy to obtain a contradiction.

THEOREM 11.7 *For any frame* $\mathcal{M}$, *if* $a \diamond A$ *is* Γ-*valid in* $\mathcal{M}$, *then* $\diamond aA$ *is valid in* $\mathcal{M}$

PROOF . Let us assume that $a \diamond A$ is Γ-valid in $\mathcal{M}$ and that $\diamond aA$ is not; then

(a) $\forall f \in \Gamma \forall m \in T \exists w \in f(m)\ w \models A,$

(b) $\forall g \in \overline{\Delta} \exists m \in T\ g(m) \not\models A.$

Let $f_0 \in \Gamma$ be such that

(c) $\forall m \in T \exists w \in f_0(m)\ w \models A.$

Using the axiom of choice as in the preceding proof, we derive a contradiction.

We omit the proof of the following

THEOREM 11.8 $\diamond aA$ *being valid in* $\mathcal{M}$ *does not imply that* $a \diamond A$ *is* Γ-*valid in* $\mathcal{M}$

12. A CONCLUDING REMARK

This investigation shows that the answer to the question as to whether (i) and (i') from Section 1 are equivalent, $1 \le i \le 4$, depends on the validity of the axiom of choice. Thus, answering a rather naive question about natural language may be involved in the consideration of a rather sophisticated question in the philosophy of mathematics.

CAUSAL PROPOSITIONS AND ESSENTIAL PROPERTIES

ŽIVAN LAZOVIĆ

In this paper I want to consider an argument recently used to justify attribution of essential properties to so-called natural kinds, such as species (for instance, tigers) or material substances (for instance, water). The argument exploits the idea that essentialism may best be understood on natural or causal interpretation of modality, and it is advanced in support of the belief in existence of natural kinds determined by appropriate essential properties. This belief – to which, by the way, I am rather disinclined – is a realist one in so far as essential properties are thought to exist independently of our prescientific or scientific theoretical assumptions. I do not intend to provide a way out of the jungle of metaphysical mysteries about essential properties. What I want is only to undermine one typical argument which is founded on the causal interpretation of the modal notions.

As we know, there can be different forms of essentialism, and they differ mainly in conditions which, according to them, essential properties must satisfy. So, for example, we may have individuating and kind essentialism. The first holds that essential properties are those properties which partly or wholly distinguish an object from the others and individuate it. Leibniz in his metaphysics and Kripke with his idea of the necessity of origin both hold such a view. But, I shall discuss only the second, kind essentialism, which defends a weaker thesis[1] and accept the possibility that different objects may have the same essential properties. Such properties determine the kind or common essence of things in question. The best known example of this form of essentialism is surely Aristotelian essentialism.

Both forms of essentialism connect essential properties with necessity and identity. An essential property F of an object a is that property which a possess by necessity in the sense that a has it in all worlds causally possible relative to a given world, and a would cease to be the same object if it ceased to have the property F. However, it is still problematic in what sense essentialists talk about necessity and identity with regard to supposed essential properties. Concerning kind or Aristotelian essentialism it is often supposed that that form of essentialism is 'firmly entrenched in the logic of causal statements', and that 'Aristotle, in his theory of essences, was, after all, concerned with some kind of natural necessity'.[2] Marcus claims that if we understand modalities as causally based, and if causal laws are necessary truths, then we are committed to the claim that kind essential

A. Pavković (ed.), Contemporary Yugoslav Philosophy: The Analytic Approach, 185–194.

properties, such as being a tiger or being water, are instantiated.

I shall try to argue that although the antecedent of that conditional may be true, the consequent simply does not follow, and to get it we need an additional metaphysical premise. In other words, I want to argue that even though the causal laws are necessary truths, this is not sufficient to determine which properties of an object are kind essential properties. So, in the first part of my paper I shall outline, following Marcus, one modal interpretation of Aristotelian essentialism. Then, I shall offer a possible account of causal propositions according to which they are necessary. Finally, I shall argue that causal propositions, so interpreted, do not imply any kind essential properties.

The basic intuition implicit in essentialism is that an essential property of an object is the property which that object necessarily possesses:

1. $\Box Fa$, or, $(\exists x)\Box Fx$

For example, to say that Aristotle is essentially a man is, according to that intuition, to take as true that Aristotle is necessarily a man. But many tautological predicates, such as 'being a man or not being a man' are also necessarily true for any object. The essentialists are not inclined to count such predicates as designating essential properties; I think that they are right because essentialism is an interesting philosophical theory only if it excludes such cases. Thus, essential properties are those that an object has necessarily and that are at the same time not necessary for some other objects:

2. $\Box Fa \bullet (\exists x){\sim}\Box Fx$, or, $(\exists x)\Box Fx \bullet (\exists x){\sim}\Box Fx$

For the sake of simplicity we may assume that F is a monadic predicate which designates any ordinary or natural property, such as being red or being water. Also we may assume for a similar reason that F is a direct predicate, i.e. that it contains no modal operator.

As we recall, Aristotle distinguished between necessary and contingent properties of an object, so that, where one object has essential properties it is also implicit that it has properties which are not necessary, and this could be represented by:

3. $(\exists x)(\Box Fx \bullet Gx \bullet {\sim}\Box Gx) \bullet (\exists x){\sim}\Box Fx$[3]

Also, one of the fundamental assumptions of Aristotelian essentialism is that if anything has any kind essential property, so called second substance, then it has that property necessarily, Or, to put in other words, kind essential properties are not properties that anything could have contingently. We can represent that assumption in the following way:

4. $(x)(Fx \supset \Box Fx)$

Conjunction of 4 with 3 gives us the modal characterization of Aristotelian essentialism:

5. $(x)(Fx \supset \Box Fx) \bullet (\exists x)(\Box Fx \bullet Gx \bullet {\sim}\Box Gx) \bullet (\exists x){\sim}\Box Fx$

In fact, the characterization of Aristotelian essentialism provided by Marcus is not complete, and if we wanted to make it complete we would have to mention at least the following two points: first, Aristotle's assumption that essential properties must be instantiated in the actual world, and second, that Aristotle spoke about identity and necessity in terms of identity over time and temporal modalities. The first point indicates that Aristotle's essentialism starts from the standpoint of realism about natural kinds; this assumption is not, as some are inclined to think, important only for epistemological reasons (in the sense that essences must be instantiated in order to be knowable).[4] Without instantiated essential properties and without temporal modalities we could not say, according to Aristotle's theory, in what sense one thing can't loose some of its properties and still continue to exist. After all, though the contemporary idea of possible worlds generalizes Aristotle's concepts of identity and necessity, Aristotle himself was concerned with intra-world identity conditions, not with cross-world identity. Anyway, these remarks only point out the incompleteness of the proposed modal characterization of Aristotelian essentialism, and do not question its correctness, especially if we have in mind kind essentialism in general.

Now when we have the modal characterization of kind essential properties, let's consider the way in which causal propositions might be conceived as necessary. Typical examples of causal propositions are expressed in singular or general conditional statements, for instance: 'If we heat this water to 100° C, then it will boil'; and: 'If we heat water to 100° C, then it will boil'. General causal statements are usually used for expressing causal laws. But the underlying idea is that when we talk about causal relations between events we presuppose some kind of necessity. Let's then assume that the connective 'if..., then...' in our examples of causal statements means something like causal implication. How could we distinguish causal implication from material implication, and also causal statements from statements of material implication? It is often thought that causal implication is different from material implication in that, causal propositions are generally used for inferring so-called contrary-to-fact propositions.[5] This is a perplexing thought, because it is not at all clear that counterfactuals could be useful in explaining the nature of causal implication. However, our ordinary usage suggests this thought so that we can say with regard to our previous examples: 'If this water were heated to 100° C, then it would boil'. Actually, many statements about dispositional properties of things imply appropriate counterfactuals, and in virtue of them appropriate causal propositions. For instance, fragility is the dispositional property, or the disposition, to fragment if subject to the impact of a certain force.

A very common use of causal implication is also to make predictions. When we know that this liquid is water and that it will be heated to 100° C, we predict that it will boil. Such a procedure can be validated by the principle that causal implication implies material implication, because when Fa causally implies Ga, we could say that if Fa then it would, in the sense of material implication, Ga. However, there are good reasons to deny that the converse holds, i.e. that material implication implies causal implication. If the converse were to hold a false proposition would causally imply any proposition, as is the case with material implication between propositions. But with regard to causality that looks unacceptable because no one would suggest that, when this water is not heated to 100° C the proposition: 'This water is heated to 100° C', causally implies both propositions:

'This water boils', and 'This water doesn't boil'.

Even if causal implication means something more than material implication, it is commonly held that general causal propositions are equivalent to the corresponding material generalizations. The idea is that 'Fa causally implies Ga' is true if and only if the corresponding general statement 'For every x Fx materially implies Gx' is true. So it seems that we can deny the equivalence between causal end material implication without denying the equivalence between causal propositions and corresponding material propositions. One objection to this idea might be that in the case of proposed equivalence we wouldn't be able to distinguish between true general causal propositions and material generalizations true by accident. Intuitively it seems that some general propositions as 'All people in this room are philosophers' are contingently true, not by any law, and that when they are true, we aren't inclined to state causal implication. This objection apparently begs the question, but regardless of that, the underlying suggestion is that causal propositions have a certain modal character. The supposed modal character clearly does not mean that causal propositions are necessary in the logical sense. In ordinary usage causal implication is not identified with strict implication. When we say that Fa causally implies Ga we don't hold that the state of affairs in which Fa obtains without Ga is logically impossible. Hence, for the causal interpretation of modalities we can introduce special symbols like " $\boxed{c}$ " for necessity and " $\Diamond_c$ " for possibility. The modal proposition " $\boxed{c}$ P" means that the proposition P is true in all worlds causally possible relative to a given world, and " $\Diamond_c$ P" means that the proposition P is true in at least one of the worlds causally possible relative to a given world. The relation between causal modalities is parallel to the relation between logical modalities:

6. $\Diamond_c P = \sim \boxed{c} \sim P$

It is also evident that the concept of causal necessity can be related to the concept of causal implication as the corresponding concept of logical necessity can be related to the concept of strict implication, and that would support the suggestion that causal propositions involving causal implication have modal character. Thus by analogy with the definition of the concept of strict implication:

7. $(P ⥽ Q) = \Box(P \supset Q)$

The concept of causal implication can be defined by:

8. $(P \rightarrow Q) = \boxed{c}\,(P \supset Q)$

where "$\rightarrow$" is the symbol for causal implication. A problem arises if we try to apply (8) to any general causal proposition, because then the modal operator for causal necessity appears inside the scope of a universal quantifier. If for every *x*, *Fx* causally implies *Gx*, then, according to 8, we can write:

9. $(x)\,\boxed{c}\,(Fx \supset Gx)$

The problem concerns the proper interpretation of this result. Intuitively it seems that to say that for every *x*, the proposition "*Fx* ⊃ *Gx*" is true on *causal* grounds, means to say that the proposition "*(x)(Fx* ⊃ *Gx)*" is true on causal grounds:

10. $(x)\,\boxed{c}\,(Fx \supset Gx) \equiv \boxed{c}\,(x)(Fx \supset Gx)$

But the status of (10) is far from unproblematic. The solution depends on our choice of the system of modal logic in which we would like to interpret necessity as causal necessity. If we choose a modal system weaker than *S5*, for instance *S4*, in which only the converse of the so called *Barcan formula* might be proved,[6] the equivalence (10) wouldn't be true, but only the weaker alternative:

11. $(\boxed{c}\,(x)(Fx \supset Gx)) \supset ((x)\,\boxed{c}\,(Fx \supset Gx))$

But if we choose *S5*, in which the *Barcan formula* and its converse both can be proved, we would have (10).

Anyway, in spite of that which choice we would prefer, once it is granted that causal propositions have a modal character, the question is: Do they commit us to kind essentialism?

Ruth Barcan Marcus offers an argument in support of a positive answer. She asks us to consider some familiar example, say, that in which we claim that if we heat (*H*) a particular liquid *a* to 100° C, it would boil (*B*). Our claim may be formulated as a causal statement expressing the singular causal proposition represented by:

12. $\boxed{c}$ (Ha ⊃ Ba)

Let's agree that (12) is true. Then a possible question might be: why does that liquid boil at 100° C? On Marcus' interpretation the answer: Because that liquid is water (W), would indicate the kind property of the liquid a. The phrase 'because', involved in the answer doesn't introduce a simple description of an event, and in that respect there is an evident difference between the first question and the following one: why does that water boil? To the latter an answer would be: because it is heated to 100° C. The phrase 'because' in the first case, contrary to the second, points to a kind property, to a property that merely characterizes a thing by specifying what kind of thing it is. In turn, apparently, that kind property provides a causal basis for the necessary connection between the being heated (Ha) and the boiling of the liquid a (Ba), which are non-essential properties of the liquid a. So, (12) must somehow follow from:

13. $\boxed{c}$ ((Wa • Ha) ⊃ Ba)

(13) with the help of (7), can be rewritten as:

14. (Wa • Ha) → Ba

This singular causal proposition is, in fact, an instance of some general law. If we choose system S4 as a model framework for causal interpretation of necessity, then (14) instantiates the following law:

15. (x)((Wx • Hx) → Bx)

The problem is now how do we actually get from (14) to (10)? What we can do, according to Marcus, is to apply the principle of exportation. But, in quantified modal logic only restricted exportation holds.[7] Hence, if we choose any system of quantified modal logic as a basis for the proposed interpretation of causal necessity, the principle of exportation can not be applied unrestrictedly. So, from (14), where two conditions, Wa and Ha, are causally sufficient for the effect Ba, it neither materially nor causally follows that if any one of them is satisfied the other would be causally sufficient for the effect. Simply, the inference:

$$\frac{\begin{array}{l}(Wa \bullet Ha) \rightarrow Ba \\ Wa\end{array}}{\therefore \quad Ha \rightarrow Ba}$$

is not valid.

If we want to get valid inference only weakened exportation should be applied to (14) and the result is:

16. Wa → (Ha⊃ Ba)

It means that when the first condition is satisfied the second would materially imply the consequent:

$$Wa \rightarrow (Ha \supset Ba)$$

$$\frac{Wa}{\therefore \quad Ha \supset Ba}$$

But, since we have started with (12), we must somehow get to it from (16). Marcus thinks that this is a place where Aristotelian essentialism is required. As we recall, the basic idea of Aristotelian kind essentialism was that if anything has any kind essential property, it has that property necessarily. To get to (12) from (16) we rely on the assumption that being water is not accidental, taking as evident that condition (2) is also fulfilled, i.e. that there are things which are not water. Thus, according to (4), from Wa we get $\boxed{c}$ Wa, which, by the valid modal principle ($\underline{P} \prec \underline{Q}) \prec (\Box\underline{P} \prec \Box\underline{Q}$) and by (16), gives us (12):

$$Wa \rightarrow (Ha \supset Ba)$$

$$\frac{\boxed{c}\, Wa}{\therefore \; \boxed{c}\, (Ha \supset Ba)}$$

Marcus' conclusion is that if causal propositions like (12) are true then we must suppose that the causal connection between external accidental properties. say, of the liquid a, is grounded on the essential property of that liquid. That essential property in turn determines the kind of the liquid a.

As we can see from Marcus' story, what makes the attribution of such a property as being water essential rather than accidental is the fact that that property has a special role in the explanation of the causal connection between the others, nonessential properties of things. In this respect essential properties are similar to dispositional properties of things. But Marcus claims also a stronger thesis that these properties are kind essential properties, in the sense that they are connected with identity conditions for the members of the kind: when an object has such a property, it is a member of the appropriate kind, and if it looses that property it also looses its identity and ceases to be the same object as before. One general objection, put forward by Paul Teller in his article 'Essential Properties: Some Problems and Conjectures'[8], shows that this Marcus' claim does not follow from the previous argument. As a matter of fact, it is not evident why liquid a, if it had the property of being water, must cease to exist when it ceases to possess that property. Really, maybe we would not reidentify the liquid a as the same liquid as before, since many of its external properties would have changed too, but that does not show that a necessarily possesses the property of being water. What is stated by the causal laws like:

15. (x)((Wx • Hx) → Bx)

is rather that the properties designated by W, H, and B, are covariant, so that whenever something is W, and is H, it must by causal necessity have B. The causal law (15) does not indicate that any of mentioned properties is essential in the Aristotelian sense, and it does not provide any justification for the belief that it is impossible for the same x to have none of those properties. According to our example, if x ceases to be water, then, even if x would not boil at a hundred degrees Centigrade, it is not necessary that x should cease to exist – it is quite conceivable that, although the properties designated by W, H and B always go together, it is possible for x to have none of them. To claim that x would also loose its identity we must rely on an additional and independent essentialist assumption that being water is the property which, when possessed by x, determines x's identity. By relying on such an assumption we simply beg the question, since our argument designed to show that there are kind essential properties, in fact assumes the existence of such properties.

There is also another problem worth mentioning. In my opinion, it shows that Marcus's argumentation fails even if we accept that causal propositions, or causal laws, when true, are necessary truths. The modal interpretation of causal propositions does not help us in sorting properties and picking out kind essential properties. To realise that we have to notice first that causal statements like:

12. Ha → Ba

are, in a sense, elliptical if 'a' has a fixed value, for instance, if it designates this liquid. Elliptical causal statements are, in fact, often used in ordinary life when we talk about causes and effects without providing all the necessary conditions which are together sufficient in the causal sense. In our example (12) may be conceived as an elliptical causal proposition because it is not mentioned that a is water. The complete proposition is in:

13. (Wa • Ha) → Ba

Anyway, this does not show that Wa has some special status, that being water is kind essential property. We could also say for a thing b that if we drop (D) this thing it would fall to the ground (F):

17. [c] (Db ⊃ Fb)

But (17) is elliptical in the same way as (12), since Db is not causally sufficient for Fb. The omitted condition is that the thing b is, or must be, heavier than air (A) . So (17) must be rewritten as:

18. [c] ((Ab • Db) → Fb)

or, according to our definition of causal implication as:

19. $(Ab \bullet Db) \rightarrow Fb$

It is easy to see that if we follow Marcus's suggestion and apply the same procedure as in our previous example, we will come to the conclusion that (17) could be derived from (19) only if the thing b has the property designated by A necessarily, that is,if the property designated by A, being heavier than air, is an essential property of b. However, it is quite clear that the property being heavier than air is not a kind essential property, and, furthermore, it is not an essential property at all – many objects, and many people too, change their weight without loosing their identity. To distinguish between such properties as having a certain weight, which are not essential, and such properties as being water, which are thought of as kind essential, it is not enough to have necessary causal propositions or causal laws; we also need some independent and basically metaphysical presupposition about what sort of properties are kind essential properties.

The conditional that forms the basis for the causal interpretation of kind essentialism after all does not hold. Even if we agree that causal propositions have modal character, and even if we have necessarily true causal committed to kind essential properties. It seems that realism about natural kinds hardly can find a way out from the jungle of its metaphysical assumptions.

NOTES

1. According to the individuating essentialism, if two objects x and y have all their essential attributes in common, then they are identical, but according to the kind essentialism, that doesn't follow.

2. Ruth Barcan Marcus, 'Essential Attribution' (The Journal of Philosophy, volume LXVIII, No 7, 1971, pp. 200, 197).

3. One form of individuating essentialism might be:

$$(\exists x)(\Box Fx \bullet Gx \bullet \sim\Box Gx) \bullet (\exists x)(Fx \bullet \sim\Box Fx)$$

4. Cf., Paul Teller, 'Essential Properties: Some Problems and Conjectures' (The Journal of Philosophy, vol. LXXII, No 9, 1975, p 235).

5. For a classic treatment of this topic compare David Lewis, 'Causation' (The Journal of Philosophy, vol. LXX, No. 17, 1973).

6. The Barcan formula is:

   $(x)\Box Fx \supset \Box(x)Fx$

7. As we know, the principle of exportation is unrestricted with regard to material implication where the deduction theorem holds.

8. Op. cit., p. 238.

INTERNALISM AND INTENTIONALITY

NENAD MIŠČEVIĆ

I

This paper is about two kinds of theories of intentionality. They will be called internalist and externalist theories. The terminology is well entrenched in epistemology, and most of the authors writing about the problem of intentionality seem to be aware that there are really two kinds of theories which can be distinguished in a way analogous to internalist/externalist distinction in epistemology. However, the issues are not always presented clearly, nor even discussed separately. It so happens that most authors which would be labeled 'internalists' by the classification also hold the doctrines that intentionality is irreducible, that functionalism is wrong and some others. They themselves present these doctrines as consequences of what here will be called the internalist stance. However, internalism is conceptually independent from these doctrines, and should be discussed as an issue in itself. I propose to do just this in the paper.

In the first part I shall briefly sketch what I think are the main varieties of internalist theories. Then I present some considerations in favour of internalism, and then discuss briefly the main consideration against it – the idea that internalism leads either to regress or to the postulation of unexplainable properties of the mental. My main aim is not to prove internalism correct, but to make clear the nature of the issue. The distinction between internalist and externalist theories is familiar from the theory of knowledge. Suppose you want to define 'knowledge' and you start by giving truth conditions for 'P knows that p'. In many theories it is required for P's belief to be knowledge that P has some justification, or some valid reason for her belief that p. Call this requirement C. Now, you may accept C, and then you have following two options:

(E) Either you rest satisfied with P's having some justification for her belief that p, without demanding that P should believe that she has it, or

(I) you impose a further requirement, call it C^+ that P should believe that requirement C is satisfied, i.e., that she should believe that her belief that p is justified.

A. Pavković (ed.), Contemporary Yugoslav Philosophy: The Analytic Approach, 195–203.

In the first case (E), you are externalist, in the second (I), you are internalist.

The distinction can be generalised to cover other domains. Take some intentional predicate, like 'thinks', 'perceives' or 'intends'. The theory will usually list some conditions a person has purportedly to fulfill in order for such a predicate to apply to her. Some of these conditions can refer to persons belief or knowledge (or intention, or whatever other attitude) that some other condition is fulfilled. In other words, a theory can require that in order that a person should have a given attitude it has to have a sort of access to the fact that she has some property, or that she satisfies some other condition mentioned in the theory. If a theory requires such a thing it is an internalist theory. Otherwise it is externalist.

One can think of the distinction in the following way: A theory Th specifying the content of a mental predicate M will have the following form:

$$\text{Th: } \underline{x} \text{ M's } \underline{y} \text{ iff } \begin{array}{c} c_o \\ \cdot \\ c_i \\ \cdot \\ c_n \end{array}$$

where $\underline{x}$ ranges over persons, $\underline{y}$ over propositions or objects, and M stands for mental predicate like 'believes that ...' 'intends that...', 'thinks of...' . 'c's' stand for conditions.

Now, among the conditions listed on the right side there may be a pair, c_i , c_j , such that c_j says that the person $\underline{x}$ has access to the fact that c_i is satisfied. (In our first example, C was c_i and C^+ was c_j). In order to specify the kind of access that $\underline{x}$ has, the theory will employ some mental predicate, call it N. If there is such a pair the theory is internalist. In short,

Int (Th):

(I) Int(Th) iff $\exists c_i \ \exists c_j \ \ (c_j + \underline{x}$ N's that $c_i)$

('$\exists$' is the existential quantifier).

Call (I) the internality condition.

A theory is externalist if it is not internalist.

The debate between internalists and externalists has tended to concentrate on the nature of the mental predicate N employed in stating the internality condition. For example, if it is necessary for George in order to know that it is raining, to again know that his reasons for his belief about the weather are valid, it becomes hard to see how George can ever begin to know anything. The externalists have been quick in spotting such weaknesses.

It is, therefore, advisable to take a closer look at the relation between M and N (the mental predicate defined by theory, and the mental predicate used in stating its internality condition), and to do this in full generality.

Suppose for any two mental predicates we can tell whether one is more complex than the other, less complex, equally complex or incomparable with the other. (Knowledge is more complex than belief, and belief and intention are either equally complex). Then we would have the following possibilities:

1. M is less complex than N
2. M is as complex as N (special case M + N)
3. M is more complex than N
4. M and N are incompatible.

Take the first alternative. A theory which operated with such a pair of predicates would explain a less complex attitude (say, believing that it is raining) by a more complex one (say, by selfascribing the property that one is such that it rains outside). It would so to speak, ascend from simple to complex:

$$\text{Asc (Th)} = \text{Int (Th)} \ . \ (1)$$

The second alternative is to stay on the same level. One can define some complex attitude in terms of an equally complex one, or in terms of an incomparable one:

$$\exists q \text{ (Th)} = \text{Int (Th)} \ . \ (2) \vee (4)$$

A special case would be 'homonymous' theory which would define, e.g., belief that p again in terms of belief (that some condition is fulfilled):

$$\text{Hom (Th)} = \text{Int (th)} \ . \ M = N$$

The third alternative is to descend from a complex M to some N which is simpler or more elementary. A theory embracing this alternative could point a way to reduction of the complex predicate:

$$\text{Desc (Th)} = \text{Int (Th)} \ . \ (3)$$

The reduction can be achieved in one step within the theory Th, or through a descending chain of theories. In the later case, we start from a couple of theories, Th1 and Th2, such that Th1 explicates M in terms of a simpler N, and Th2 the explicates N in terms of some N′ simpler than N. In the ideal case we can then build chains of the required length.

A rough classification of theories according to alternatives just mentioned would take the following form:

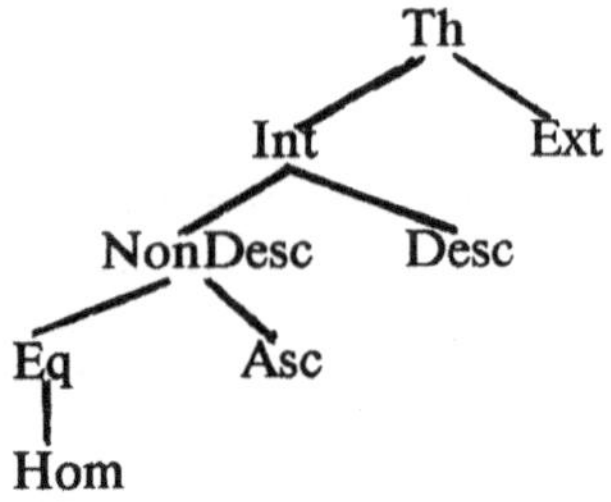

II.

The usual format for defining intentional notions in recent philosophical work is dictated by the assumption that there are representational states, that these states are somehow linked to the 'world', and that having 'intentionality' consists in occupying such states. In this paper, I shall accept that format. Accordingly, I shall consider theories of the following form: For a person, say John, to believe that it is raining consists in John being in some functionally specified mental state S, such that S is suitably related to the fact (situation, proposition) that it rains, and perhaps in some further conditions being satisfied. In more schematic form:

Th: x M's y iff (1) x is in S
(2) S is related to y (correlation condition)
(3) ?

I shall assume that condition (2) should eventually refer to some kind of causally sustained correlation between x's mental state S and proposition or object y, so that I call it 'correlation condition'. In order to be more specific I shall provisorily accept Dretske's characterisation of correlation, and suppose that S carries information (in an objective sense) that Y (in case y is proposition, or alternatively about y if y stands for an object). The same argument will apply to other correlational theories.

Now, the internalist (working within our format) will typically claim that the correlation condition is necessary but is far from being sufficient. He will claim that a further condition should be added to the effect that our thinker x should have some cognitive access to the fact that his state S is related to the outside world. He will say that it is not enough for John simply to be in a state which carries information about rain. So, he will insist that an internalist condition be added to the list:

ThInt: X M's y iff (1) x is in S

(2) S is related to y (corr. cond.)

(3) x has access to (2) (internality condition)

The externalist will rest content with (1) and (2). So his theory within the same format will differ from the internalist one only in rejecting the internality condition.

Now we have our first problem: which theory is better, the internalist or externalist one? Should we accept the internalist condition or not?

I would like to present three arguments in favour of internalism.

The first argument simply points out what is intuitively plausible and what seems to be absurd.

First, it seems natural to suppose that the following is true: If John thinks about apples then he believes that he thinks about apples. Moreover, if John has no idea about apples it will not be the case that John thinks about apples. Now, this seems to imply that in order to think about apples John has to have access to the content of his apple-thoughts.

Second, suppose that John is in a state S which plays the functional role of a belief. The state S is suitably correlated with apples. To be more precise, suppose that state S is actually a token of a mental word A. Suppose further that A is causaly or informationally related to apples. It is still possible that John has no inkling that S has to do with apples. He is not aware of the relation, and when in S he is disposed to say things like 'The Democrats are certainly going to loose'. The externalist will still claim that when he is in S his belief is about apples.

There are two strategies the externalist can use to avoid the absurdity. One is to define the functional role so as to exclude the deviant cases (either by explicit definition or by appeal to 'normality', proper functioning and the like). However, in this case it is the functional role which fixes the content, and we have to deal with an entirely different theory. The other is to bite the bullet and to deny that conscious access is important. However, the internalist is not forced to postulate the conscious access. He can accept that there are cases of unconsciously structured beliefs, and insist that the believer has in this case an unconscious cognitive access to her state. Take the example of a skilled driver who is not paying attention to the road because he is absorbed in a conversation, but is still avoiding obstacles. His behaviour seems to show that his cognitive apparatus has processed the input information to the point of giving it a distal interpretation - interpretation in terms of features of the road, or even presumed intentions of the driver in the approaching vehicle. So his unconscious beliefs should be of the form 'There is this drunken madman approaching me, I better be careful'. It is difficult not to conclude that his cognitive apparatus has access to the content of such belief, to what the belief is about.

The second argument for internalism aims to show that cognitive access is an important advantage for organisms that possess it. This, if successful, does not in itself show that internalist condition should be built into the definition of intentional states, but it does put the internalist condition in a favourable light for naturalist philosophers.

Consider two living systems, Ext and Int who are very much alike except for the following difference: Int is capable of monitoring its internal states, and of interpreting them in terms of their distal objects. Ext is simply hardwired in such a way that certain input states trigger appropriate central processes which result in favourable output – Ext behaves quite successfully in its normal environment. Now, Ext can triangulate upon an object in the sense that it is capable of coupling together several states S_1, S_2 and S_3 which as a matter of fact 'denote' the same object O. However, Int will be able to learn much more quickly, to construct more powerful 'hypotheses' about its environments, and consequently to adapt to changing circumstances which are far beyond Ext's powers.

The same point could be put more bluntly the following way:

The externalist typically claims two things

a) that 'information' which S carries about some object O is objective, independent of organisms <u>x</u> which is in S.

b) that organisms are typically information gatherers and information users. (Substitute 'reliable indication' for information if you prefer).

Now, what evolutionary reason could there be for gathering information which is 'objectively' encoded in <u>x</u>'s state S, if the information is not available to the organisms? Or, to come closer to the internalist standpoint, suppose a state S carries (objectively) information I about a predator. Of what use could that information be for an organism which does not somehow access the fact that I has to do with predators?

To sum up: the availability of the fact that S is suitably (informationally, indicatingly) correlated with O is an important advantage, and it is plausible to suppose that organisms will be selected for it. The non-availability of such facts renders information less useful (or more difficult to use) and will not be favoured by evolution.

The third argument concerns the distinction between original (or primary) and derived (or secondary) intentionality. A printed page or a book is about an event - say the Battle for Britain - because it is written in a language spoken by a population and the printed words in the book mean something for these language speakers. The 'intentionality' of books and pictures is derived or secondary, whereas the intentionality of our mental states is primary.

It might seem that the secondary character of books' 'intentionality' depends on the fact that the very same printed characters are open to different interpretations, and that this is the distinguishing mark of secondary intentionality. This, however, is not so. Suppose that the meaning of printed text is unique and fixed, e.g., by the fact that the text is written in an 'ideal' formulaic language used only by members of some super-Vienna Circle. Its 'intentionality' would still be derived from the intentionality of the minds of the Super-Viennese. So, the secondary character does not depend on interpretational multiplicity. It might be that there is a converse relationship - that the fact that books' intentionality is secondary allows for the multiplicity of interpretations.

Now, the distinction between primary and secondary intentionality is well entrenched, natural amd moreover supported by those contemporary externalists who are at the same time realists about intentionality. So, let us suppose that everybody (except the anti-realist) wants to maintain the distinction.

Now the argument takes the form of <u>modus tollens</u>:

1) If you are externalist you cannot have a principled way to draw the distinction. (The incompatibility premiss).

2) The distinction is to be drawn (common opinion).

---

3) So, you better not be externalist.

This is a case of practical inference about a very unpractical topic.

The burden of the argument rests on the incompatibility premiss (1). To put it bluntly, the premiss states that the externalist has no principled way to distinguish the intentionality of the book about the Battle for Britain, from the intentionality of John's thoughts about the same battle. (Daniel Dennett has argued that most contemporary theories about intentionality are powerless to do the distinguishing, and has argued against the distinction. I think that his starting point is right, but that it shows exactly the contrary of what Dennett takes it to show - it shows that something is wrong with externalist theories).

The internalist can always claim that the relevant difference between John and the book in relation to the Battle is that the book is about the Battle but that it has no access to what it is about whereas John's thoughts are related to the Battle in such a way that John is aware that they are about the Battle.

What can the externalist say about the distinction? It would not do simply to claim that there is a difference in the number of steps from John, viz., the book to the Battle. It would not do to claim that John's thoughts about the Battle are somehow more direct - maybe he learned about the Battle from the book. There is a correlation between the book and the Battle, and there is a correlation between John's mental state and the Battle. The book can carry more information, and it can be more reliable than John's mental state.

The only intuitively plausible way for the externalist to draw the distinction is to appeal to the functional role - but this is out of question for a book or a painting. But if this role is totally independent from the content of the state, nothing is achieved (the state plays its role so to speak by its brute physical force). If the state plays its role at least partly in virtue of its content, then the real distinguishing feature is the fact that in the case of John the system (John, his cognitive aparatus, his consciousness or whatever) has access to the content, and this is precisely what the internalist claims.

So the incompatibility premiss is plausible and probably correct. If you are an externalist you cannot distinguish the primary from secondary intentionality. If you care for the distinction you have to give up externalism.

The foregoing was a sketch of a case for internalism. The defense concludes by stressing that internalism does not rely upon an overestimation of the role of consciousness, that internalism accounts for the important fact we are information seekers and information users and that it fits into the naturalistic picture of humans interacting with their environment.

Now, if we have succeeded in showing that internalist theories are more promising than externalist ones, it remains to choose the right kind of internalist theory. It is here that we must take care to avoid the dreaded regress which seems to threaten most internalist theories. In the case of a theory of intentionality, the regress may arise in specifying the internalist condition - x has access to the fact that S stands for y. It seems that x must be in some other state S_1 in order to have independent access to the object y, so that it may represent that S stands for y. But then, does it have to access the fact that S_1 also stands for y? If no, why do we need the internality condition at all? If yes, the regress arises. We might after all have to weaken the internality condition, in order to avoid the regress. This would mean choosing some specification of the access and of its object which would be more primitive than our definiendum - in the vocabulary of the first section: choosing an N less complex than M. Then our theory would be a 'descending' one, an Desc (Th).

Some people doubt that this could be done, and therefore base their internalist stance on an appeal to the mysterious properties of the mental. This is certainly a wrong way. If internalism cannot be formulated in a suitable 'descending' theory, it will have to be given up. However, I hope to have shown that this would be a desperate move indeed.

SCIENTIFIC PERSUASION

STANIŠA NOVAKOVIĆ

> 'You cannot force anybody by arguments to take arguments seriously, or to respect his own reason'.[1]

I will try in this paper to tackle in greater detail the problem of scientific persuasion, i.e. the problem of arguments scientists use in trying to persuade other scientists to reject one theory and accept another. Therefore, I am going to discuss the arguments scientists use in order to convert members of a scientific community and make them believe that a particular hypothesis is -- for the moment -- the best available theoretical solution to a certain scientific problem.

A relatively comprehensive but brief outline of a viewpoint on the problem of scientific persuasion is found in the famous book by Thomas Kuhn, The Structure of Scientific Revolutions. It is well known that the normal practice of a scientific community takes place within an accepted paradigm, i.e. within a dominating general theory in a certain domain of science. A change of paradigms constitutes a scientific revolution, while, at the same time, the decision to abandon one paradigm always means the decision to accept another, to accept a new World View. But, the process of changing world views means struggle, so that the parties engaged in the conflict have to use the technique of mass persuasion.

One has to bear in mind, of course, that, as Kuhn says, 'the proponents of competing paradigms are always at least slightly at cross-purposes. Neither side will grant all the non-empirical assumptions that the other needs in order to make its case...Though each may hope to convert the other to his way of seeing his science and its problems, neither may hope to prove his case'.[2]

The transition from one paradigm to another, Kuhn believes, cannot be made one step at a time, forced by logic and neutral experience. It must occur -- after a shorter or longer-lasting clash of arguments -- all at once (though not necessarily in an instant) or not at all.[3] And, although a paradigm change cannot be justified by proof, it may not be said that no arguments are relevant nor that scientists cannot be persuaded to change their mind albeit the process may take a long time and be accompanied by considerable difficulties.[4]

A. Pavković (ed.), Contemporary Yugoslav Philosophy: The Analytic Approach, 205–228.

As we discuss the techniques of persuasion, or argument and counterargument in a situation where there can be no proof, a new question arises, necessitating a kind of study which has not been undertaken preiously. Certain scientists embrace a new paradigm for all sorts of reasons, and, usually, for several reasons at once. Some of these reasons, as Kuhn remarks, may lie entirely outside the apparent sphere of science.[5]

Now, another question emerges: is it, nevertheless, possible to single out some sort of argument which will prove to be particularly effective in the struggle for the paradigm change?

Kuhn remarks on this topic: 'Probably the single most prevalent claim advanced by the proponents of a new paradigm is that they can solve the problems that have led the old one to a crisis. When it can legitimately be made, this claim is often the most effective one possible...'[6] It will be far more effective if the problem of the old paradigm was repeatedly explored and attempts to remove it proved futile again and again. Besides, the claim mentioned is particularly likely to succeed if the new paradigm displayed a quantitative precision strikingly greater than its older competitor.

But, when it happens that the candidate for a paradigm initially helps not at all with the problems which brought the crisis about, evidence must be drawn from other parts of the field. In such a case, particularly persuasive arguments can be developed if the new paradigm permits successful prediction of phenomena which were entirely unsuspected while the old one prevailed.[7]

Another kind of consideration which can lead scientists towards rejecting an old paradigm in favour of a new one is of aesthetic nature (the new theory is said to be 'neater', 'more suitable', or 'simpler' than the old one), but such arguments are less effective in the field of empirical sciences than of mathematics.[8]

In any case, Kuhn wants to stress that these arguments are neither individually nor collectively compelling. The key moment which determines whether there is a persuasive argument or not, is, of course, the manifestation of a crisis in science, the fact that the old paradigm met certain difficulties or that the number of these difficulties is increasing constantly.

However, the question of the origin and existence of persuasive arguments does not have any bearing on the question of power of these arguments. So, let us now see what Kuhn has to say concerning this.

First of all, Kuhn points out the fact that the defenders of the traditional theory and procedure can almost always indicate the problems which its new competitor does not solve and which, from their viewpoint, are no problems at all.

Besides, paradigm debates do not really treat of their relative problem-solving ability, although they are for good reasons usually couched in those terms. 'Instead', says Kuhn, 'the issue is which paradigm should in the future guide research on problems many of which neither competitor can yet claim to resolve completely. A decision between alternate ways of practicing science is called for, and in the circumstances that decision must be based less on past achievement than on future promise...A decision of that kind can only be made on faith.'[9]

So, it is obvious that, to Kuhn, all the persuasion over rejection of an old and acceptance of a new paradigm boils down to arguments of irrational nature. He is evidently ascribing a very low value to the rational power of persuasive arguments in science, and this, I think, is one of the greatest deficiencies of Kuhn's position.

Bearing in mind his views on science in their entirety, one can certainly understand this particular irrationalist tendency of Kuhn's thought. In fact, I think that this persuasion topic does not quite fit into his general position, especially since he holds the view that scientific revolutions (i.e. the changes of paradigms) do not bring any progress in science; that view is based, of course, on his doctrine of incommensurability of paradigms.

The doctrine holding that the change of paradigms amounts to pure gestalt-switch, which includes the incommensurability, does not really require any kind of rational persuasive argument! The reason why Kuhn discusses, nevertheless, this kind of argument, lies, I presume, in the simple fact that Kuhn could not ignore something which is obviously an aspect of the everyday practice of theoretical scientists. It is true, of course, that persuasive arguments cannot be taken as a proof; but, this is so, among other things, because there is no actual case where both the old and new paradigms refer to the identical scope of features (not to mention different levels of reality). It usually happens that the scope of the new paradigm is wider than the scope of the old one; but, even in this case, it does not mean that the old paradigm (perceived, of course, completely in the light of the new) should become part of the new one. In fact, it may happen quite often that some ingredients of the old paradigm are not included in the new one; this explains the possibility of the old theory pointing to some problems which cannot be solved by the new one, though no impediment to the former. Consequently, such a possibility represents also the main basis of persistence for the defenders of the old theory.

Finally, Kuhn does not point in his original text to yet another essential reason why persuasive arguments cannot be taken as proof: the existence of value judgement in science and, especially, in persuasive argumentation. This issue will be discussed more fully later on.

Now, let us finish the discussion of Kuhn and his ideas by finding out what kind of problems he encounters when speaking of persuasion and, at the same time, of incommensurability of competing paradigms. He became aware of this problem only in his Postscript to The Structure of Scientific Revolutions.

Let us recall the basic characteristics of scientific revolutions: awareness of anomalies, which can last for quite a long time and deeply penetrate the ruling paradigm, brings about a state of crisis ending with the emergence of a new candidate for the paradigm and the ensuing battle over its acceptance. However, in Kuhn's opinion, the transition between competing paradigms cannot be made one step at a time, forced by logic and neutral experience; as the gestalt switch, it must occur all at once (though not necessarily in an instant) or not at all. Here then comes Kuhn's concept of the incommensurability of competing paradigms, since after the gestalt switch the proponents of competing paradigms practise their trade in different worlds.[10]

Now, since the vocabulary used by the parties in discussing certain experimental or observational situations during a debate over theory-choice consists predominantly of the same terms, they must be assigning some of those terms to nature differently, so that, inevitably, their communication is only partial.

'As a result, the superiority of one theory to another is something that cannot be proved in the debate. Instead, ...each party must try, by persuasion, to convert the other...Nothing about this relatively familiar thesis implies either that there are no good reasons for being persuaded or that those reasons are not ultimately decisive for the group. Nor does it even imply that the reasons for choice are different from those usually listed by philosophers of science: accuracy, simplicity, fruitfulness, and the like. What it should suggest, however, is that such reasons function as values and that they can thus be differently applied, individually and collectively, by men who concur in honoring them. If two men disagree, for example, about the relative fruitfulness of their theories, or if they agree about that but disagree about the relative importance of fruitfulness and, say, scope in reaching a choice, neither can be convicted of a mistake. Nor is either being unscientific. There is no neutral algorithm for theory-choice, no systematic decision procedure which, properly applied, must lead each individual in the group to the same decision... To understand why science develops as it does, values interacts with the particular experiences shared by a community of specialists to ensure that most members of the group will ultimately find one set of arguments rather than another decisive.'[11]

Kuhn calls this process persuasion. But, taking his concept of a scientific revolution into consideration, Kuhn admits that there exists a greater problem. Two men who perceive the same situation differently (out of two paradigms) and nevertheless employ the same vocabulary in its discussion, must be using words differently. How can they even hope to talk together, much less be persuasive?

In order to come up with at least a preliminary answer to that question, Kuhn offers further specifications of the nature of the difficulty.

Kuhn writes: 'The practice of normal science depends on the ability, acquired from exemplars, to group objects and situations into similarity sets which are primitive in the sense that the grouping is done without an answer to the question 'Similar with respect to what?'. One central aspect of any revolution is, than, that some of the similarity relations change. Objects that were grouped in the same set before are grouped in different ones afterward and vice versa. Think of the Sun, Moon, Mars, and Earth before and after Copernicus; of the free fall, pendular and planetary motion before and after Galileo; or of salts, alloys, and a sulphur-iron filing mix before and after Dalton... Therefore, when such redistributions occur, two men whose discourse had previously proceeded with apparently full understanding, may suddenly find themselves responding to the same stimulus with incompatible descriptions and generalizations. Those difficulties will not be felt in all areas of even their scientific discourse, but they will arise and will then cluster most densely about the phenomena upon which the choice of theory most centrally depends.[12] Such problems, though they first become evident in communication, are not merely linguistic, and they cannot be resolved simply by stipulating the definitions of troublesome terms... Part of the difference is prior to the application of the languages in which it is nevertheless reflected.'[13]

As a matter of fact, their everyday and most of their scientific world and language are shared. But, it is in the smaller remaining part that they differ, so what they should do in such a situation is to recognize each other as members of different language communities and then become translators:

'Since translation, if pursued, allows the participants in a communication breakdown to experience vicariously something of the merits and defects of each other's points of view, it is a potent tool both for persuasion and for conversion. But even persuasion need not succeed, and, if it does, it need not be accompanied or followed by conversion. The two experiences are not the same, an important distinction that I have only recently fully recognized'.[14]

Consequently, persuasion means convincing somebody that one's view is superior to his own and therefore ought to replace it. That much, Kuhn notes, is occasionally achieved without recourse to anything like translation, although arguments statable in the vocabulary both groups use in the same way are not usually decisive, at least not until a very late stage in the evolution of the opposing views. In contrast to those who are new in a profession, few of those with a considerable length of service will be persuaded without some recourse to the more extended comparisons permitted by translation.

And now, in Kuhn's opinion, here lies the essential difference between persuasion and conversion -- to translate a theory or a world view into one's own language is not to make it one's own. 'For that, one must go native, discover that one is thinking and working in, not simply translating out of, a language that was previously foreign. That transition is not, however, one that an individual may make or refrain from making by deliberation and choice, however good his reasons for wishing to do so. Instead, at some point in the process of learning to translate, he finds that the transition has occurred, that he has slipped into the new language without a decision having been made'. In order that the acceptance of a new theory be effective, conversion is essential. 'The conversion experience that I have likened to a gestalt switch remains, therefore, at the heart of the revolutionary process'. In that sense persuasion, good reasons for choice, provide only the motives for conversion, or a climate in which it is more likely to happen, i.e. in which a scientific revolution (in Kuhn's sense) is more likely to occur.[15]

In the Postscript to his idea of scientific revolutions and the role of persuasion in it, Kuhn rightly notes that scientists share their everyday as well as most of their scientific world and language, but that in the process of persuasion it may be of great importance, and of great help, if certain terms and expressions of the new theory were translated into the language of the old one. But, if that is so, one may ask: What really survives of Kuhn's idea of incommensurability among the paradigms?

A very good point is Kuhn's warning also that translation of a new theory into the language of the old does not necessarily mean that the new theory will be adopted as one's own, but only that all the necessar conditions for a conversion are provided. This, now, means that the process of revolution is not as simple as it appeared to be: articulation of a paradigm (the normalscience) -- crisis -- instant revolutionary change of paradigms. In addition to the previously observed fact that the state of crisis may last quite a long time, it becomes obvious that the process of persuasion (which is preceding conversion, i.e. revolutionary transition) may last quite a long time too. But, this calls in question another of Kuhn's assertions: that two paradigims cannot exist at the same time. Kuhn believes that if the new viewpoint endures for a time and continues to be fruitful, the research results expressed in this way are likely to grow in number; so they can play a decisive role in the process of persuasion, and Kuhn even speaks of a 'very late stage in the evolution of the opposing views'. Hence, it is obvious that we are dealing here with a long-lasting parallel existence of two paradigms. However, since the new theory may not continue to be fruitful, why not admit the possibility of a third view-point competing for the position of a paradigm which will overcome the others. This line of thought leads, unfortunately, to a position which Kuhn rejects, e.g., the theory of permanent revolutions, or to the acceptance of Feyerabend's idea of the proliferation of theories!

But, even if we do not pursue this line of thought and stay close to the possibility of parallel existence of the two paradigms, one may ask: Why should we connect the concept of scientific revolution with the moment of conversion only; why not with the moment of the first emergence of the new theory? Or, one may ask further: Does the revolution occur at the moment when the new theory becomes so fruitful that it gets its first converts, or only when the majority of the community of scientists gets converted, as well as how to determine that moment?

It seems more proper to connect our notion of scieitific revolution with the entire process which starts with the emergence of the new theory and lasts until it becomes the ruling paradigm having pushed out the old one. In any case, it seems highly tendentious to insist, as far as Kuhn is concerned, on connecting a scientific revolution merely with the moment of conversion!

In addition, when one speaks of understanding new terms and expressions which the new theory brings along, then it is not inappropriate to speak of 'translation' as well, but only in a very figurative way. When Kuhn himself admits, though only in the Postscript, that the greatest part of the language is common to all scientists, and points to the fact that they share a history, except, perhaps, the immediate past, then it is not so easy to understand what really has to be translated in order to be apprehended and what is the justification for using radical expressions such as 'communication break-down' and the like.

As far as comparison of statements coming from competing paradigms is concerned, one can notice that in all actual cases, e.g. in physics, there exist quite a few neutral statements and expressions which have the same extensional meaning or reference in different theories. In C.R. Kordig's opinion, these common language expressions may be of two sorts: statements of logic and/or mathematics, and statements of empirical or observational character. For example, some expressions in microphysics may be specific for that theory, i.e. may have extensional meaning preventing their use in the ordinary experience on the macro-level; nevertheless, all other expressions in microphysics are taken from the ordinary language.[16]

Hanson, who, like Kuhn, supported the opinion that scientific theories replacing one another are incommensurable, when discussing the transition from the theory of Tycho Brahe to that of Johann Kepler, noticed that there was enough observational uniformity in these two theories that the following statement may appear in both: 'Our sense observation shows only that in the morning the distance between horizon and sun is increasing'. This statement, of course, does not tell us whether the sun is ascending or the horizon is descending.[17] In this case, ordinary language does not look neutral; namely, we usually speak of the sun which is ascending. Here, though, Kordig's remark is relevant: he believes that, perhaps, it is more correct to say that our beliefs are not neutral, rather than saying that for our languages. In any case, the observation statement quoted above shows that we are able to describe in a theoretically neutral language an invariable state of affairs discovered by observation. Therefore, when we speak of an experiment or an observation, there is no reason to maintain (as Feyerabend, Hanson, Kuhn and Toulmin sometimes do) that such statements are completely saturated with theory - that they wholly depend on the particular theory which was put to test. If that were the case, it would have been impossible for Kepler to use observational reports of Brahe (which he did use), since Brahe upheld a completely different theory.

Or, take the relation between the physical theories of Galileo and Newton. It will be agreed that the Galilean theory concerned itself solely with the terrestrial phenomena, while the Newtonian included the celestial ones as well. So, although it is possible to distinguish the subject matters of these two theories, certain objects referred to in terms and expressions of the Galilean physics (theory T_1), are the same as those referred to in certain terms and expressions of the Newtonian physics (theory T_2). Ergo, an observational invariance occurred in the transition from T_1 to T_2, and hence an extensional invariance of meanings. As Nagel, discussing the problem of reduction of theories in natural sciences, observed very well, "the subject matters (of Galilean and Newtonian physics) are in an obvious sense homogenous and continuous; for it is the motions of bodies and the determinants of such motions that are under the investigation in each case, and in each case inquiry is directed toward discovering relations between physical traits that are the common concern of both disciplines'.[18]

The invariance of meanings is present, Kordig says, not only in connection with the coextensive objects the competing theories refer to, but also takes into consideration the abstract objects such as mathematical statements. These statements are usually neutral in relation to the rival scientific theories.[19]

But, since the presence of value-judgements in science is also one of the main reasons why persuasive arguments cannot be considered as proof, it will be appropriate to discuss briefly the role of value-judgements in the process of acceptance of hypotheses. Philosophers differ very much on this issue.

Nowadays, I guess it is almost generally accepted that no scientific hypothesis, or theory, can be verified or confirmed completely. However, in spite of that fact, scientists always decide to accept a particular hypothesis. If this is so, scientists are making, consciously or unconsciously, a value-judgement. Here we have a value-judgement since, as noticed by R. Rudner[20], scientists' decision to consider a particular hypothesis as acceptable, i.e. to believe the evidence in favour of the hypothesis to be strong enough, is made, generally, in the light of the seriousness of the mistake they would have made if they did not accept that hypothesis. Scientists make value-judgements as well in the sense that it is very difficult to exclude completely the influence of one's own subjective beliefs and prejudices, whatever their origin may be.

In The Structure of Science, Nagel defended the view that in sciences, and in social sciences particularly, it might be possible to neutralize the considerable pressure of value-judgements which obviously exists. R.C. Jeffrey, however, attempted to present a view that there was really no sense in speaking of scientists making value-judgements, as they do not simply either accept or reject hypotheses, but provide them only with certain degrees of probability, thus creating the possibility for a rational decision to act on the basis of a certain hypothesis; this, however, has to be separated clearly from the acceptance of that hypothesis.[21] One is certainly justified in making this kind of distinction. It really exists; namely, scientists may even act on the basis of a hypothesis they do not yet accept, as well as decide not to act in accordance with a hypothesis they consider to be true. But, it does not entail that anyone has to deny that the greatest part of the normal scientific practice is done on the basis of hypotheses consciously accepted and not easily abandoned.

In any case, certain methodological or meta-scientific values are normally considered (by most of the contemporary philosophers of science) to be the reasons which may play a decisive role in the process of persuasion. I can mention quite a few such values: (1) withstanding a great number of serious tests aimed at falsification of a particular hypothesis; (2) extension of comprehensiveness of the concept-model, i.e. providing greater universality, or, more precisely, making broader connections between different and unconnected fields of investigation, which, in fact, amounts to increasing the explanatory power of a hypothesis; (3) greater possibility of anticipating new facts, or greater possibility of deducing various consequences; (4) greater simplicity; (5) greater applicability, etc.

In the preliminary considerations, one may feel it is appropriate to mention also the possible sources of resistance of scientists to scientific discovery. In that respect, one can hardly add anything new to the sources discussed by American sociologist of knowledge, Bernard Barber[22]: resistence is coming from (1) substantive concepts and theories held by scientists at any given time; (2) methodological conceptions scientists entertain at any given time; (3) religious ideas of scientists; (4) models of social interaction: a) professional standing of a discoverer, b) pattern of specialization, c) official scientific organizations or publications, d) rivalry between different 'schools', e) opposition of older scientists to the younger ones. But, although the influence of these sources of resistance to scientific discovery cannot be willingly avoided, in science (in contrast to the majority of other human intellectual activities), certain norms are accepted which are a rather reliable counterweight to resistance: open-mindedness, criticism and self-criticism, authority of reason and argumentation, objective tests... Finally, some forms of resistance, as the extreme form of disagreement, may even lead to elucidation and further growth of certain viewpoints.

But, let us go back to the reasons which are not the source of resistance, but of acceptance of a hypothesis. I have already mentioned -- in that respect -- that a very important role may rightly be attached to some basic methodological values. Some authors, like I. Lakatos, point out that we should not forget that in the process of persuasion an irrational element is always present. Namely, criticizing demarcationists (whose methodology, or appraisal criterion, demarcates better from worse knowledge, defines progress and degeneration) for their firm support of the view that rationality resides only in what is articulated in human knowledge, Lakatos indicates that this means the following: they presume that they can lay down the statute law for rational appraisal which can direct a lay jury in passing judgement so that the philosophy of science becomes the watchdog of scientific standards, applied by great scientists either sub- or semi-consciously.

Now, if we look into certain episodes of the growth of scientific knowledge, Lakatos goes on to say, it is not possible to tell for certain in every case that they fit, and to what degree they fit, into our rational reconstructions. In such cases, militant demarcationism is ready to throw away all the efforts invested into defence of 'old' theoretical ideas which do not fulfil standard requirements and reasons, but originate from affinity, authority, imagination and the like. As examples of such militant demarcationism, one can take Newton's well-known fight against hypotheses not deduced from phenomena, or Popper's rejection of psycho-analysis, etc.

Strict demarcationism is obviously not in accord with the everyday scientific practice which has never been obliged nor will ever be obliged to follow any universal or standard procedures. Moreover, if we allow the existence of some general standards which scientists apply sub- or semi-consciously, one cannot deny that it often happens that more or less irrational elements do play a significant role in the acceptance of a new theory.

Yet, it would be a mistake, in fact, it would ruin any possibility of searching for sense, i.e. intellectual development aspect in the history of science, to end up opposing strict demarcationism. The other extreme is called 'elitism' by Lakatos. Representatives of this view (Polanyi or Oakeshott, for example) hold that only a privileged elite has the craft of science; a large part of science belongs to the 'tacit dimension' (i.e. is inexpressible) which is shared and understood by the elite only; only they can judge their own work.

This kind of irrationalist approach represents another extreme practice, since the history of science contains any number of examples testifying that normal scientific practice consists of completely rational arguments understood by anyone.

In addition to those two, Lakatos warns us that there is also a third extreme position --`scepticism'. 'Scepticism', Lakatos says,'regards scientific theories as just one family of beliefs which rank equal, epistemologically, with the thousands of other families of beliefs. One belief-system is no more 'right' than any other belief-system; although some have more might than others. There may be changes in belief-systems but no progress'. This school of thought, which can be traced back to the Greek tradition of Pyrrhonian scepticism, temporarily silenced by the stunning success of Newtonian science, is regaining momentum today, as noticed by Lakatos, particularly in the anti-scientific circles of the New Left; its most original and colourful version is Feyerabend's 'epistemological anarchism'. Feyerabend grants that philosophy of science is a perfectly legitimate activity; it may even influence science. Nevertheless, any belief-system -- including those of his opponents -- is free to grow and influence any other belief-system; not one of them has epistemological superiority.[23]

This school of thought obviously denies, in principle, the possibility of considering any reason that may persuade any scientist to abandon his theory in favour of any other one. However, I would like to emphasize that the possibility of discussing persuasive reasons depends not only on how we approach some methodological questions (the problem of which also is not in harmony with the actual scientific appraising scientific theories), but also on our general concept of scientific knowledge. As opposed to a realistic concept of science within which it makes good sense to talk of persuasive argumentation in science, within an instrumentalist view there simply is no need to discuss this subject. Let us investigate it more closely.

Consistent instrumentalism does not have any need to enter into discussion of the reasons which might be fundamental in the process of scientific persuasion. Since the instrumentalists' sole interest is whether a hypothesis can 'save the phenomena', and not whether it is true or false, any hypothesis that 'works' or might be used as a good instrument for prediction is welcome. Therefore, any theory which is not senseless, or useless, and 'saves the phenomena', may be considered equally good; strict instrumentalism, consequently, admits co-existence of any number of hypothetical explanations of the same set of phenomena without even trying to establish any kind of priority among them.

It could not be otherwise, if scientific theories are considered only as convenient instruments for summarizing and systematizing certain sets of facts, as well as for deducing consequences which can be used in man's practical activity. From this standpoint, strictly understood, it is not even important whether a certain hypothesis is simpler than the other one, or appraised to be a better one by any other theoretical criteria. Scientists will use the hypothesis (or theory) that they consider most convenient in a particular situation. According to a strict instrumentalist, nothing can prevent anyone, in principle, from taking the position that it is possible to tolerate co-existence (and simultaneous practical use) of several different theories relative to the same set of facts. Preference may be given only to the theory which is 'simpler to use'.

It is clear that, with this concept of science, the process of persuasion and of the acceptance of scientific theories, is radically different from the one when we take the realist concept. The instrumentalist position, of course, is not always strict and consistent. For example, P. Duhem speaks of some criteria for theoretical appraisal of scientific theories, and of simplicity in particular, which may turn out to be an argument, too, even in the process of persuasion, regardless of the degree of restriction in the sense of persuasion we are dealing with in this case.

In the history of science, cases have been noted in which scientific hypotheses, especially in the initial stages of their development (particularly before a metaphysical theory -- in Popper's sense of the term -- becomes scientific) were accepted primarily as instruments, and it was only considerably later that they were accepted as statements which pretended to tell us something (true) about reality. The case with atomism was the classical example. It took quite a while for this theory to become a scientific one, one which might be empirically tested. Details on gradual transition from the instrumentalist to the realist approach in the atomic theory can be found in the article of M.R. Gardner, 'Realism and Instrumentalism in 19th Century Atomism'.[24] The case was similar with the Copernican theory, despite the fact that Copernicus himself was a realist.

The realist position also has its difficulties which impede the attempt to formulate a precise and complete conception, when speaking of persuasion in science. One of the difficulties consists in the almost universally accepted view that scientific hypotheses (or theories) more or less always transcend the class of facts available (and that even the reports of direct observations are not out of the domain which is subject to criticism and new formulation).

Another difficulty is evident in the disagreement that may exist between the new hypothesis and the previously accepted theories (even accepted observations); it should not influence our decision concerning the fate of a new hypothesis. Naturally, this discord gives rise to a new problem which calls for a new investigation with a view to reexamining the situation and searching for the cause of the discord, since it may not be contained in the new hypothesis.

Finally, the difficulty may lie in the fact that the question whether a particular hypothesis is successful or not cannot be answered easily, quickly or decisively. The decision is reached slowly and gradually, and every step in that direction is susceptible again to revision. As Feyerabend noticed very well -- his evidence being a number of the newest scientific theories whose characteristics not only included unexpectedness, but also reached beyond boldest human imagination -- even the absurdity of a particular hypothesis cannot count as a general argument against that hypothesis. It is a reasonable consideration for the choice of one's own theories to demand that they seem plausible to oneself. This is one's private affair, Feyerabend says; but, to declare that only plausible theories should be considered is to go too far.[25] Nevertheless, if we start from a theory that looks absurd (and in all other cases, too), it is important that later on a athorough investigation be carried out of the difficulties the theory meets, as well as the objections raised against it.

I will now discuss the arguments (mentioned in the recent literature) which may be used, or which are used to characterise the process of persuading scientists to give up an old theory and accept a new one.

(1) Withstanding greater number of serious tests intended to falsify a hypothesis.

This Popperian argument requires a detailed theoretical clarification. To begin with, it seems rather obvious that, in the process of persuasion, withstanding the tests performed with the intention to achieve possible falsification of a hypothesis (i.e. testing of deduced consequences which are new and unexpected, even such that if they prove wrong, they would become crucial evidence against that hypothesis) has greater weight than any other kind of test.

Let us consider, for example, the case of Einstein's general theory of relativity. Since Newton's formulation of the laws of gravitation is valid only in the inertial co-ordinate system of classical physics (defined by the condition that the laws of mechanics must be valid in it) and does not agree with the special theory of relativity (distance is not invariant with respect to the Lorentz transformation), Einstein's general theory of relativity constitutes the first serious effort, since Newton's time, to reformulate the law of gravitation in the sense of making it even more general. But, as told by Einstein himself,[26] after he succeeded in formulating his general theory of relativity, he was faced with the following question: Are we not in danger of carrying speculation too far from reality? He believed, of course, that there must be a possibility of bridging the new theory and observation, that the new theory of gravitation had to stand the test of experiment. Namely, Einstein thought that all hypotheses, disregarding their attractiveness, had to be rejected if they did not correspond to reality. Therefore, Einstein himself was interested more than anyone else in seeing how his new theory of gravitation would withstand experimental tests. If some deduced consequences did not come up, Einstein stated, he would have been ready to admit that his theory was not good and would have abandoned it.

Now then, since the old theory of gravitation is the special limiting case for the new one (if the gravitational forces are comparatively weak, the old Newtonian law turns out to be a good approximation to the new laws of gravitation), all the observations which support the classical theory support the general theory of relativity as well. So, even if no additional observation could be quoted in favour of the new theory, Einstein and Infeld say, if its explanation were only just as good as the old one, given a free choice between the two theories, we should have to decide in favour of the new one. If someone were to wonder why, the answer would be the following: From the formal point of view, the equations of the new theory are more complicated but, from the point of view of fundamental principles, their assumptions are much simpler; the two frightening ghosts, absolute time and inertial system, disappeared; no assumption concerning gravitational forces and their dependence on distance is needed; the gravitational equations have the form of structure laws, the form required of all physical laws since the great achievements of the field theory.

Still, that is not enough and cannot be a decisive argument in our acceptance of the new theory. We need certain new deductions not contained in Newton's law of gravitation. One such is the bending of light rays in a gravitational field.

According to the classical theory, a beam of light is weightless and it will not be affected by the gravitational field. The new theory, however, holds that a beam of light carries energy and energy has mass. But each inertial mass is attracted by the gravitational field as inertial and gravitational masses are equivalent. A beam of light will bend in a gravitational field exactly as a body would if thrown horizontally at a velocity equal to the velocity of light. Earth's gravitational field is too weak, of course, for the bending of light rays in it to be proved directly, by experiment. However, solar gravitational field is strong and famous experiments performed during the solar eclipses showed conclusively, although indirectly, the influence of a gravitational field on the path of a light ray.

The next test carried out confirmed the prediction that there ought to be deviations in the elliptical paths of planets moving around the sun (the ellipses should rotate very slowly, relative to the co-ordinate system connected with the sun). The general theory of relativity predicts the magnitude of this effect. Since the attraction between the sun and Mercury is stronger than the attraction between the sun and any other planet (the distance is shortest), the biggest chance, if any, to find a deviation from Newton's law lay with Mercury. In fact, the deviation of the motion of the planet Mercury from the ellipse was observed before the general theory of relativity was formulated, and no explanation could be found. The general theory of relativity developed without paying any attention to this special problem and it was only later that the conclusion was drawn from the new gravitational equations about the rotation of ellipse in the motion of a planet around the sun.

The third experimental test was based on the prediction that a clock placed on the sun would have a different rhythm from the rhythm of a clock placed on Earth, since the influence of the gravitational field is much stronger on the sun than it is on Earth. Now, since atoms represent, so to speak, clocks which emit waves in definite rhythms, it was found out, by measuring the wave-length of light emitted by a sodium atom placed on the sun, that it was slightly greater than the wave length of light emitted by the same atom on Earth.

In connection with the first kind of persuasive arguments, we should not forget that there have always been scientists unready to accept this kind of argument. Francis Bacon, long ago, described these scientists very nicely; he called them 'dogmatists'. As their characteristics indicate, first of all, they are simply not able to see, they are blind to the facts that conflict with their theory; and, even when forced by others to notice the facts, they usually deny the truth of those same facts or belittle their importance with the aid of a 'frivolous distinction'.

How and why some scientists behave like this is primarily a socio-psychological question I will not try to discuss here. Nevertheless, there is another, methodologically very interesting question: Are there situations in which one is justified, or, better still, in which it is advisable or even necessary to ignore the evidence against a hypothesis? Trying to answer this question in a nutshell, I have been inspired by the ideas of J. Agassi, put forward in his article 'When Should We Ignore Evidence in Favour of a Hypothesis?'[27]

First of all, I will mention some views which may ignore, as a matter of course, evidence against a hypothesis. I have again in mind, primarily, instrumentalists such as Pierre Duhem. Instrumentalists point out that the reports of scientific observations, i.e. our empirical evidence, are subject to alterations parallel to the changes of our theoretical framework within which these reports are made. Other theoreticians emphasized (Freud, for example) that it was not only permissible but also completely appropriate to overlook some facts for a while. Others, still (P.A.M. Dirac, for example), stressed that, in their work, they favoured aesthetic considerations rather than close agreement between theory and facts. Finally, there were theoreticians such as M. Polanyi who were philosophically opposed to formulating any rules for scientific procedure; they thought that scientific method was ineffable and its practice a matter of the intuition of leading scientists.

As for instrumentãlism (or conventionalism), P. Duhem, not a strict instrumentalist himself, pointed out that the facts need not change or disappear, while the continuity is maintained by translating from the language of one theory into the language of the other. However, if one were to support the extreme opinion that there is no translatability, it would imply that one is denying the possibility of argumentation against a scientific theory, or in its favour alike. In any case, the instrumentalist view may be methodologically suitable only to those interested in the history of science, but not at all to creative scientists aiming at the future.

Freud's remark, as I understand it, is a kind of warning against hasty rejection of a theory, i.e. it is yet another way of supporting the principle of tenacity. Anyway, it is advisable to give enough time to each new hypothesis -- even if it encounters conflicting evidence -- in order to let it develop all its explanatory power.

The opinion of Dirac is rather uncommon among physicists; of course, it is more popular among mathematicians. If the scientists interested in empirical sciences were to turn completely to non-empirical criteria, they would not be able to practice empirical sciences any more.

Nevertheless, it is clear and widely accepted nowadays that there may be cases when scientists will reject the observation report which conflicts with their hypothesis, just as there will be those when they will reject a hypothesis. Moreover, one is also forced to conclude that it would be wrong to impose any rigid methodological rule such as the one according to which scientists should immediately, resolutely and forever reject a hypothesis, whenever it encounters a single phenomenon which contradicts it. This is not denying that a critically orientated scientist should be able to realize a discrepancy between his hypothetical view and the counter-example, and, if he accepts the counter-example, to begin an immediate search for another theoretical solution.

It is obvious, consequently, that Popperian rigid view, or, as Lakatos put it, 'dogmatic falsificationism', has to be replaced by a more refined conception in which the Popperian basic rule will be weakened. Nevertheless, it does not mean that one has to deny the appropriateness of Whewell's thought that a true scientist should cherish 'candid and simple love of truth, which makes him willing to suppress the most favourite production of his own ingenuity as soon as it appears to be at variance with realities'.[28] It also does not mean any belittling of the great persuasive power of the argument that one ought to accept those hypotheses which withstood hard and serious empirical tests.

(2) The next persuasive argument may be characterized as the 'explanatory power' of a hypothesis (or a theory).

Now, if we were to try to explicate what we have in mind when using this syntagm ('explanatory power'), I would like, first of all, to mention the capacity of a hypothesis for solving more problems than the competing one, as well as its capacity for relating the formerly different and unrelated segments of human knowledge (or fields of investigation); it leads hence usually not only to a greater universality of a hypothesis (or a theory), but also to greater possibilities in anticipating new facts; i.e., in explaining as well the phenomena the old one was not capable of explaining.

This kind of argumentation (2) is older than (1) and had appeared in several different forms and under several different designations. Whewell's 'consilience of inductions' was the first significant treatment of this sort of argumentation. Whewell had in mind a few possible cases: first, when the hypothesis can explain two or more known but different classes of facts (or laws), requiring until then separate and independent explanatory hypotheses; one can achieve reduction of 'theoretical baggage' in this manner --- if nothing else. The second case, when acceptance of a new hypothesis makes it possible to anticipate previously unknown (or, at least, undiscussed) phenomena; these phenomena may be completely different and totally unexpected. This fecundity of explanation is surely an important feature in favour of the hypothesis possessing it. Therefore, Whewell concludes: '...When the hypothesis, of itself and without adjustment for the purpose, gives us the rule and reason of a class of facts not contemplated in its construction, we have a criterion of its reality, which has never yet been produced in favour of falsehood'.[29] Finally, the third case, when a new hypothesis can predict or successfully explain the occurrence of phenomena which, according to the previous experience, we would not expect to happen. The second and the third cases obviously represent an expansion of the explanatory power of our hypotheses. In contrast to the second case, the third one gives us the possibility of testing our hypothesis (or theory) in the most effective and convincing way. Therefore, one can say in all certainty that the 'consilience of inductions' is one of the most powerful persuasive arguments in favour of a hypothesis.

Whewell says that in his extensive study of the history of science he did not find any example in which consilience of inductions gave testimony in favour of a hypothesis discovered afterwards to be false.

'The instance in which this (consilience of inductions, S.N.) has occurred, indeed, impress us with a conviction that the truth of our hypothesis is certain. No accident could give rise to such an extraordinary coincidence. No false supposition could, after being adjusted to our class of phenomena, exactly represent a different class, where the agreement was unforeseen and uncontemplated. That rules springing from remote and unconnected quarters should thus leap to the same point, can only arise from <u>that</u> being the point where truth resides'.[30]

This argument which ought to give us criteria for the choice of appropriate hypothesis, has been lately called the 'inference to the best explanation' (Gilbert Harman), while a few philosophers have been working out the concept of 'the best explanation' (Keith Lehrer, Paul Thagard). Those who follow this line of thought claim to give more adequate description of the process of evaluating theories in comparison with the hypothetic-deductive model. Relying on Whewell and Peirce as their precursors, these philosophers are concerned with the analysis of the reasons which may be used in persuading scientists that a particular hypothesis offers a better explanation of the evidence we have than any other hypothesis. The main criterion they use is 'consilience', usually adding simplicity and analogy.

Thagard considers a theory to be 'consilient' when it explains facts of different sorts, or explains laws in different fields. Such a theory unites and systematizes. Therefore, to say that a theory is 'consilient' is to say more than that it 'fits the facts', namely, it is to say, first, that the theory explains the facts and, second, that the facts it explains are taken from more than one area. However, one may get the impression that the most 'consilient' hypothesis or theory is the one which explains any fact whatsoever. This could be achieved only by adding a whole set of auxiliary hypotheses which would make possible for any phenomenon to fall under the theory, but such an approach would be unsatisfactory as it would conflict with the other criteria, namely, precision and simplicity.[31]

Newton's mechanics can be taken as an outstanding example of the kind of reasons used to persuade us that we ought to choose a particular theory as the best explanation of a set of facts at the moment. It abolishes the dichotomy established by Aristotle between earthly and heavenly regions which had persisted for more than a thousand years. Newton made it possible with his theory of gravitation to apply the mechanics concerned with motions on Earth (Galileo's laws) to the motions of planets (Kepler's laws) and show that all these laws were simply a consequence of general mechanics. It also made possible to explain motions of comets, tides and so on. Thus Newton's scheme offered a theory of an obviously greater power of explanation and unification.

Or, let us take quantum mechanics, which, by giving explanation of the previously unconnected phenomena, exceeds by far any competitor; the theory provides explanation of spectral frequencies of certain atoms, phenomena of magnetism, solid state of matter and various other perplexing phenomena such as the photoelectric effect and the Compton effect.

But, if we would like to avoid accusation of discussing the examples in physics only, let us also consider Darwin's theory of evolution. I think of Darwin's lengthy argument for his theory of evolution of species by means of natural selection. In his book The Origin of Species he cites a vast array of facts explained by his theory of evolution but inexplicable by the then accepted view that species were independently created by God. Darwin provides explanation of facts relative to the geographical distribution of species, the existence of atrophied organs in animals and many other phenomena. He concludes: 'It can hardly be supposed that a false theory would explain, in so satisfactory a manner as does the theory of natural selection, the several large classes of facts above specified'.[32]

(3) Simplicity of a hypothesis or a theory.

Simplicity is often cited as a criterion of theoretical appraisal of rival hypotheses before they are put to test. Having that in mind, the first question to be considered is the following: Do we deal here with two different concepts of simplicity or just with different uses of the same concept?

It seems to me that simplicity, as the reason for opting for a theory which has already been tested, has an essentially different content when compared with the concept of simplicity with whose aid we make a theoretical evaluation of a hypothesis immediately after it is introduced and before putting it to test. As for this difference in content between the two concepts of simplicity, it will suffice to point out that the recently introduced hypothesis cannot have any associated auxiliary hypotheses, whereas the hypothesis which encountered criticism and had been tested already, rarely remains without auxiliary hypotheses. Now the kind of simplicity we are looking for is primarily associated with the requirement of a minimal number of auxiliary and ad hoc hypotheses in conjunction with the hypothesis or theory we opt for. This kind of simplicity can only be used for evaluating those hypotheses or theories which have already been critically tested. This is the concept of simplicity relevant in the persuasion procedure.

This concept of simplicity requires, of course, further clarification. The explanation of facts F by a theory T requires a set of given conditions C, but it also may have a set of auxiliary hypotheses A, i.e., such hypotheses which have not been part of the original theory. Set A were assumed in order to help explain some elements of F which -- as it subsequently turned out -- could not be explained by T itself.

This can even be better understood if we consider a few examples from the aforementioned article by P. Thagard. Huygens' theory of light, includes statements such as those that light consists of waves in the ether and that light waves are propagated according to his principle that a wave is formed around each particle in the medium, the particle being its centre. To explain the laws of refraction and reflection and other phenomena, Huygens assumed that waves were spherical. But to explain the irregular refraction in Iceland crystal, Huygens surmised that some waves were spheriodal. The latter assumption, restricted in use to one class of facts, is an illustration of the auxiliary hypothesis. Similarly, to explain Snell's law of refraction, Huygens assumed that the speed of light was lesser in denser media. These two assumptions were not independently acceptable at the time of Huygens, so they could not belong to C.

Lavoisier obviously had this concept of simplicity in mind when he criticized the phlogiston theory because it required a number of inconsistent assumptions to explain facts easily explained by his theory. These examples show, as P. Thagard has pointed out, how simplicity puts constraint on consilience: a simple consilient theory must not only explain a range of facts only; it must also explain these facts without making a host of assumptions with limited application.[33]

Now, in the case of the kind of simplicity that is connected with persuasion in accepting an already tested theory, there is also no place for syntactic or semantic simplicity, or notational or pragmatic simplicity which, of course, may be relevant when evaluating the theoretical acceptability of a recently introduced hypothesis or theory. There is also a sense of the concept of simplicity which may be shared in both cases and used in both situations, for both purposes. That is simplicity in the Popperian sense, the sense of a system characterized by the paucity of parameters.

(4) Degree of fitting of a hypothesis into a system.

Several authors mention, as a separate reason, the degree of theoretical support by more universal theories, or analogy (P. Thagard), or even 'many-sided connection' (C. Kordig). I think -- when we speak of persuasion -- that all these meanings are covered by the formulation of the fourth argument. Someone may even say that fitting of a hypothesis into a system is too broad and that it is not easy to guess at the real content of this formulation. Using Laudan's syntagm, we may say concisely that we have in mind such a theory which does not create any 'conceptual problems'.

In any case, this criterion is a complex one. In that sense, when speaking of fitting a hypothesis into a system, one may think of several different aspects.

The first and most general of aspects is the following: acceptance or non-acceptance of a radically new scientific theory, or scientific research programme, is not a purely empirical question; our metaphysical beliefs, for example, may play an important role. But, since it is not customary to cite explicitely our most general metaphysical presuppositions about the world, it is not always easy to identify them.

The next aspect (of lower generality) is often cited by scientists as an important argument speaking in favour of choosing their hypothesis or theory and it holds that the theory is in accordance with, or at least, does not contradict other, already accepted hypotheses or theories, especially those concerned with some related problems.

The third aspect of this criterion, is of a special kind; it would be a simple indication that a particular hypothesis or theory has fulfilled the condition of being not only maintainable from the inside, but also able to correlate well with those theories which make the organized totality or system of knowledge in a particular scientific discipline.

Finally, it is certain that analogy is not only an important tool for generating scientific hypotheses, it may also be used as a heuristic factor directing scientists towards a particular kind of hypotheses. Likewise, it may be used on the level of persuasion, when scientists try to persuade their colleagues that a hypothesis or a theory ought to be considered good and appropriate.

Thus, for example, Darwin used the analogy between the artificial and natural selection for heuristic purposes, but he also claimed analogy to be one of the bases for belief in his theory. Huygens, Young and Fresnel each used the analogy between the phenomena of sound and the phenomena of light to support the wave theory of light.

But, it has to be shown in what way analogy is connected with explanation, or how it is used in persuasive arguments. (Namely, simple indication that there is certain analogy cannot suffice since one can always find some disanalogies). P. Thagard also tried to do this, connecting analogical inference with explanation in the following way:

> 'Suppose A and B are similar in respect to P, A and R, and suppose we know that A's having S explains why it has P, Q and R. Then we may conclude that B has S is a promising explanation of why B has P, Q and R. We are not actually able to conclude that B has S; the evidence is not sufficient and the disanalogies are too threatening. But, the analogies between A and B increase the value of the explanation of P, Q and R in A by S'.[34]

As regards the way in which analogy works as a persuasive argument, it is obvious that we get increased understanding of one set of phenomena if the kind of explanation used -- the kind of model -- is similar to the kind already used. Now it will be appropriate to warn that this formulation does not mean that an explanation is a reduction to the familiar. Scientific explanations often employ unfamiliar notions and introduce entities as peculiar as positrons and black holes. However, other things being equal, the explanations afforded by a theory are better if the theory is familiar, that is, if it employs mechanisms, entities or concepts which are already used in established explanations. The use of familiar models is not essential to explanation, but it helps.[35]

If we consider theories at the lower level of generality, then a frequently used argument in favour of such theories is simply pointing out that those theories are linked with the general theory or paradigm which has proved to be highly successful. Nevertheless, this argument can often be very unreliable. Laudan, for example, cites the case of theories of mechanistic physiology in the late seventeenth century (such as those of Borelli and Pitcairn) which were regarded highly in many circles where the mechanistic research tradition flourished, even though, judged entirely on the basis of their own merits, they were significantly inferior to certain theories in other, less successful research traditions.[36]

Finally, it has to be mentioned that all four arguments may be formulated negatively; in that case, they may be used to talk scientists out of certain (usually old) theories. Namely, negative formulations point out essential methodological deficiencies of a theory: (1) it did not withstand sufficient number of serious tests, (2) it has small explanatory power, (3) it is less simple, and (4) it fits poorly into a system of theories, i.e. creates conceptual problems.

Nevertheless, when speaking of arguments directed against the acceptance of certain scientific hypotheses, we must take heed that in such a process, one of the most powerful arguments -- apart from calling attention to various logical deficiencies -- is drawing attention to the anomalies in the sense that certain theories are saturated with 'empirical problems' of which they cannot be rid.

*

*   *

Apart from the four arguments discussed, it is certainly possible to find several more. Still I have restricted myself to these four, since I believe that the additional ones which may be traced in the literature (increasing the scope of the conceptual model, greater generality, greater economy of thought, manifold connections, more precise formulation, etc.) can be designated as special cases or sub-species of these four. If that is not possible, then we are dealing with arguments which have not been accepted by the majority of scientists (greater possibility of being operational, greater possibility of being mathematical, greater elegance, etc.)

Even if I have succeeded, more or less, in drawing attention to the basic persuasive arguments which are used among scientists, I would not like to suggest that any scientific hypothesis or theory will be immediately accepted if and when we present some of these arguments in its favour! The question of how many arguments and which of their kind we need in order to make sure that a hypothesis or theory will be accepted by a community of scientists is not the subject of our discussion. In any case, it goes without saying that the process of persuading a group of scientists to favour a particular hypothesis or theory, usually proceeds hand in hand with the process of talking the same group out of another.

Finally, from all that we have said, it can be seen that these four arguments are not independent of each other. During actual episodes in the growth of science, where the persuasion procedure is becoming apparent, these arguments are usually intertwined. We have to analyze them separately, of course, to enable scientists to make a choice and use the most adequate and most effective methods of persuasion. Conscious use of any of these arguments may contribute to their change and improvement, which may also be of some significance both for the methodology of science and actual practice of science.

NOTES

1. Popper, K.R., The Open Society and Its Enemies, vol. 2, London, Routledge & Kegan Paul, 1962, p. 391.

2. Kuhn, T.S., The Structure of Scientific Revolutions, Chicago & London, The University of Chicago Press, 1962, p. 147.

3. v. Ibid., p. 149.

4. v. Ibid., p. 151.

5. v. Ibid., pp. 151–152. The Belgian philosopher H. Perelman, in his treatises on general theory of argumentation, differentiates between 'persuasion' (or 'convincing'), which requires rational means, and 'talking into' (or 'animating'), which relies on irrational means.

6. Ibid., p. 152.

7. v. Ibid., p. 153.

8. v. Ibid., pp. 154–155.

9. Ibid., pp. 156–157.

10. v. Ibid., pp. 67, 84 & 149.

11. Ibid., pp. 198–200.

12. Ibid., pp. 200–201.

13. Ibid., p. 201.

14. Ibid., pp. 202–203.

15. v. Ibid., p. 204.

16. v. Kordig, C.R., The Justification of Scientific Change, Dordrecht, D. Reidel, 1971, p. 93.

17. v. Hanson, N.R., Patterns of Discovery, Cambridge, University Press, 1959, p. 182.

18. Nagel, E., 'The Meaning of Reduction in the Natural Sciences', in Philosophy of Science, ed. by A. Danto & S. Morgenbesser, Cleveland, World Publ. Co., 1960, p. 291.

19. v. Kordig, C.R., Op. cit., p. 97.

20. v. Rudner, R., 'The Scientist Qua Scientist Makes Value Judgements', Philosophy of Science, 1953, vol. 20, No. 1, p. 3.

21. v. Jeffrey, R.C., 'Valuation and Acceptance of Scientific Hypotheses', Philosophy of Science, 1956, vol. 23, No. 3, p. 245.

22. v. Barber, B., 'Resistance of Scientists to Scientific Discovery', Science, 1961, vol. 134, No. 3479.

23. v. Lakatos, I., Philosophical Papers, vol. 2, ed. by J. Worrall & G. Currie, Cambridge, University Press, 1978, pp. 107–112.

24. Philosophy of Science, 1979, vol. 46, No. 1, pp. 1–32.

25. v. Fayerabend, P.K., 'Realism and Instrumentalism: Comments on the Logic of Factual Support', in: The Critical Approach to Science and Philosophy, ed. by M. Bunge, Glencoe, Ill., The Free Press, 1964, p. 305.

26. v. Einstein, A. & Infeld, L., The Evolution of Physics, Cambridge, University Press, 1961, pp. 235–240.

27. v. Ratio, 1973, vol. 15, No. 2, pp. 183–205.

28. Whewell, W., Novum Organon Renovatum, London, 1858, p. 81.

29. Whewell, W., Philosophy of the Inductive Sciences., vol. II London, 1847, pp. 67–68.

30. Ibid., p. 64.

31. v. Thagard, P.R., 'The Best Explanation: Criteria for Theory Choice', The Journal of Philosophy, 1978, vol. LXXV, No. 2, pp. 81 & 85.

32. Darwin, Ch., The Origin of Species, New York, Collier, 1962, p. 476.

33. v. Thagard, P.R.., Op. cit., p. 87.

34. Ibid., p. 90.

35. v. Ibid., p. 91.

36. v. Laudan, L., Progress and Its Problems, Towards a Theory of Scientific Growth, Berkeley, University of California Press, 1977, pp. 83–84.

IS THE EVIL DAEMON A SCEPTICAL DEVICE?

ALEKSANDAR PAVKOVIĆ

Some contemporary sceptics – as well as many anti-sceptics – assume that the Cartesian Evil Daemon or any of its modern equivalents is the most powerful device at the sceptic's disposal. So far as I know, no one – except perhaps Gassendi – has ever asked whether the device is a sceptical one at all. In the fifth Objections to the Meditations Gassendi accuses Descartes of unnecessarily introducing novelties into philosophical arguments; according to him, in order to chase away one's old prejudices it is not necessary to invent a deceiving God or a malicious daemon. By inventing these, one is merely replacing old prejudices with new ones. The old sceptic Gassendi wants to suggest, I think, that the Cartesian Evil Daemon hypothesis is not only superflous but contrary to the sceptic's purposes. I shall not try to argue that this is so. I shall only point our how poor a reason the Evil Daemon hypothesis is for any doubt and how it thus differs from some standard sceptical reasons for doubt. And I shall try to show that Nozick's argument from a latter-day Evil Daemon hypothesis attributes to the sceptic concerns which are not his. These are the principal reasons for my doubting that the Evil Daemon hypothesis or its latter-day equivalents are sceptical devices.

What is a sceptical device ? A sceptical device is any device or mechanism deployed in a sceptical argument; and a sceptical argument, I take it, is an argument stating reasons for doubt or for suspension of belief concerning matters of interest to philosophers. Arguments such as Nozick's which purports to deny that we possess knowledge or justified belief are sceptical only in so far as they state such reasons. Our ignorance or lack of justified beliefs may be – and often is – a good reson for suspending belief or for doubt. But sometimes we lack adequate justification for a belief or knowledge and yet the belief we have seems more plausible than any alternative beliefs we may have on a particular mattter. In such a case, the lack of justification or knowledge is not a sufficient reason for suspending belief or for doubt. And in such a case a sceptic would have to offer additional reasons if he is to justify his suspension of belief or his doubt. Since the sceptic traditionally aims at doubt or suspension of belief, I take it that arguments for ignorance or lack of justified belief need not necessarily be sceptical.

Even if this view of sceptical arguments seems too restrictive, nothing in what I am going to say hinges on it. I am not going to argue that the Evil Daemon hypothesis is not a sceptical hypothesis because it is not used in

A. Pavković (ed.), Contemporary Yugoslav Philosophy: The Analytic Approach, 229–240.

arguments stating reasons for doubt or suspension of belief. In fact, I think that one can use the hypothesis in an argument for suspension of belief.[1] I shall only maintain that the two arguments I shall discuss do not follow the pattern of traditional sceptical arguments and do not exhibit traditional sceptical concerns.

I

Let us then start with the original Evil Daemon hypothesis. At the end of the first Meditation, Descartes is struggling to keep his mind free from his old beliefs which still force themselves on him in spite of his previous arguments showing that they are uncertain and doubtful. They crowd his mind because he has grown accustomed to them; thus he has to struggle against the strong force of habit. But that is not all: the old beliefs are still highly probable - and in spite of his previous arguments - he has more reasons to accept than to reject them. To counter both the force of habit and stronger reasons for acceptance, Descartes decides to feign or pretend that all these beliefs are "false and imaginary".

This decision is Descartes' first - and I think crucial - step in his attempt to banish his old beliefs from his mind. The second is his invention of the Evil Daemon; he supposes that there is a powerful but evil daemon bent on systematically deceiving him in any of his beliefs. So having decided to pretend that all his beliefs are false, he supposes that there is such a daemon who causes him to hold these false beliefs. This causal hypothesis should then explain how it could come about that the beliefs he pretends to be false are false as he pretends. The hypothesis should then causally explain the imagined or pretended falsity of his beliefs; if so, the hypothesis should be stated as follows:

> For any belief which I pretend to be false, suppose that there is an evil daemon which causes its falsity.

In this form the hypothesis is not a modal one: it does not state that it is logically or epistemically possible that an evil daemon should cause a belief to be false. Modern philosophers often state the hypothesis as the follows:

> For any belief I hold, there may be, for all I know, an evil daemon who causes it to be false.

In this form the hypothesis states an epistemic possibility: that it is epistemically possible that an evil daemon causes a belief to be false. And from such a statement of epistemic possibility it is easy to infer the corresponding uncertainty, viz. that for any belief I hold, it is not certain that the belief is not false. But since the original hypothesis does not state an epistemic possibility, originally – in Descartes' text – it is not used as a sole premiss in an inference of this kind.

But if not used in this way, in which way does Descartes use the hypothesis ? Having already decided to pretend that his old beliefs are false, of what use is this causal hypothesis to Descartes? Well, this is how he explains its use:

> I shall obstinately hold on to this thought [of the Evil Daemon, A.P.] and if by its means I shall not be able to get to know any truth, at least it is in my power to suspend my judgment.[2]

The hypothesis is a device for suspending one's judgment and keeping it suspended. As I understand it, it functions in the following way: whenever I (the inquirer) feel any inclination to accept any of my old beliefs – e.g. that I see a square tower or that I have a body – I shall imagine that an evil daemon wants to deceive me by causing me to have this false belief. This thought should counter – and hopefully override – any inclination I have to hold this old belief to be true. It should counter it by bringing to my mind very vividly a possible causal explanation of its falsity. By imagining an evil daemon – with or without its more picturesque paraphenalia – I should let myself become something of an epistemic paranoid: the evil daemon is, at least in my imagination, lurking everywhere. This, if anything, should cause me to become somewhat suspicious about any of my old beliefs and should disincline me to regard them as true.

But apart from causing me to be suspicious, does it give me any reason to become coy about my old beliefs ? Well, a causal explanation of a belief's truth or falsity often provides reasons for accepting or rejecting it. For example, a causal explanation of our sense-perception may show how some of our sense-impressions are caused by bodies outside us; this would – as Descartes suggests in the sixth Meditation – provide us with reasons for accepting at least some of our perceptual beliefs in external bodies. By analogy, Descartes might have thought that the causal explanation by an evil daemon would help him persuade himself, a rational inquirer, in the falsity of his old beliefs. But the analogy here has clear limits: a causal explanation of sense-perception is an explanation of how external bodies cause various mental events whereas the explanation by an evil daemon is an explanation of how an imagined or supposed evil daemon causes the imagined falsity of our beliefs. The former is thus an attempt to explain what actually takes place; the latter is an attempt to explain what an inquirer imagines or pretends is taking place. It seems then that for a mock-rejection of beliefs Descartes provides a mock-reason.

But is a mock-reason enough for Descartes' grand scheme of getting rid of all his old beliefs ? Well, Descartes has decided, by his own will, to pretend or feign that his beliefs are false. And there seems to be nothing to constrain the means he could use to prop up or to explain the pretense: for any means he can always pretend that these support or explain his earlier pretense that his beliefs are false. So his mock-reason is as good as Descartes himself takes it to be: it is as good a reason for pretending that all beliefs are false or for imagining them to be false as Descartes believes it to be.

Suppose now that his Evil Daemon hypothesis is to serve as a real counterweight to the reasons we all have for the highly probable beliefs which Descartes wants to banish. So, whenever I believe that I have a body, to counter this I should assume or imagine that my belief is false and that both the belief and the grounds I have for it are caused by an evil daemon. In this way I am pitching the grounds I have - what I feel at the moment to be doing by my own will, e.g. moving my body - against an assumption. And a natural question would be : What support, if any, have I for the assumption ? In other words, why do I think that there is an evil daemon which makes me feel the way I do ? And not only does Descartes fail to discuss any such question, but it seems to be self-defeating to try to find out any evidence or support for this assumption: on the assumption, the daemon is more powerful than I am and is intent to deceive me systematically; if so, he is able and willing to make sure that I never find any evidence for his existence. If he were to fail to do so, he would obviously fail to deceive me; for if I found any evidence of his existence and doings, I would have good reasons not to accept many beliefs I have and he could not deceive me any longer.

The Evil Daemon hypothesis thus implies that a human inquirer can find no evidence nor ground for or against it. In Descartes' view this implication is fatal to the hypothesis: his attack on the hypothesis starts with a refutation of this very implication. In the Second Meditation he argues that the imagined daemon cannot falsify Descartes' belief in his own existence and so even if there is such a daemon it is not sufficiently powerful to decieve a thinker such as Descartes. Descartes' cogito should thus show how a human inquirer can find some initial evidence against the hypothesis and Descartes in other Meditations seeks to find some more.

Undeterred, some contemporary philosophers - for example, at one stage of his development Keith Lehrer[3] - find in this implication a most fruitful source of allegedly sceptical arguments. If the hypothesis is so constructed as to avoid any confirmation or disconfirmation, humans can find no conclusive reasons to claim that the hypothesis is false. If there are no conclusive reasons to be found against the hypothesis, there is no way to conclusively defeat the claim that the hypothesis is true. And therefore, there is no way humans can rule out the possibility that the hypothesis is true; for all we can know, the hypothesis may be true and there may be good evidence for it which is inaccessible to us. It is logically and physically possible that no evidence be accessible to us showing that we are in fact being deceived.

Moreover, for any belief one holds, one can construct a hypothesis which would both falisfy this belief and imply that no evidence against it is accessible to us. Suppose that our belief that the posting boxes we see around us are red is false; suppose further that the postboxes are yellow but that our visual apparatus is so constructed that they appear to us red; and, finally, suppose that no human scientist can discover this quirk in our visual apparatus. Our hypothesis thus falisfies a rather widespread belief and rules out a possibility of the human discovery of evidence for or against it. But would this implication of the hypothesis increase its plausibility ? And would it increase its plausibility so much as to enable the hypothesis to override the evidence of our sight and the scientific evidence - if there is any - that postboxes we see are red?

In order to argue that so constructed a hypothesis overrides the present evidence, one would have to show that the existence of evidence that is in principle inaccessible to humans, is as good reason for accepting it - or holding it to be true - as is the evidence of, say, one's senses. We see red postboxes and have further reasons to believe that our sight is not deceiving us in such cases. One would now have to show that our hypothetical inability to find out that our sight is in fact deceiving us overrides or defeats all those reasons for our belief; or if it does not overide them it is at least as good reasons as they are. I shall not argue now that it is impossible to show this. I shall only point out an analogy which I hope suggests that there is no urgent need to argue for this. Suppose you see your colleague take some money out of your desk; you have otherwise excellent reasons to believe that you are not being deceived (e.g. your eyes are in order, you notice various features of the thief that only your colleague has etc.). Now suppose another allegedly sceptical colleague of yours puts it to you that everyone in your department gets deluded or illuded from time time in a similar way but that there is no way anyone is going to find out that one succumbs to illusions or delusions of this kind. On the basis of this hypothesis your allegedly sceptical colleague then urges you to suspend judgment in the case of your thieving colleague and to drop the whole matter. If you follow his advice and drop the matter, you have decided to prefer a good story to a well-grounded one. It is a good story in so far in it attempts to explain in an ingenious or humorous way why the evidence or grounds one actually has is of no use. But no support or evidence is offered or even suggested in its favour. So the story you accept is totally unsupported while the story you rejected is as well supported as such a story can be. Note too that the story you have accepted does not offer any evidence for rejecting the well-grounded evidence: it only re-classifies - without any grounds for doing so - your present evidence as illusory or delusory appearance. So we are back where we started from when discussing the Evil Daemon hypothesis: in accepting such a hypothesis we are opting for a good story as opposed to a well-grounded one.

I shall not argue that a decision to opt for a good story is in such or any other case irrational; for Descartes seems to tenatively accept a good story only to turn it against itself in a highly rational way. I only want to suggest that this decision is not characteristic of the sceptic's strategy.

Consider the sceptic's strategy in the arguments from illusion and from perceptual relativity. Sometimes a tower at a distance visually appears to me round but when I come closer it then appears to me square; and an oar immersed in water appears to me bent while the same oar outside of water appears straight. Assume that a tower is either square or round and an oar either bent or straight. On this assumption, one of the pair of perceptual beliefs I formed on each of these occasions is false. And yet for each of the beliefs I had the same kind of evidence – the evidence of sense. From this it follows that evidence of sense at least sometimes supports false perceptual beliefs. So far there is nothing sceptical nor particularly controversial in this line of argument. But the sceptic now infers that since the evidence of sense sometimes supports false beliefs, this gives us reasons to doubt the reliability of the evidence of sense on any occasion on which we rely solely on this kind of evidence. And from this, the sceptic finally concludes that we have good reasons to doubt the truth or justification of any perceptual belief based on this kind of evidence.

The principle of inference on which the sceptic relies here seems to be this:

> If evidence of a certain kind sometimes supports false beliefs, then one is entitled to doubt the reliability of such evidence.

Or, in much simpler terms, past mistakes give us reasons to doubt what we believe now because both present and past beliefs are based on evidence of the same kind. In this form, the principle may appear quite plausible. One can, of course, ask whether such an innocuous looking principle can warrant the final sceptical conclusion that we have good reasons to doubt any perceptual belief based on the evidence of sense alone. For the principle seems to license the inference to the doubt in the reliability of evidence but not to a much more general doubt concerning truth or justification of perceptual beliefs. Both Descartes and Hume questioned the inference to such a general doubt; Descartes thought that mistakes concerning "distant and hardly perceptible" objects do not justify a general doubt in any perceptual belief based on the evidence of sense; there were some beliefs of this kind – e.g. that the perceiver has a body – for the doubt in which one needs an argument from past dreams. And Hume in his first Enquiry thought that such a sceptical argument shows only that the evidence of sense alone cannot be fully trusted.

But whatever the shortcomings of this undoubtedly sceptical argument, its strategy is clear: one starts from mistakes or false beliefs recorded in the past and argues that such mistakes justify doubt concerning any present belief of the same kind. Thus the argument is an amplification – and perhaps an exaggeration – of past recorded errors. Arguments guided by similar amplification were advanced by sceptics concerning inductive and deductive reasoning, memory-beliefs and mathematical beliefs. The argument from the Evil Daemon hypothesis is not an argument of this kind. This argument relies on the principle which could be put as follows:

> Any ingenious story which both explains how a belief may be caused to be false and is so constructed as to avoid any human disconfirmation, provides good reasons for doubting the truth of the belief.

Relying on such a principle, the arguer here searches not for past mistakes but for a good story; and, arguably, his task is much easier than that of a sceptic. For some beliefs - such as beliefs about one's own body and mind - it is not always easy to find mistakes of the kind which would allow the amplification favoured by the sceptic. For not only are such mistakes rare, but when they occur one can easily explain them as effects either of inattention or malfunction; and such an explanation may prevent the amplification from these cases of mistaken belief which the sceptic favours. Namely, cases of malfunction and inattention are simply not of the kind concerning which the sceptic would want to generalize. He would certainly not want to argue that since I mistake the place at which I feel a slight pain because I was not attending to it, I may be mistaken about my pains in general.

No such impediment blocks the arguer from the Evil Daemon hypothesis; his concern is not with past mistakes and the generalization from such mistakes. He only has to be inventive and to find a good story. The philosophers however do not offer much of a variety - there is the one with an evil daemon and and a couple of others with scientists to do his work.

However poor story-tellers philosophers seem to be, the principle on which the Evil Daemon argument relies raises the question mentioned before: do stories without any foundation - however ingenious they may be - offer reasons for doubt ? Such a question does not arise in the case of the sceptic's principle: past mistakes are reasons for, or sources of, doubt. If I repeatedly fail to remember correctly my telephone number, I have good reasons to doubt that I shall remember a number given to me by a new acquaintance; and in consequence, I may write it down. And if my kind friends notice that I repeatedly confuse types of red wines, they would be entitled to doubt, however kind they are, my latest stories of a Beaujolais that I drank.

In short, the traditional sceptical arguments exploit our tendency to learn from mistakes; this tendency of ours makes us cautious about the matters concerning which we have noticed to be mistaken. For if we are to learn from our mistakes we should not rush and continue making the same mistakes over and over again. So in such a case caution is both natural and rational; and an expression of this caution is the doubt with which one sometimes approaches a subject-matter of one's past mistakes. In saying that this tendency or attitude is natural, I am not saying that it is universal; I am only saying that such an attitude is not considered to be out of place at least not among those who are not overly self-confident. Overly self-confident people often do not learn from their mistakes and thus what I have said so so far does not apply to them.

One can, of course, try to undermine a well-founded belief by inventing a good story which explains how such a belief may become false. Such an attempt does not exploit one's natural tendency to learn from past mistakes but rather a tendency which only Hume would perhaps consider natural - a tendency to prefer a good story to a well-founded one. Whatever tendency one exploits here, there is always the old unwanted question: why believe the ingenious story at all and on its basis reject our well-founded belief ?

II

In his Philosophical Explanations Robert Nozick seems to suggest an answer to this question. According to him, the principle which the argument from an Evil Daemon appeals to presupposes that knowledge is closed under logical implication or entailment. In his view, this is a wrong but widely held assumption and by constructing an argument from the evil daemon the sceptic exploits it in order to show that we know nothing or almost nothing. So the good story of an evil damon or its latter day doppelgänger is also an instructive tale: it tells us how coupled with a wrong assumption about knowledge such a story can lead to an incredible and false conclusion.

Roughly, Nozick maintains that since the truth of an evil daemon hypothesis is incompatible with the claim that we know anything, from the hypothesis it follows that this claim of ours is false. The old story of an evil daemon Nozick replaces with a new story of powerful scientists on a far away planet who by thoroughly scientific means cause a subject - in this story Nozick himself - to have false beliefs about his life on Earth. I shall present the argument in the following simplified form. A far away planet is here replaced by a far away city - New York -and Nozick's Jerusalem is replaced by Brisbane, the place where this paper was read first.

1. I know that if I am in Brisbane then I am not in New York where scientists cause me to believe - as it happens, falsely - that I am in Brisbane.

2. I do not know that I am not in New York where scientists cause me to believe, falsely, that I am now in Brisbane.

Therefore, by tollendo tollens,

C. I do not know that I am now in Brisbane.

In this argument the Evil Daemon hypothesis i.e. its equivalent is only a denied consequent of the conditional and not an independently stated hypothesis. Therefore, the old unpleasant question : Why believe in the hypothesis? does not arise here. And in the second premiss I - that is, Nozick - admit his ignorance of the falsity of this hypothesis. The second premiss does not seem to be very controversial: even some non-sceptics would admit that they have no knowledge of the falsity of the hypothesis although they think they have good reasons to believe that it is false.

The first premiss seems to aspire to be an instance of knowledge of general conceptual truths: everyone knows that one cannot be at two places at once. And so I should know that if I am in Brisbane, I am not somewhere else, e.g. in New York. Now if one does not accept the first premiss on the ground that it is an instance of knowledge of this kind, it is unclear on what ground one could accept it: Nozick for one offers no other justification for it. Now I would like to suggest that it is far from obvious that the first premiss is such an instance and that, therefore, its acceptance by the sceptic – or more likely by his alleged opponent – is unwarranted.

Let us then examine our first premiss. There is no doubt that I know that if I am now in Brisbane, then I am not in New York; that is, I know that if *p* then not *q* given that *p* and *q* are logically incompatible. But do I know that if I am now in Brisbane then I am not in New York where powerful scientists cause me to believe, – as it happens, falsely, – that I am in Brisbane ? Well, if the consequent of this conditional is to be analysed as a conjunction of two independent statements, it is no longer clear that I do know this. For in such a case I would have to know that if I am now in Brisbane, then I am not in New York *and* the scientists do not cause me to believe that I am in Brisbane. This is to say, I know that if *p* then not *q* and not *r*. Now *p* here does not imply not *q* *and* not *r*: it is quite possible that I am in Brisbane and not in New York and that the scientists still cause me to believe that I am in Brisbane. If he is to hold the first premiss as an instance of our knowledge of conceptual truths, the sceptic or his opponent cannot accept such an analysis of the consequent. What he needs is an analysis in which the two statements *q* and *r* are not independent. The second, that scientists do not cause me to believe that I am in Brisbane is clearly dependent on the first that I am not in New York; therefore, the arguer here seems to be saying that if I am not in New York, then scientists do not cause me to believe that I am in Brisbane. This, of course, is equivalent to the conditional that that if scientists cause me to believe that I am in Brisbane, then I am in New York. So, arguer here seems to be committed to saying that I know that if I am now in Brisbane, then if scientists cause me to believe that I am in Brisbane, I am in New York. Now this is an extraordinary feat of knowledge and mobility: I know that if I am in Brisbane then on some further condition I am in New York.

The third possibility is to forbid any truth-functional analysis of the consequent. But such a prohibition would in effect stop any explanation of what I know when and if I know the truth of the conditional. For, as has been pointed out, it is uncontroversial that I know that if I am in Brisbane, then I am not somewhere else; the question is do I, *in virtue of knowing this simple truth*, also know that some other quite specific things are not happening to me. And to answer this question we would have to find out how my not being somewhere else is related to the things that are allegedly not happening to me. Perhaps the relation is not truth-functional; but if so, one would have to explain what the conceptual truth is the knowledge of which is asserted in the first premiss. Until one explains this, one has no grounds to accept the first premiss.

Of course, one can argue that it is not the sceptic's worry why someone - e.g. his opponent - endorses the premiss; it is enough that someone does. Well, this won't do for three reasons. First, in this case, it is not clear that anyone endorses the first premiss; and so one would have to indicate some reasons why one would accept it. Second, unless one understands why one endorses the first premiss, one cannot explain what the sceptic is aiming at with such an argument. And, finally, the sceptic's worry is - contra Nozick - why his opponents believe what they do; for his aim is to undermine his opponents' reasons and to do so he should find out what they are.

But suppose all these worries about why's and how's and opponents are just a lot of quibbling. The sceptic simply takes a plausible claim about one's knowledge and from this deduces that there is no knowledge. How does he do it ? First, he takes a plausible candidate for knowledge e.g. the statement that I am in now in Brisbane. Second, he finds a simple statement incompatible with this, e.g. that I am in New York. Third, he somehow fuses the second statement with a claim that scientists are causing me to believe, falsely, the first statement, e.g. that I am now in Brisbane. In this fusion the work of an evil daemon is exemplified; this is then an equivalent to the original Evil Daemon hypothesis. Finally, the sceptic claims that his subject knows that if the first statement is true, then the second - now incorporating an evil daemon equivalent - is not true. That is, I know that if I am now in Brisbane, I am not in New York and that being so the scientists are not causing me to believe that I am in Brisbane. And the rest is as in Nozick's argument: he denies knowledge of the falsity of the second statement (incorporating the evil daemon equivalent), and deduces the denial of the knowledge of the first statement.

It is the third step - as I suggested above - that does the trick; for it yields the second statement that I not in New York and that being so the scientists are causing me to believe, falsely, that I am now in Brisbane. But for the trick to work it is necessary both that the first statement should imply the second and that I know that it does; that is, it is necessary both that "I am now in Brisbane" should imply " I am not in New York and that not being so, scientists are not causing me to believe that I am now in Brisbane" and that I should know this.

It is certainly possible to argue that both of these necessary conditions hold; but not only has no one - not even Nozick - produced any argument for it, but it is highly implausible to expect the sceptic to do so. For he is not - traditionally or non-traditionally - concerned with implications among statements nor with the conditions under which we know them to hold. If the argument requires to show that we know a certain implication to hold, then it is not a sceptical argument – or at least not obviously so. If on the other hand we think that the sceptic uses the argument with no intention of clarifying how the argument works, we are assuming either that he is rather simple-minded or that he is arguing in bad faith.

The latter of course is an old charge against the sceptic. But for the charge to be substantiated one would have to show that the sceptic does use an argument of the above kind. And since I find no reason to believe that he does, I shall not consider this charge. The former view, that the sceptic in his simplicity ignores the difficulties which his argument encounters is compatible with another old charge against the sceptic: that he raises the standards of knowledge so high that we, humans, simply cannot attain them. Let us see how the above argument could be used to support such a charge.

The charge is this: the sceptic always demands that in order to know anything, one should rule <u>any</u> possibility of a mistake in any belief or candidate for knowledge. The possibility of mistake envisaged by an evil daemon hypothesis or its equivalent is that I may be deceived by scientists or daemons in my belief that my name is Sasha. If I am to know that my name is Sasha, I should be able to rule out this very possibility. And one – perhaps the best – way of doing this is to know that if my name is Sasha this implies that scientists are not causing me <u>falsely</u> to believe that my name is Sasha. Therefore, the sceptic's demand that any possibility of mistake be ruled out here exemplified in the requirement that I should know the implication. And since the sceptic's general demand is often implicit, he feels no obligation to clarify the particular instance of it. In his eagerness, he simply ignores the need to provide such an explanation; and those less eager than he can gleefully point out how simple-minded he is in his approach to these complex issues of knowledge.

But does the old charge stick ? Does the sceptic require that our knowing something rule out the possibility of any mistake ? There is no indication that any sceptic ever required that we should know that if <u>p</u> then nothing is causing us to believe, falsely, that <u>p</u>. Moreover, none of the arguments advanced in the Ten Tropes as reported by Sextus Empiricus require such a wholesale ruling-out of mistakes; and none of Hume's arguments rest on such a requirement; Gassendi's doubts concerning Descartes' procedure we reported above. These are examples from the works of self-confessed sceptics. But I do not propose to count the heads on this issue. For even if found a self-confessed sceptic who tacitly presupposes that any possibility of mistake should be ruled out, this alone would not make it a characteristic of sceptical arguments. Sceptical arguments traditionally attempt to undermine specific grounds or evidence in which the sceptic's opponent puts his trust. For the purposes of such an argument, there is no point in requiring that the possibility of <u>any</u> mistake be ruled out. Such arguments at most require that mistakes of a certain kind be ruled out. If the evidence of sense is to present good reasons for perceptual beliefs, it should not lead to mistaken beliefs of the kind all of us have encountered at one time or another. According to this requirement, in order for us to trust our senses, it is not necessary that we rule out the possibility of deception by an evil daemon or its latter-day doppelgänger. This is not what the sceptic is tacitly suggesting or proposing.

But why not extend this cautious requirement to any possibility of deception including the one by an evil daemon ? Indeed, why not? There is nothing to stop us there . But if we do, we should be clear - as Descartes seems to be - what the purpose of such an extention and an argument based on it should be . It is of no use to produce such an argument, attribute it to the sceptic, and then easily refute it. For in this way one has not refuted the sceptic but only an argument attributed - in my opinion wrongly - to the sceptic.

NOTES

1. I discuss a version of such argument in the third essay in my book of essays Razlozi za sumnju (Reasons for Doubt), Beograd, forthcoming.

2. The Philosophical Works of Descartes. translated by J. Cottinghamn et al. Cambridge University Press, 1984, Vol I.p.16.

3. In his "Why Not Scepticism?" in the Philosophical Forum, 1971. Reprinted in Pappas and Swain (eds.) Essays in Knowledge and Justification.

WHY WON'T SYNTACTIC NATURALIZATION OF BELIEF DO?

MATJAŽ POTRČ

In this paper I shall talk about beliefs, but I shall take 'belief' to be shorthand for all intentional mental states (even desires, wants, etc.) Further, I take naturalization in its broadest sense of getting a concept in such a shape that it could be usefully studied by some established science. Since psychology is the science which explains intentional mental states, a question arises as to how to conform the philosophical account of them to psychological practice.

Stich (1983) is particularly sceptical as to the possibility that belief be naturalized. According to him cognitive science has simply to explain it away. But clearly he thinks that in doing so naturalizing of the intentional is not explained away as well. I take his view to be important because he pushes to the extreme an otherwise hidden trend in cognitive science: the tendency to naturalize beliefs via syntactization. The criteria for syntactization in the theory of mind surely deserves clarification.[1] I take the important contrast within the question of naturalization of belief states to be between naturalization via syntax and naturalization via content.

Although naturalization via content might change our intuitive notion of content considerably, it is the only one that preserves the concept of belief. Naturalization via syntax leaves us without any viable notion of content at all.

There are at present two theories of mind which view the content of mental states as compatible with, or even a pre-condition for, cognitive science: Strong and Weak Representational Theory of Mind (RTM). Strong RTM claims that mental states have contents and that generalizations among them are accounted for in terms of the content. The Syntactic Theory of Mind (STM) counters this by claiming that mental states are relations to syntactic tokens and that interactions among mental states are to be explained in terms of their formal or syntactic properties. Weak RTM, agrees with both those statements of the STM, but it embraces Strong RTM as well, in viewing mental states as having content or truth conditions.

If such is the picture we are faced with, it is clear that something will have to go. The STM itself shows very clearly what the syntactization of content of intentional mental states means. It ends up with the claim that

A. Pavković (ed.), Contemporary Yugoslav Philosophy: The Analytic Approach, 241–250.

there can be no beliefs ordinarily or scientifically construed, no intentional mental states which would depend on the content for their characterization. In short, the STM is useful because it eliminates the semantical content altogether. This is the first way out of the difficulty of handling content in psychology and in cognitive science. I do not accept this solution, but, as I said, it is useful because it makes explicit the tendency which has been implicit in the proposals to reduce the content, and thus the semantics of beliefs to their syntax. This is what I take to be the main obstacle for cognitive science, and the major source of difficulty for an appropriate naturalization of intentional mental states such as beliefs.

I see no easy way out of this difficulty, but nevertheless I think that something could be gained by pointing to its systematical appearance. There is thus, in my opinion, a tendency in cognitive science to systematically syntacticize the content of beliefs. I think that, on the contrary, sources external to the organism have a strong influence on the content of mental states such as beliefs, and that with their help the beliefs might be explained. My claim shall be that the naturalization of belief cannot proceed predominantly via a syntactic route.

I shall point out to one of the sources of the tendency to syntactize the content of beliefs, which tends to eliminate external distal stimulus as the main explanatory source of the content. Although I think that such a source is predominant, it does not influence directly all the contents of intentional mental states. All I am arguing is that without such a source in the background, there is no actual belief.

My remarks will be sketchy and programmatic. I hope to elaborate on them in the future.

Troubles with STM

I want to consider briefly how the STM takes the extreme strategy of syntactization to be the right one for naturalizartion of intentional mental states like beliefs. In Stich's book beliefs are studied according to their content first, and it is later claimed that they should be dispensed with, at least in cognitive science. The proper domain of the cognitive science and thus the ground for naturalization of beliefs would be a pure syntactic idiom. Such a syntactic idiom would dispense with beliefs insofar as they are conceptual hybrids, about which only folk psychology can talk. If we try to account for them scientifically, their distinctions are found to be traced too coarse-grainedly or too fine-grainedly. Psychology and cognitive science, in studying intentional mental states, should restrict their attention to the part that leaves out their content. The naturalization of beliefs should thus proceed without reference to contents.

The first claim made by the STM is that the contents are folk psychological constructs, and that they are introduced by postulating the similarity of the subject to the one who ascribes the content. The reference, indexicality of content, and the stimuli would count only insofar as the ascriber of content can hold the belief of the organism to be similar to his own. But this clearly will not do as a criterion, since infants, animals and mentally handicapped do not allow any such comparison. Moreover, many important areas of psychological research would be put out of reach by this move (e.g. child and developmental psychology). The complete banishment of content, restricting psychology to current internal syntactic states, without regard to their external sources and history, is what is left. Here I presume that there might be one other way out - naturalizing the content and stressing the importance of the external distal stimulus that plays a role in determining it.

There is an argument proposed in the defence of the STM -the replacement argument. It is designed to prove that psychology should concentrate exclusively on the study of current internal mental states of the organism.

Syntactical theory, of course, concentrates on the study of mental states without content. But are such states still intentional? If it turns out that mental states studied by the STM are not intentional, and if intentionality is the feature which allows for causal explanation of mental states, as well as for the explanation of their interactions, it is quite clear that syntactic stance by itself will not do and this could be proof that content is important for the cognitive science. But, what is the replacement argument?

Take me, and take my replica, whose intentional mental states are exactly like mine. Perhaps some gangster substitutes this replica of me. Whether the current actual internal states of me or of the replica are studied there is no difference for psychology. It might be the case that my replica and I do not share all externally attributable properties, and thus that we do not behave identically, even if our current internal states are the same. For example, if my replica sells my car, his behaviour can not be explained by "selling", because he does not own it, even if he is in exactly the same syntactic state as I would be on this occasion. Such conclusion seems dubious to the STM because the facts external to the replica and me do not count in the psychological explanation of our behaviour. It might seem that they count, it is argued, but there are many externally attributable properties which can not interest cognitive science. The STM wisely restricts the psychological study to the syntactic element only, without taking semantics into account; semantics studies the content of mental states, and this study can be connected with everything in the world, thus with all the sciences. The current syntactical internal mental states will do without reference that introduces similarity, and without historical considerations. Otherwise, it would be like chemistry trying to explain lowered price of boiled French wine on the market.

There seem to me to be some difficulties in arguments of this kind offered in support of the STM. First, I would like to claim that there can be no replica that has exactly the same mental states as I do. Further, I would like to add that the STM by restricting itself to the study of mental states without semantic content, leaves out what it otherwise proposes to be the science of: the intentionality, i.e. the directedness of mental states to something, to some content, which further depends on external stimulus in some way. If this is so then naturalization via syntactization will not do, and content should be necessarily presupposed in the study of beliefs.

The first claim is that there can be no such replica. For such a replica with the mental states exactly identical to my own to be possible, there should exist completely identical environments for both of us to inhabit. Remember that the story told us how I have been replaced by a replica. But for identical mental narrow states to be there for both of us, this should happen to my replica as well. After being abducted by gangsters, either I can be in exactly identical environment as my replica, or again, this cannot be the case. In that case I would not be in the same environment, there could hardly be any internally equivalent states in us. If there would be such equivalent states, the internal adjusting mechanism of one of us to the environment should be differently shaped in the way that it would fit a different environment producing, as a result, the same actual internal current mental states. But this would imply that the very internal states would not be the same for the two fo us. The second possibility is that of identical environments. But then there would be just no point in replacing me by my replica. This point might also be stated in the following way. It might be that my replica and I are two, or that we are not. If we are two, we cannot be an exactly (physically) identical environment. And if our current states at least partially reflect differences in environment, then we cannot be in exactly the same mental states. Take a clear difference in the environment: if my replica is lying in a quiet corner of the beach, and if I am on the busiest corner of the city at the same moment, our actual mental states could hardly be the same. Further, take the case where I cease to exist: from this very moment there is no replica there. Typically the STM does not like Twin Earth examples, it wants to stay on the same Earth. But staying here will hardly bring two identical environments and replicas into play.

Moreover, banishing memory from the study of belief will hardly leave room for any belief explanation. In that case it is miraculous how even completely syntactical intentional states could be formed. And, restricting the study of mental states to the ones that actually take place would presumably not leave intentionality with us. And if intentionality is the proper object of cognitive science, we are left wondering what the STM could be a science of. Take the following example: either my actual mental states are connected to the typewriter, which I am now using, or they are not. If they are, they are intentional. But how can a mental state of mine count as intentional in such a case, if it is not about the typewriter?

Here we seem to have touched the very source of STM: its desire to get rid of the ambiguity concerning referential and semantic parts of intentional states, present in any cognitive science. How to achieve this? If we leave out the content, we seem to obtain a pure syntactical account of mental states. Only the internal states of individuals count. I doubt again that such a proposal could be appropriate for naturalizing of any kind. Let me for a moment indulge in a speculative critique, based only in part on Stich's explicit wording.

The STM proposes a bifurcation in the very notion of belief. There is an autonomous syntactic part that fits well the purposes of the cognitive science, and there is another part which belongs to the other sciences. The cognitive science part of content varies according to the theory that brings it in. For the Weak or Strong RTM, "the shirt is crumpled" would be a good start, but for the STM the only part of interest is the sentence's syntactical equivalent. The part that is the subject matter of other sciences varies. The STM's goal is to get rid of this part, the content, because this would mean escaping from a conceptual hybrid. But, getting rid of the content, of the nonautonomous part of belief, we are simply left without belief as such.

Let us now have a look at the folk psychological content "The shirt is crumpled". Suppose that it takes the place of "p" in "S believes that p". We encounter the following situation: For the Weak RTM, the fact that the shirt is crumpled cannot be properly explained by any standard science, like physics, chemistry, and so it should be accounted for by some special science that takes care of the intentional idiom (Fodor). The trouble for the STM comes from the fact that the criteria are here provided syntactically. But the case of the crumpled shirt is exactly the case where the STM cannot settle only by explaining the abstract structure of a sentence stating that "p". In the case of the Weak RTM we are left wondering by which miracle the syntactic constraints could explain the content, and in the case of STM whether the mere syntax can save the intentional. To see this let us first grant the syntactic part is autonomous (independent from content). Then, if there is some intentional state left for psychology, it should have some content. Still worse there might be a hierarchy of contents suitable for the STM. "7+5=12" might be fine as a syntactically defined content, because there would be not much left over for other sciences. "Water is H_2O" would be "trespassing" into domain of a particular science, chemistry. So content would have a consequence, after all, on the shaping of the cognitive science. But this would be at variance with the claim of the STM that content should be left out of science. The strategy for the STM is to say that content should be dispensed with because it could only be reached by ascription, by similarity and thus by homo mensura. The ascription of content would take the mental states of the organisms to be similar to ours. No place would be left for realistically construed identities of contents.

From Syntax to Semantics

The goal of psychological practice is to explain behaviour. It is widely agreed that the internal mental states and their connections which cause other mental states and behaviour are important in explaining the behaviour of an organism. The view that internal causal connections and their proximal stimuli are important has been useful in rejecting the behaviouristic approach, and in explaining the behaviour of different organisms.

The standard way to introduce this view is by explaining different features of an organism's behavior with the help of what psychology prespupposes is in the organism - representations.

The peculiarity of representational states, in whichever way they might be interpreted, is that they form a system, which allows them to act on each other causally. Now, if representational states act on each other, cause each other and then cause behaviour, it is tempting to suppose that such a system of representations might do the whole job. Or at least it might do the job that is of interest to the practice of psychology as a natural science. So, one idea that arose naturally from such an approach has been that it suffices to explain beliefs syntactically, because the syntax can explain all the interactions among the representation as well as their connections to the world.[2]

The problem - everybody admits that it is a difficult one - concerns the content of mental states. The causal interplay among representations is not possible without the content. Such a content could be introduced by the English sentence "The house is on fire" or "I want to save myself". It is by their contents that the belief that the house is on fire and the desire that I want to save myself explain my behaviour, e.g. my running out of the house as fast as possible.

Without an appeal to the content, it seems, we would miss some important generalizations concerning causal connections among mental states. In some ineliminable way, semanticity has to be introduced into the picture perhaps even to explain the very syntax. Semanticity might depend on many things, but most obviously it depends on the relation between the organism and the world. This connection is preserved even in the solipsistic manifesto RTM provides, because the content is there to stay if the behaviour of an organism is to be explained. The very relation to the external stimulus has been in part forgotten because of the momentum of the internalist reaction against behaviourism.

If there is a crucial aspect of syntax that depends on the external world, psychology should be concerned with the environment the organism is facing. But this might seem plausible because, in this case psychology should introduce many sciences into the picture. But surely this is implausible: although everything in the world can become the content of beliefs, psychology is not supposed to care about everything in the world.

The trouble is that there are many levels in an interpretation at which semantic content of representations begins to be of interest to a psychologist. This question is critical, because it influences directly our picture of the borders of psychology as a science and thus of the naturalization of such intentional mental states as beliefs. Let me sketch an example. At one extreme there would be the solipsistic claim that the contents of mental states should be narrowly construed and thus that psychology should be interested in them exclusively: it should be interested in that which roughly does not transcend the borders of any organism. Although those states might be caused by distal stimuli, the claim continues, this could be none of the psychologist's business. The other variant, which seems quite different, is to explain the content as the distal informational input, depending in a crucial way on the external stimulus. There is a shift concerning this crucial point in Fodor (Loewer, 1987): his "strongly" solipsistic position precede his psychosemantics, the latter being closer to the Dretskean informational content theory.

I shall not try to explain these interesting differences here; I only claim that what is crucial in both views is that they ascribe a causal role to the semantic content of beliefs. The position of early Fodor, I think, is quite mysterious, and Stich has correctly accused it of inconsistency (although Stich's proposed solution is opposed to mine which opens windows to the universe around).

The second position which relies on the external sources of the stimulus, on its information, in order to explain the content of mental states, seems to be more acceptable. Let me provide just a hint, pointing in this direction. The frame problem, encountered in AI, might be quite useful as an illustration. How can a computer mimick the success of a living organism in its environment, in its choice of the information it needs to survive? To process information, an organism has to choose it first. But it could not exercise that choice were it not to have a genuine access to the external information. Even the most solipsistically minded theory of semantics should thus include a description of the way semantics is extracted and selected from the external stimuli. So the access to those stimuli should be accounted for.

Some Remarks on the Twin Earth Examples

The identity of content of beliefs does indeed generate some puzzles. The Twin Earth examples have brought out this point. The strategy of those examples has been to show how the narrow (or syntactic) "side" of a belief does or does not vary, in a case where the external stimulus varies. The fact that on our planet we have H_2O in the lakes, and the imagined state of affairs on the Twin Earth, namely that there is some XYZ, similar to our water there is paralleled by the possibility that my twin and I are in different mental states.

Although the connection is not direct, because it is mediated by the formality constraint on syntax, I think that the source of such examples is to be traced to Frege's example fo the Morning Star and the Evening Star. Let us take these names in a Fregean fashion – the first is assimilated with the star's property of showing up in the morning and the second in the evening.

Take my belief that this object in front of me is the Morning Star. Despite the fact that "Morning Star = Evening Star", I do not need to believe that this is the Evening Star as well, because I might not know the already stated equivalence (or some other description of it).

The syntactization I take to begin where someone points out the syntactic frame in such a way that the external stimulus ceases to determine the content of mental states like beliefs.

My idea is that the syntactization is supported by an oclusion of the importance of external source of information. this source of information is decisive in acting back on the belief, as it is the crucial part of this belief's content. There is a tendency to start with various descriptions of one referent. But although it is right that various aspects of referent are introduced by descriptions ("Evening Star" can be viewed as another descriptive name of the Morning Star), it is sometimes forgotten that differences in descriptions themselves depend on influences of external stimuli, or on the systematic differences of the mutually dependent, even holistic influence of external stimuli.

Beliefs are usually produced by the action of stimuli, which leave traces on the content of intentional mental states. There is always at least a trace of a stimulus of this kind, and belief is usually produced by stereotype, arrived at by repetition of such traces. If I believe that snow is white, I have a content that can wildly vary individually, according to my particular input in experience (my content and the content of someone in Africa who, has never seen snow are different, because we had different kinds of information available). The disposition-forming of my belief is different from his. But activating of the disposition depends on the outer stimuli again. For this, there should be one internal component available, the stereotype both of us form about snow.

Let us stay with the Morning Star – Evening Star example. One might ask: how can one defend the view that the external stimulus is predominantly determining the content of beliefs concerning it, when it seems that the descriptions associated with the names do the job of determining the content? My answer is that there is simply no identity supporting my belief that this is the Evening Star until the external environment causes the belief that this is the Evening Star; and so this belief is introduced by the surrounding environment. Compare the desire of Oedipus to marry Iocasta. When the feedback information arrives from the external, objectual sources and from the community, the identity "Iocasta = Oedipus's mother" becomes important, not earlier. Thus the additional external bodies and events cause the content of this information.

This is the response to the claim that believing "under description" does not allow us to take the external source to be predominant in belief forming, as to the content of beliefs.

Of course, many intentional mental states, like plans or goals, do not seem to depend directly on external stimuli. But plans and goals are based on generalizations concerning external objects and situations. Each of them includes various traces of external stimuli in belief formation. And informational mechanisms leading to such traces are to be studied to account for such belief formation.

It might be argued that the Twin Earth cases as sketched above already point to external stimuli as predominant, because it is the difference in the stuff, H_2O or XYZ, that matters for the contents of beliefs involved. But there are two further possibilities. The first is that difference of stuff does not have any influence on the content of beliefs, and this is thus a very clear example of the predominance of syntactic element. But even if in such examples the stuff matters for the difference in the content of intentional states concerning it, still it might not be the content as such that counts, rather the identity among contents introduced by the examples; and this identity again supports the view that the syntactic element is what really counts. The second possibility is to claim that there are simply no beliefs, and thus no contents of beliefs: all that is interesting is some syntactic equivalents of beliefs.

To the first claim, that the difference of stuff does not influence the content of belief, I would reply that it would then be miraculous that beliefs could be formed at all. It is obvious that the information concerning the difference in various stuffs is what matters to forming of belief. Thus the cases in which there is no specific information whose source is the stuff present, might be viewed as deviant. It is the difference of beliefs that counts at this stage, not their identity. And as the beliefs are formed, it is obvious that refining them depends on the external information. As for the claim that these are belief-like syntactic states not genuine beliefs, one might wonder on what grounds one is to assert that they are like beliefs. For those actual current mental states of interest to the STM are hardly arrived at or passed from one to the other without external information available to the organism. But this means that the very principle of syntactization, the banishing of content, had better be dispensed with syntactic theory is based on syntactic transformations; but how can we have those transformations without the constant external information inflow and elaboration? There might be another way out in saying that all mental states are innately given to the organism and that their transformation are brought about by fiat. But then how can they be current? A theory of mind should not allow such a crucial precondition of itself to remain unexplained.

NOTES

1. I think that even some theories which apparently take a negative stance towards it (Churchland, 1986) could be put under the very same label.

2. I do not doubt that syntax plays an important part in the explanation of the behaviour and of the representations that lead to it.

BIBLIOGRAPHY

Churchland, P.S., 1986, Neurophilosophy, Bradford Books/MIT Press, Cambridge, Mass.

Loewer, B., 1987, 'From Information to Intentionality", Synthese 70, 287–317

Stich, S., 1983, From Folk Psychology to Cognitive Science, The Case Against Belief, Bradford Books/MIT Press, Cambridge, Mass.

AN ARGUMENT AGAINST THEISM

ALEKSANDAR PRAŽIĆ

Is the concept of an omnipotent being a contradictio in adjecto? Is the statement that God is an omnipotent being analytically false? Is it possible to produce an "ontological proof" of God's non-existence? These questions strike one as a bizarre mixture of contemporary and outmoded issues, but this is surely not a crucial objection against them. Many people, including some of those who are particularly interested in the issues that generate these questions would say that they are misplaced and that their consideration is only a remarkable waste of time, because the formulation of these questions leaves aside the most important issues in the field and, therefore, nothing hinges on any solution that we might give to them. An atheist could say that the scientific Weltanschauung supplies him with a much more important and reliable basis for his attitude towards religion than any logical analysis of propositions concerning God's attributes and his existence could offer, while a theist could say that logical considerations turn out to be irrelevant when compared with the truth found in the depth of one's heart and with the authority of revelation, which form the unshakable basis for his beliefs. In this manner, the real controversy between atheism and theism can be seen as the controversy between scientific conceptions and the content of the revelation, or as the controversy between, on the one hand, an interpretation of religion as a symptom of human inferiority in the circumstances determined by nature or history and, on the other hand, an interpretation of religion as an irreplaceable basis for any true humanism. Only this sort of controversy could be a fruitful one, and could lead to a change in beliefs, which is the only aim controversies between theists and atheists have. A mere logical analysis never converted anyone to any religion, nor has anyone stopped believing in God on the basis of it.

Up to a point we must agree with the above mentioned claims. It is true that the psychological effect of logical analysis may be negligible, or even non-existent. However, this doesn't mean that analyses of this sort are irrelevant for all kinds of controversies. The least we can do is to distinguish between controversies in which a variety of persuasive methods are employed and controversies in which participants limit themselves to various methods of argumentation. In persuasion, all means are permitted, the value of each of them being determined on the basis of psychological effectiveness, while the value of argumentation depends on the rules which can be stated

A. Pavković (ed.), Contemporary Yugoslav Philosophy: The Analytic Approach, 251–262.

independently of the knowledge concerning the aim of the debate, so that the psychological effect on the opponent need not be taken into account in assessing it.

It is also necessary to expose as naive the atheist's belief that his scientific Weltanschauung is scientifically founded and proven, or, in other words, his belief that the reliability of the results of scientific inquiry is transferable to the scientific world-view. It must be said that the part of the scientific Weltanschauung which falls beyond a sum or a summary of scientific results lies, as a matter of fact, beyond the boundaries of science and has no scientific foundation whatsoever. Just those feature which distinguish the scientific Weltanschauung from science constitute its non-scientific part. It is possible to move from the fact that statements about God belong neither to the fruitful hypotheses which determine the direction of scientific inquiry, nor to well-confirmed scientific theories, to the conclusion that God does not exist and that there is no other reality but the one from which science starts and to which it always returns; but this move is not a scientifically justified move and science never makes it. For instance, when Charles Sanders Pierce writes that reality is only what a considerable scientific progress brings along, or that reality is only the content of beliefs acquired by means of scientific methods,[1] then this is neither a scientific nor a methodological claim, but only an expression of Pierce's beliefs and attitudes which do not spring from science. The extra-scientific origin of the basic claims of the "scientific Weltanschauung" shows that its reliance on scientific conceptions is not sufficient for its scientific character.

On the other hand, we must underline that it is indeed naive of a theist to believe the mere subjective feeling of certainty or the authority of revelation are sufficient to refute all logical objections. For instance, the question whether the authority of revelation has any firm basis independently of the content of the revelation, is a perfectly reasonable question. An excellent illustration for this can be found in Saint John's Gospel. The Pharisees cannot accept that Jesus is the son of God , for he has no one but himself to testify for him.[2] Jesus answers: "I am not alone, but I and the father that sent me. It is also written in your law, that the testimony of two men is true".[3] And when they ask him: "Where is thy father?", he answers: "If ye had known me, ye should have known my Father also".[4] Even if we leave aside the confusion which is made here between the legal acceptance of the testimony and its truth, we can object that the quoted passage reveals a circularity which seriously undermines the authority of the founder of Christianity.

We also cannot accept the claim that the foundation of religious feelings "in heart" or in some other higher inspiration is enough to make it inappropriate to approach its contents rationally. The question concerning the origins of knowledge about something should not be confused with the question concerning its logical possibility. When St. Paul writes that God gave the insight into the holiest secrets to the uneducated and the "poor in spirit" just in order to show how insignificant human wisdom is and to expose the vanity of those who consider themselves wise, "for wisdom of this world is foolishness with God",[5] then he thereby discredits human reason only as a source of knowledge – he claims it is unable to gain insight into the most important religious truths without God's help and guidance. But this claim doesn't eliminate the question concerning the logical possibility of these religious truths. Even if reason is really insufficient for certain knowledge to be gained, that doesn't mean that the possibility of conceiving the object of that knowledge without contradiction is not one of the necessary conditions of the object's existence. A theist could reply by saying that we cannot imagine God because we lack the necessary conceptual apparatus, and therefore the question concerning God's logical possibility is a futile one. But, if no idea of God is possible, then we also have no idea of what is claimed to exist when it is said that God exists, and the belief in God becomes in this manner a belief in anything that cannot be conceived of – a consequence that no believer in any religion would be prepared to accept. Given the fact that the belief in God is always tied to some conception of God, it is unclear why logical analysis of such a conception should be inappropriate.

In this way, we have already blocked the way for the theist's possible objection that the discovery of contradictions in a concept of God disqualifies only that particular concept, but that it cannot demonstrate that God doesn't exist. At this point we should immediately ask the theist what conception of God he has in mind, forcing him to admit that there are characteristics of the God he believes in which he is unable to give up if he wants to go on believing in the very same God. Logical analysis or empirical investigations can point out that the "holy scriptures" contain contradictory or factually false statements and passages; a believer can then reject, modify or re-interpret them without any serious consequences to his basic religious beliefs. For instance, the claim that the world was created in six days, which is at odds with most scientific theories, can be interpreted by a theist as a metaphorical way of speaking, or it can be even rejected without the rejection seriously challenging important parts of his viewpoint. But, if it were demonstrated that God is not the creator of the world and man and that he is not the just judge who distributes adequate punishments and rewards to all men, namely, if it were demonstrated that God is not, in Leibniz's phrase, the architect of the cosmic machine and the ruler of the Kingdom of Souls[6], then the concept of God would be robbed of its essence; that is, there would be nothing left of any importance in the name of which a believer could reject the above characteristics and yet remain a believer.

An atheist who wants to enter a debate with a theist in order to refute the latter's position should concentrate on the claims which contain the decisive and the irreplaceable propositions like the creation of the world by God, God's judgment, etc., and he should carefully investigate all the implications of these claims. What kind of argument should he advance here? If he is expecting to find a final solution in empirical sciences, then he fails to perceive that not only these propositions but even their remotest consequences have no relation whatsoever to empirical facts, and, accordingly, fails to see that the results of scientific investigations bear no relevance for the determination of the truth value of the propositions in question. Therefore, he can only hope that logical analysis will demonstrate unavoidable contradictions in his opponents' beliefs. If a concept of God is a self-contradictory one, we can conclude that the necessary conditions for God's existence are not satisfied; however, if a concept of an "extra-sensory" object is a non-contradictory one, no empirical science can demonstrate that, in fact, it doesn't exist. So, if the atheist doesn't want to limit himself only to persuasive methods in his debate with a theist, then there is only one way open to him - the way of logical analysis. If a proof of God's non-existence is possible - a proof in the proper sense of the word, as opposed to the mere expression of atheistic beliefs - then it is only an "ontological proof"; namely, the one which is based upon the analysis of the very concept of God. The expression "ontological proof" is used to refer to a particular sort of proof of God's existence. But, although it has been shown that the move from an analysis of a concept to the existence of the object the concept covers is problematic, the same cannot be said about the move from a self-contradictory concept to the non-existence of its object. The analyses of the concepts of circle and of square do not imply the existence of any real circle or square, but the analysis of the concept of round square does imply the impossibility of any such object ever existing in reality.

All that has been said thus far shows that the questions with which this essay began can be dismissed only by those theists and atheists who do not trouble themselves very much with thinking and who misjudge the nature of the evidence they offer in support of their views. In these questions a particular stress is put on the demand for an analysis of the concept of omnipotence, which is supposed to be an essential feature of God's nature. Of course, if we were able to show that a notion of an omnipotent being is an contradictio in adjecto we would not thereby prove that God does not exist, but only that an omnipotent God does not. But this is not insignificant, since omnipotence is an attribute no theist is ready to give up, because the attribute in question is closely connected with God's deeds such as the creation of the world out of nothing (creatio ex nihilo) and granting men eternal beatitude or eternal damnation. Such a proof would certainly challenge the existence of God as we are used to conceive him. Another reason for emphasizing omnipotence is in fact that there is already an argument which purports to demonstrate that God doesn't exist by showing that the concept of an omnipotent being is a self-contradictory one.

We shall now analyze this argument.

Many readers will probably feel disappointed when, after such a long introduction, they finally see what it is all about. Atheists will exclaim: "Is this triviality really considered to be more important for the refutation of religion than scientific explanations of natural and social phenomena?", while the exclamations of the theists will probably run along the following lines: "Is this triviality considered to be an equal opponent to the depth and seriousness of religious life and to thousands of years of tradition?". All I have written so far should serve to prevent such exclamations. Enough has been said already about the usefulness of involving both scientific theories and religious feelings in the debate. As far as the tradition is concerned the following should be stressed: there is no tradition of such greatness that logic should bend before it. But what if analysis shows that the argument in question is nothing but a sophism? Even then the analysis will be very useful. On the one side, it will put in their place the arrogant ones who believe that they can resolve the problems which have troubled the most excellent minds over the centuries by means of mere conjuring tricks, and, on the other side, every exposure of a sophism for what it is prevents a misuse of logic; and this eliminates the misuse of the most reliable procedures we have for grounding our beliefs.

Let us now turn to the argument against God's omnipotence. The argument consists of two questions which have to be answered by an affirmation or a denial (i.e. a simple "Yes" or "No"). The first question is: "Is God omnipotent?". The answer is "Yes". The next question follows: "Can God create a stone which would be so heavy that God himself could not lift it?. If the answer is "Yes", then it follows that God would be unable to lift the stone, and if the answer is "No", it follows that there is something God is unable to create. Both answers imply that there is something God cannot do, which obviously contradicts the claim that he is omnipotent. The argument thus claims to be a reductio ad absurdum of the thesis concerning God's omnipotence, i.e. the apagogical proof of the thesis that God is not omnipotent. Indirectly, it is a proof of God's non-existence, for it seems unjustifiable to go on calling a being of limited powers by the name of God. I do not know who first discovered this argument, but it is usually advanced by people without philosophical education or a particular interest in theological questions. They do not take it as the main ground on which their atheism is built, but they certainly consider it as a very useful means to confuse a believer and put him in a situation in which he doesn't know what to say - for both options at his disposal could be interpreted as denying a very important part of his faith. "Credo, quia absurdum est" cannot be treated as an answer in the proper sense of the word; it is just an expression of a bizarre attitude by which adoption our opponent willingly excluded himself from the realm of rational and meaningful discourse.

I now want to claim that the argument in question is a mere sophism, and I will justify my claim by analyzing it.

The first thing the analysis shows us is that the assumption of God's omnipotence is the basis of the whole argument. An affirmative answer to the question concerning omnipotence is not only given by a theist, but by an atheist too. The affirmative answer does not imply that God in fact exists, but only that the attribute of omnipotence is an essential characteristic of the concept of God. This concept must be shared by those who believe that it does refer to something "out there" and those who believe that it does not. But if a certain assumption is accepted and agreed upon by both sides in a controversy, then the implications of that assumption must be accepted too. In other words, once there is an agreement concerning an assumption, all propositions logically incompatible with it are ipso facto excluded. In the argument under consideration, it means the following thing: anything incompatible with God's omnipotence must be abandoned. If God is omnipotent, then it is impossible to conceive, without contradiction, of a stone so heavy that God himself would be unable to lift it, or, in other words, a stone un-liftable by God is a logical impossibility under the assumption of God's omnipotence. Since God is omnipotent per definitionem, the very concept of such a being excludes the possibility of the creation of such a stone - and that creates no contradictions at all. To put it simply, precisely because he is omnipotent, God cannot create a stone which he could not lift. Saying that he cannot create it doesn't imply the lack in his power of creation, but only the logical impossibility of a so-defined stone. Accordingly, the sophism concerning God's omnipotence is relying upon the confusion between his power and logical possibility. If a given thing is logically impossible, under the generally accepted definition of God, then we cannot accuse God of being powerless in respect to the thing in question. In that case even the necessary condition for God's demonstration of his power has not been fulfilled.

So, the argument against God's omnipotence turns out to be a sophism because of the logical mistake known as quaternio terminorum. In order to demonstrate this, it is necessary to put the argument in syllogistic form. This could be done in the following manner:

The major premise:	If there is even a single thing a being cannot do, then that being is not an omnipotent one.
The minor premise:	God is a being who cannot create a stone so heavy that he himself cannot lift it.
Conclusion:	God is not omnipotent.

This is a Celarent mode of the first syllogistic figure, where the minor premise and conclusion are treated as universal judgments, as prescribed by traditional logic which used to treat singular judgments as a species of universal ones.[7] Therefore, the deduction is a valid one from a formal point of view. However, the analysis of the meaning of the terms involved reveals that, due to the lack of precision, the premises only give an illusion of containing the terminus medius. Namely, the verb, "can" ("cannot") in the major premise refers to the power to do something, while the same verb in its first appearance in the minor premise refers to the logical (im)possibility. So, the premises contain not three but four concepts, and therefore no conclusion follows from them. This becomes obvious when the minor premise is precisely stated: God is a being who, per definitionem, cannot create a stone so heavy that he himself cannot lift it, i.e. a stone so heavy that God cannot lift it is a logical impossibility.

In order to clarify this point further, I will construct an argument which has an easily detectable sophistic character, while, at the same time, it has a form identical to the one which denies God's omnipotence. The new argument starts from the assumption of God's omniscience – an assumption which is by no means an arbitrary one, since no believer would accept the notion of God which doesn't include the property of omniscience, because knowing everything is closely related to the ability to judge justly. It is necessary to be omniscient in order to take into account all the circumstances which might be relevant in judging the degree of responsibility of each man. An atheist would also agree that this characteristic is a part of the very concept of God, adding that it, in fact, represents a hypostasis of an ideal towards which humanity has always strived.

Let us turn to the argument. If God is omniscient, then there is no type of knowledge which he lacks. But knowledge concerning one's ignorance is definitely a type of knowledge. Human beings very often have no more reliable knowledge than the knowledge that they are ignorant in certain respects. However, God does not have that type of knowledge, and therefore, he is not an omniscient being. Alternatively, one could say that, given that God is an omniscient being, he does have that type of knowledge too, but that means that there is something he does not know.

It is clear even at first glance that this is a sophistical argument, based on the neglect of the fact that any ignorance is per definitionem excluded in God's case. It is alleged that there is a contradiction between God's omniscience and his lack of the type of knowledge which it is impossible to have for an omniscient being. A being, assumed omniscient, is required to have a knowledge which presupposes some ignorance; when such a contradiction occurs at the beginning of the argument, it is not surprising that the conclusion is going to be contradictory too. A reader can recognise the same objections as those already invoked against the argument concerning God's omnipotence. Neither of these arguments succeeds, i.e. neither proves that omnipotence or omniscience are the properties which cannot be conceived of without contradiction, so that it would follow that an omnipotent or an omniscient being is a logically impossible one. However, in spite of the fact that these sophisms are invalid as arguments, they are not totally worthless. They do point to the only reliable method at an atheist's disposal, if he wants to argue with the proponents of different religions. This is the method of logical analysis and the analysis of meaning, i.e. the method by which one can find contradictions in the concepts religious people form of the highest being. Of course,religion is not of cognitive origin, and therefore the discovery of such contradictions is not a sufficient condition for the conversion of theists to atheism. But it would be unjustifiable to expect surely intellectual debates to achieve such far-reaching consequences as the abandonment of religion. It is enough for an atheist to bring his opponents to the point at which they, if they still want to disagree with him, have no other choice but to hide behind Tertulian's maxim "Credo, quiaabsurdum est" and thereby exclude themselves from the class of rational beings.

Given that I can describe so nicely the way an atheist should follow in his argumentation, if I am asked whether I have tried to follow my own advice, I would reply that I am not in possession of any definitive solution, but that I do have some ideas which may be of interest. The analyzed arguments prove neither that an omnipotent being cannot be conceived without contradiction, nor that an omniscient being cannot be conceived without contradiction. But, it might be possible to prove that it is impossible to conceive of a being who is omnipotent and omniscient at the same time. Such an argument would not be an unimportant blow to theism, since not very many believers are ready to deprive their God of any of these two properties.

So, the question is now, whether an omnipotent and omniscient God is logically possible. If it is shown that such a concept is not a self-contradictory one, that does not constitute a proof that such a concept refers to something real; but, it is shown that such a being is logically impossible, then invoking the authority of revelation does not help, nor are the feelings in our hearts and our religious conscience of any consequence.

Leibniz was right in his insistence that demonstration of logical consistency of the concept of God must precede any attempt to prove his existence. Therefore, before proceeding any further, it is necessary to establish that the concept of God doesn't contain any contradictions, i.e. to demonstrate that essential properties attributed to him are compatible with each other. Leibniz tried to demonstrate this on the basis of a definition of God as the most perfect being. If God is the most perfect being, then his properties are various perfections, i.e. absolute properties. "I call every simple quality which is positive and absolute, or expresses whatever it expresses without any limits, a perfection".[8] He believed that the very definition of perfections excludes the possibility of their mutual incompatibility. For A and B to be incompatible properties, B must be a negation of A or a negation of some part of A. But, it the property A is a perfection, then B cannot be its negation, since A and B, per definitionem, express all that they express without any limits; B cannot be a negation of some part of A either, since a perfection is, per definitionem, a simple property which cannot be analyzed.[9]

If we want to prove that the concept of an omnipotent and omniscient being is a self-contradictory one, then first we must show that omnipotence and omniscience are not perfections in Leibniz'z sense, i.e. that the statement of a being's omnipotence and a statement of a being's omniscience are not logically independent statements. G.H. von Wright defines the lack of logical independence in the following manner: "In order to show that two singular propositions, p and q, are not logically independent, one has to show that at least one of the four combinations - p&q, p&¬q, ¬p&q and ¬p&¬q - is a logical impossibility".[10] In our consideration, p stands for "God is omnipotent" and q stands for "God is omniscient". Of course, we are not interested in showing that there is just any sort of logical dependence between these two propositions, but only the one named "logical incompatibility" and expressed by "not: p and q. In other words, we have to show that the conjunction of p and q is false if we suppose that both p and q are true. We are not challenging the logical possibility of a being who is neither omnipotent nor omniscient, nor the possibility of an omnipotent but not omniscient being, nor the possibility of an omniscient but not omnipotent being - we are trying to deny the possibility of a being who is both omnipotent and omniscient. This can be achieved only by an analysis of omnipotence and omniscience, namely, by an investigation of the relationship between the implications of the propositions attributing these properties to a certain being. Leibniz denies the very possibility of such an analysis by claiming that perfections are simple qualities, i.e. that a perfection capable of further analysis would not be a simple quality, but an aggregate. It should be noted that this is a very simplistic view of logical analysis. The discovery and investigation of different implications of the assumption that some being has a "simple quality" should be distinguished from the process of division of a whole into its parts (e.g. analyzing a chemical compound into its elements). And besides, what gives us the right to claim that omnipotence and omniscience are absolute properties, until we have investigated all the possible consequences of possessing these properties?

I shall now argue that it is possible to show that some consequences of omnipotence contradict omniscience, and that an omnipotent and omniscient being, even if he is a God, cannot be conceived of without contradictions.

If God is an omnipotent being, then he can do everything that it is logically possible for him to do. Therefore, he can do everything that is conceivable without contradiction under the assumption of his omnipotence. Is temptation such a thing? It is difficult to see why the answer to this question could not be an affirmative one, i.e. it is difficult to see why it would be logically impossible for an omnipotent being to bring another being into a situation in which it can be determined whether another being does or does not possess certain characteristics. The fact that the ability to tempt is indeed attributed to God can be seen by quoting several passages from the Old Testament's famous story of Isaac's sacrifice: "It came to pass... that God did tempt Abraham, and said unto him ...: Take now thy son, thine only son Isaac, whom thou lovest, and get thee into the land of Moriah; and offer him there for a burnt offering upon one of the mountains which I will tell thee of".[11] When Abraham brandished a knife to slaughter his son, "the Angel of God" appeared from the sky and prevented him from doing so by delivering to him the following passage: "Now I know that thou fearest God, seeing thou hast not withheld thy son, thine only son from me ... By myself have I sworn, saith the Lord, for because thou hast done his things, and has not withheld thy son, thine only son: that in blessing I will bless thee, and in multiplying I will multiply thy seed".[12] Expressions "now I know that thou fearest God" and "for because thou hast done his things, I will bless thee" undoubtedly show that God did not know in advance what Abraham will do in a situation in which he had to choose between such values as the life of his only son and the devotion to his God. God could not have known that, if words "it came to pass ... that God did tempt Abraham" are to have any meaning at all. Ignorance is a necessary condition of any temptation. Of course, this is not a complete ignorance, but quite a special brand of it which follows from the definition of "temptation". If a tempter knows what the tempted will do, then we cannot speak of temptation any more. Therefore, if God has tempted Abraham, he could not have been omniscient. On the other hand, if God is omniscient, then he cannot be omnipotent, since he is unable to tempt. In order words, it is impossible to conceive without contradiction of an omnipotent and omniscient being, which is what we wanted to prove.

It is important to note that this argument is not subject to the same objection as the sophism about divine omnipotence. Namely, it cannot be said that the very assumption of an omnipotent and omniscient being makes temptation impossible for that being. The fact of the matter is that the assumption itself is logically untenable, so that we have no right to make an appeal to it. If we were to claim that for an omnipotent and omniscient being anything that contradicts its omniscience is logically impossible, even if compatible with his omnipotence, that would mean that the power of such a being is limited by his omniscience, and, since limited, should not be called omnipotence anymore. It would be like supposing that there was a round polygon and then claiming that is impossible for such a geometrical figure to have all its parts at the same distance from its centre, since that would contradict the notion of many-sidedness. As properties of circularity and many-sidedness are determined independently of the assumption of a round polygon and since they do make that assumption a self-contradictory one, so the analysis of properties of omnipotence and omniscience, which are defined independently of a conception of an omnipotent and omniscient being, do make that conception a contradictory one.

Besides the argument based on the notion of temptation, it is possible to think of other arguments which make the logical impossibility of an omnipotent and omniscient being even more obvious. For instance, it is unclear why the power of forgetting would contradict omnipotence. We humans do not have that power, although we would be only too happy to erase some of the contents of our memories. But, if that power doesn't contradict the notion of omnipotence, then it must be, <u>per definitionem</u>, attributed to a being who is credited with it. Since forgetfulness contradicts omniscience, it is clear that an omnipotent being cannot be an omniscient one at the same time.

Let us take a look at another example of the incompatibility between omnipotence and omniscience. If a being is an omnipotent one, then anything that is logically possible for him is also really possible for him. But then such a being cannot be an omniscient one, since omniscience means a lack of secrets not only in the present and the past, but also in the future. Omniscience does not only imply the knowledge of what is logically possible in a situation, but also the knowledge of what is actually going to happen. It is unclear how such knowledge could be possessed by a being who possesses an absolute power of choice. Therefore, if God is per definitionem an omnipotent and omniscient being, then the concept of God is a contradictio in adjecto.

This conclusion returns us to the opening sentence of this paper. Let us take a concluding look at the main results of our previous considerations. The revelation contains propositions which empirical science can easily refute, but also propositions which are beyond the realm of empirical science. The right method of refutation of the latter claims doesn't consist of dismissing the revelation by labeling it a non-empirical and an irrational source of knowledge, for it is possible to dismiss the objection that the difference between a theist and an atheist is exactly the same as the difference between a man with a normal capacity for colour discrimination and a colour blind person. The complete refutation of the above mentioned propositions is possible only via the discovery of contradictions in them, i.e. by means of logical analysis. It should be born in mind that proofs of God's non-existence don't establish anything over and above an inconsistency in the consistency in the concepts which are used by some people in a given culture, in order to give form to their religious feelings and beliefs.

Translated from Serbo-Croat by
Bojana Mladenović

NOTES

1. Philosophical Writings of Peirce, selected and edited by Justin Buchler, New York, Dover Publication, pp 36–41.

2. The Gospel According to Saint John, 8–13.

3. Ibid., 8.16,17.

4. Ibid., 8.19.

5. The First Epistle of Paul the Apostle to the Corinthians 3.19.

6. G.W. Leibniz, Philosophical Papers and Letters, A selection translated and edited by L. E. Loemker, D. Reidel Publishing Co., Dordrecht–Holland/Boston–USA, 1976, p 652.

7. M. Cohen and E. Nagel, An Introduction to Logic and Scientific Method, London, G. Routledge and Sons Ltd., Il, para 2.

8. The Ontological Argument, ed. by A. Plantinga, Doubleday & Co. Inc., Garden City, New York, 1965, p 55.

9. Ibid., p 56.

10. G.H. von Wright, Explanation and Understanding, Routledge and Kegan Paul, London 1971, p 195.

11. The First Book of Moses, 22.1, 2.

12. Ibid., 22.12, 16, 17.

A RELATIVISTIC CRITICISM OF REALISM

SVETOZAR SINDJELIĆ

The convergent realists Boyd, Newton-Smith, Putnam, Shimony, Niiniluoto and others[1], claim that realism can be empirically confirmed by abduction. They consider that convergent realism presents the only possible explanation of the observational success of science, and consequently, that it is necessarily valid. I want to show that their assessment is not correct and that at least one variant of anti-realism (which we shall denote as <u>interactionistic divergent relativism</u>[2]) also explains the success of science; moreover since it has more acceptable consequences, it is epistemically superior to realism. In other words, if realism and this kind of relativism are seen as rival meta-theoretical hypotheses, relativism is to be preferred.[3] My argument (in favour of relativism) will be based on the following:

(1) Divergent relativism accounts for the empirical success of science at least equally as well as convergent realism.

(2) If realism is plausible, then, on the one hand, the so-called <u>relation of correspondence</u>[4] among successive theories must hold absolutely, whereas, on the other hand, <u>non-referential</u> theories can have no empirical success, and the <u>empirical underdetermination</u> of theories in a more radical sense is not possible. However, (a) it is impossible to ascertain the manner in which the relation of correspondence may hold under radical scientific change, (b) non-referential theories <u>very often</u> had convincing empirical success, and (c) empirical underdetermination seems to be always possible. Thus, methodological facts (a), (b) and (c) are in total contradiction with realism, whereas divergent relativism explains them without any problem.

(3) There are no methodological facts which are better explainable in the context of realism than in the relativistic frame.

(4) Empirical, metaphysical and epistemological assumptions underlying realism are less plausible and more awkward than those underlying divergent relativism.

I will first outline the basic tenets of convergent realism and its abductive argument, as well as the main theses of interactionistic divergent relativism.

A. Pavković (ed.), Contemporary Yugoslav Philosophy: The Analytic Approach, 263–281.

Convergent Realism and Divergent Relativism.

In general, for a mature science, convergent realism asserts the following:

(1) The theoretical statements of science are approximately true in terms of their correspondence to reality. This means that:

(a) the entities postulated by theories exist approximately as postulated,[5] and

(b) stand in approximately such causal (and other) relations as ascribed to them by theories.

(2) In the historical development of science more recent theories are closer to the truth than older theories in the same domain.

(3) More recent theories preserve the truth of their predecessors in one way or another. They are usually thought of as preserving the terms and basic laws and generalisations. Thus, earlier theories represent limiting cases of the later ones. It is the so-called principle of correspondence, convergence or retention.

(4) In principle, it is always possible to find evidence that a theory is true or false; or more precisely, if we are confronted by two rival theories, one can always decide which one is closer to the truth. There is a criterion for truthlikeness; and most often, empirical success is assumed to be such a criterion.

(5) The scientific language becomes more and more naturalistic, i.e., it develops a terminology which quantizes reality into its real constituent elements. In the same manner, methods used by science become increasingly co-ordinated with reality, and as a result, a naturalisation of epistemology ensues. The progressive accumulation of theories, scientific language, methods and epistemic criteria leads to an unambiguous and true description of reality. The convergence of science takes place through its frontal advance towards the truth in a dialectical manner: a modification of every single theory provokes modifications of the other existing collateral theories; these modifications affect language and methods; language and methods in turn affect theories; these together influence epistemological standards; standards repeatedly work upon the rest, and so on and so forth.[6]

(6) Science investigates a mind-independent reality. Scientific theories are true or false as a result of something strictly autonomous of the human mind.

(7) As Hellman has pointed out, there is no impenetrable veil of perception: "perception is a complex process involving physiological response to physical input from the environment in virtue of which we perceive parts and aspects of that environment".[7] Since our senses are helped by true theories, there are no inaccessible levels of reality.

One of the most interesting points of convergent realism concerns its own empirical confirmation. It is argued that convergent realism explains the observational success of science whereas anti-realism cannot explain it. As Putnam concisely expresses this thesis: "The positive argument for realism is that it is the only philosophy that doesn't make the success of science a miracle".[8] Shimony is slightly more extensive: "both the evidence concerning the correspondence principle and that concerning the progress of science could only be regarded as fantastic networks of coincidences if the Hypothesis of Versimilitude were false...Kuhn's thesis that the truth plays no role in the progress of science is analogous to maintaining that progressive coherent decipherment could occur even though no such things as the initial message and the rules of encoding exist.[9] This argument is called by Niiniluoto the "no-miracle" argument.[10]

On the other hand, the relativism I am referring to can be roughly described by the following set of statements:

(1) The role of the referential framework[11] (called hereafter simply the framework) is of paramount importance to knowledge. There is no knowledge without it; nor reasoning, perception, formulation of facts, etc. The way in which we see, understand and describe reality depends on such an apparatus; and its change modifies our description of reality. A framework, in accordance with its schemes, orders our experience – by segmentation and classification of given raw material, and by the organisation of the so obtained elements into perception and other cognitive structures.

(2) A framework is a priori to experience. It is arbitrary (as is the case with different geometrical systems). (Nevertheless, it may be the result of adaptive processes). Its action is predominately unconscious, to a large extent exterior to the critical mind of the subject of knowledge, and is to a great extent tacit and undiscursive(i.e., it can be only partly expressed in linguistic assertions).

(3) The structure of a framework is extremely complex, having different and relatively separate levels. Some basic patterns are incorporated to our physiology; others are a part of our fundamental psychological expectations, language, background knowledge as well as of scientific paradigms and scientific theories. This structure, or more precisely, these structures, can be changed, but their changability is inversely proportional to level depth. In fact, the structure of a framework on the physiological level is changeable only in the long run.[12] The same

rule holds for discursiveness; the less level depth, the more discursive structures.

(4) What common sense sees as aspects or patterns of reality, are actually a product of the interaction of two systems: a framework and reality (or from a biological point of view, the organism and the environment). The patterns of these two systems, taken separately, are not only unknown, they are unknowable in principle by the subject of knowledge.[13] Hence, regarding our knowledge, we always deal with a system which is the result of the interaction of two different systems, whose structures are unknown to us. So, we can never identify the "real" patterns of reality.

(5) Framework development is evolutionary. Knowledge (with all its cognitive instruments) plays an adaptive role. It is a result of dynamic interaction between the above-mentioned two systems. Frameworks can be developed, as the evolution seems to show, in completely different ways.

(6) There is no preferred framework. The interaction of two systems allows a whole variety of alternative organisations of experience (or description of reality) which can be equally successful from the adaptational point of view.[14]

(7) Frameworks, both in the narrower sense and in the broader sense, are incommensurable.

(8) It is clear from statements (1) - (7) that scientific theories cannot be true in the sense of corresponding to reality. The world, as we see it, represents to a great extent the objectivised patterns of our framework. One implication of this position is that one cannot differentiate between the a priori elements of perception and the physical inputs from the environment.

It is necessary to point out that some other forms of relativism, including variants of interactionistic relativism, are not incompatible with realism. For example, interactionistic relativist Konrad Lorentz holds that certain natural laws are so all-pervading that organisms necessarily have to incorporate the ability to take them into account.[15] But, a variant of relativism I call divergent interactionistic relativism (which is outlined here) is not reconcilable with realism. This incompatibility is mainly due to the fact that divergent relativism disputes the existence of a preferred framework.

The Explanation of Empirical Success

As we have seen convergent realists assert that if convergent realism (together with the thesis of versimilitude which is its integral part) is not plausible – then the empirical success of science cannot be explained in any way. Newton-Smith expounds this in the following way:

> We can offer a positive argument for realism as an explanatory hypothesis. Consider the fact that theories are increasing in predictive power. What is it about contemporary theories that make them better able to provide this? On the assumption of the thesis of versimilitude, we have an answer: Theories provide better predictions because they have uncovered more about the world. It would be a total mystery that they should do this unless they have uncovered more truth at the theoretical level.[16]

In the continuation of this discussion typical for a convergent realist, Newton-Smith points out that there is also, one other kind of success of science which is, in his opinion, inexplicable on an instrumental construal of theories. Simply, a certain consequence of some theoretical solution can prove itself to be true (successful) in some distant field of science, and in solving problems which do not seem related to those it was formulated for. Newton-Smith gives the example of Planck's introduction of the quantum of action in solving the black body radiation problem. The quantum of action thus introduced turned out to be applicable to the apparently unrelated problem of the specific heat of solids. This kind of success is not surprising for realists at all; Planck merely discovered "the underlying mechanics of science".[17] Sometimes, even undesirable side-effects of scientific theories show themselves to be successful.[18]

Newton-Smith thus shows that there are two levels of the realist "no-miracle" argument:

(a) explanation of the empirical success of science, and

(b) explanation of what might be called the anticipation of novel facts.

Let us assume that both points are explicable in the context of convergent realism, although this may be disputed, because only the truth can explain the success completely (not the approximate truth).[19] It seems to me that classical instrumentalism[20] cannot easily explain point (b), whereas it can explain point (a) without too much trouble. But in the framework of divergent relativism both (a) and (b) can be explained at least as well as in the framework of convergent realism. I will start with point (a).

The instrumentalist explanation would be as follows: When a scientist constructs a theory, he already possesses the kinds of facts relevant to it. By the operational linking of these kinds of facts, he <u>calibrates</u> his model in such a way that it further serves to predict new facts which are within the system determined by the above-mentioned kinds of facts. When, over the time, the field of experience expands (as a result of operating with new kinds of facts), the predictive power of the system diminishes; a theory ceases to be empirically successful; anomalies appear. The new situation demands the construction of a new theory which encompasses more kinds of facts, and, very often, facts of a different kind than those within the old theory. Thus the success of science may be ascertained in advance for a given domain. The explanation of increased predictive power does not at present lie in increasing versimilitude. It is due to the fact that a greater spectrum of facts is taken into consideration and the new operational co-ordination of a theory and these facts.

The problems of instrumentalism begin with point (b). How is it possible for a theoretical system which is calibrated in advance to anticipate novel facts? Within instrumentalism there is no convincing answer. Nevertheless, this does not mean that a plausible anti-realistic answer does not exist. The explanation which divergent relativism offers is at least as good as the realistic one. It even works in a similar manner.

The realistic explanation that theories are empirically successful because they are approximately true, presupposes that successive theories in the same domain have to have an expanding common core, a common core which may be called versimilitude, and which will grow over the time, becoming truer and truer. The problem with the realistic explanation is that this kind of a common core has never been identified.

On the other hand, the relativistic explanation of the success of science is also dependent on the statement that successive theories have something in common. This common core would be the successful interaction between a framework and reality, or, in other words, a good correlation of patterns originating in organisms and the relevant patterns of reality. These patterns can be organised in a whole variety of ways. Every combination leading to operational success represents a set of good adaptational schemes. Thus, in this context, the productive link between framework patterns and reality patterns explains the empirical success of science.

Scientific theories (as human knowledge on the whole) have a predominantly adaptive role. They are the result of the interaction of two complex systems whose patterns one cannot isolate and separate. We cannot identify the characteristics of either of these systems as they are independent of one another, because we cannot consider nor imagine one without the other. Successful theories can correspond only to successful adaptive schemes, and by no means to reality itself. A whole variety of successful responses exists for every adaptational problem. To question which one is true is just as to raise the question of whether fish-gills or dolphin's lungs reflect the truth about the world. Successful adaptive schemes are not unequivocally determined.

Thus a significant objective component exists in scientific theories; but this objective component is never unequivocally determined and discursively expressed; also, it does not cumulatively grow with the development of science. Nevertheless, the anticipation of novel facts is a consequence of this objective component.

In brief, an explanation of the anticipation of novel facts is as follows: every framework organises reality in a certain manner. It is not disputable that theories formulated in a given paradigm represent some kind of echo of real regularities mediated by paradigm. Relativism only denies that a scientific theory represents a true, or approximately true description of real facts, i.e., that it describes things as they really are and that it is meaningful to speak of things really existing in one and strictly determined way. Although a successful framework orders relevant real regularities in one of an infinite many operationally adequate ways, an anticipation of novel facts occurs because every such ordering preserves some consequences of these regularities (which are unknown to us in its real sense). To use a metaphor from physics: a change of a geometrical system demands a new conception of physical laws so that the resulting theory would be brought into operational accordance with experience. But in this change many known and unknown predictions will be preserved, although they would acquire new meanings.

This explanation may, I admit, seem relatively obscure. But I also think that the mentioned mysterious mixture of the referential patterns and the schemes of reality explains the success of science equally well (or equally unsuccessfully) as a strange mixture of the truth and non-truth in the realistic explanation.

Also, the objection may be raised to this relativistic explanation of the empirical success of science that it is developed from the majority of relativistic assumption and due to this, circular to a certain extent. But the very same objection may be raised against the realistic explanation, and as a result such objections are a matter of small weight in this context.

Problems Which Realism Cannot Explain

If realism is a valid position, there are three consequences which must absolutely hold:

(1) The thesis of versimilitude has to be true - what automatically means that the principle of correspondence, in some of its aspects, must hold.

(2) Non-referential theories cannot be empirically successful - simply because they cannot be approximately true.

(3) The underdetermination of theories by data is not a real phenomenon, at least in the more radical forms.

If at least one of the given statements is not true, realism cannot be a plausible position. I will try to show that not one of them is true, and furthermore, that this fact is perfectly explicable in the context of relativism.

(1) The trouble with the thesis of versimilitude

The validity of the thesis of versimilitude is the central pivot of convergent realism. From this thesis follows the validity of the principle of correspondence. The increase of versimilitude should always be accompanied by an expansion of something in the theory or closely connected with the theory. Realists take for granted that even in the most revolutionary scientific change, a new theory preserves some part of the old one. However, to date, nobody has succeeded in offering even a satisfactory semantic definition of versimilitude. Some realists have indicated what the above something could be, but not one of these guesses withstood the ensuing criticism.

The most famous attempt to characterise versimilitude was made by Karl Popper. He maintained that the constantly increasing quantity of truth in a succession of theories consists in the retention of all the true empirical consequences of earlier theories. Every latter theory explains everything that the former explains, expands in relation to the former to explain the facts on which the former failed, and the last includes additional confirmed novel predictions.

Unfortunately, Popper's attempted specification of versimilitude turned out to be unsuccessful. Miller and Tichy have demonstrated that it is impossible for a theory to explain the failure of the earlier one and at the same time to preserve all its true consequences. Simply the assertion that T_2 retains all the true consequences of T_1 and at the same time explains its anomalies leads to a contradiction.[21]

Of course, realists may reply that Popper's definition of versimilitude is only one of many possible definitions – and that some other will certainly be more successful.[22] However, if one scans the list of entities conceived by realists as candidates for what new theories preserve, it seems that such a position cannot be defended. Laudan gives the following list of these entities: T_2 entails T_1 (Whewell), the true consequence of truth content of T_1 (Popper), the "confirmed" portions of T_1 (Post, Koertge), the theoretical laws and mechanisms of T_1 (Boyd, McMullian, Putnam), T_1 as a limiting case (Watkins, Putnam, Krajewski), T_2 explains why T_1 succeeded insofar as T_1 succeeded (Sellars), reference for the central terms of T_1 (Putnam, Boyd).[23]

Every more detailed analysis of scientific change (in the case of more radical shifts) indicates that nothing mentioned above is present in the new theory.

Fayerabend's and Kuhn's conclusion known as the incommensurability thesis – are extremely relevant to this problem.[24] Very roughly outlined, on the occasion of theory change, the conceptual framework must be replaced also – thus it is no longer possible to speak of retention, limiting cases, preserved portions and mechanisms of the earlier theory, and the same true empirical consequences as well. Simply, one theory has one kind of consequences conceived as true, whereas the latter theory has a different kind of such consequences. Theories organise reality in different ways, and thus they do not speak literally about the same things.

It seems clear from the presented argument that realism copes badly with the probable invalidity of the correspondence principle. On the other hand, relativism is not only perfectly compatible with this invalidity, it is also compatible with the incommensurability thesis.

(2) Success of Non-referential Theories

A direct deductive consequence of the statement that the approximate truth (together with its essential constituent – adequate reference) represents the only possible explanation of the success of science – is that theories which are not approximately true cannot have empirical success. However, a whole majority of theories, undoubtedly non-referential by realist criteria, had great empirical success in their time.

Laudan cites a number of such successful theories which employed non-referring central terms: the crystalline spheres of old astronomy, the humoral theory of medicine, the effluvial theory of static electricity, "catastrophistic" geology, the phlogiston theory of chemistry, the caloric theory of heat, the vital force theories of physiology, the electromagnetic aether, the optical aether, the theory of circular inertia, and theories of spontaneous generation.[25]

Several cases from the history of physics may illustrate this point. Newton's optics treated light as particles and was very successful. Later, it was replaced by the empirically superior Frensel's theory which conceived light as a pure wave-like phenomenon. After it had been confronted with some anomalies, Frensel's theory was replaced by new Planck's and Einstein's photon theory which again operates with particles. Also, as Kuhn has pointed out, in the succession of Aristotle's, Newton's and Einstein's mechanics one cannot see coherent direction of ontological development. "On the contrary, in some important respect, though by no means in all, Einstein's general theory of relativity is closer to Aristotle's than either of them is to Newton's".[26]

Of course, relativism encounters no problem with the genuine reference of theoretical terms. It explicitly asserts that theoretical terms do not refer to real entities. Actually, they are different modes of the organisation of the given.

(3) The underdetermination of theory by data

The thesis of the underdetermination of theory by data (or the evidential indistinguishability thesis) consists in the following: if we have a theory T, we always can construct an alternative theory T′ which will be empirically Equivalent to T (having the same observational consequences) and will be logically incompatible with T in the non-trivial sense (i.e., it is not only a different formulation of T apparently incompatible with it. This thesis contradicts realism; if it holds, realism does not hold and vice versa.

The empirical underdetermination thesis is supported by two arguments. The first of them originates in physics, whereas the second one is epistemological.

The first argument runs as follows: In physics, it is very difficult to imagine the unequivocal and full determination of a theory by facts, i.e., all observational evidence explained by a theory, may be explained in a different theoretical manner as well. It holds also for the most general theory of the world. A good example of such a reasoning is that by Poincare, Einstein, Reichenbach and others concerning the possibility of a real geometry of the world. They asserted that one can operate only with whole conjunctions (G_e & F_1) and (G_{non-e} & F_2) – where F_1 and F_2 are differently specified physical laws, G_e is Euclidian geometry whereas G_{non-e} is the non-Euclidian geometry.[27] These conjunctions may be in accord with experience only as such (as conjunctions). It is all the same which conjunction one chooses. The relevant evidence support any of them equally. We can proceed from any conjunct and bring it into accord with the evidence. No facts that may be later obtained can change this – when we commence from one alternative, we will see all later facts through this alternative.

The epistemological argument is outlined very simply by Quine:

> We may expect this because of how scientists work. For they do not rest with mere inductive generalisations of their observations: mere extrapolations to observable events from similar observed events. Scientists invent hypotheses that talk of things beyond the reach of observation. The hypotheses are related to observation only by a kind of one-way implication; namely, the events we observe are what a belief in the hypotheses would have led us to expect. These observable consequences of the hypotheses do not, conversely, imply the hypotheses. Surely there are alternative hypothetical substructures that would surface in the same observable ways.[28]

Realists have raised a range of objections to the thesis of underdetermination. Many of them state that it is not possible to have criteria by which one can know whether we deal with really incompatible theories or only notational variants of one and the same theory. It is interesting that a realist, Newton-Smith, argues that the thesis of underdetermination holds, but perhaps not in the radical sense. He constructs several examples of equivalent but incompatible theories with a tendency to prove the weak thesis of underdetermination - the thesis that it is only sometimes possible to construct evidentially equivalent but incompatible theories for the same evidence.[29] His argumentation seems to be cogent. But, Newton-Smith claims that his argument does not falsify nor attack realism. This would be possible only on the basis of the proof of the radical form of the thesis - and, in the opinion of Newton-Smith, there is no such a proof. Nevertheless, I think that Newton-Smith's result attacks the realistic position. If nothing else, it refutes the argument that the underdetermination is fictitious, actually being only a set of notational variants of a single theory.

Of course, the question may be raised whether the relativist who accepts the incommensurability thesis can consequently accept the underdetermination thesis. Namely, if one accepts the incommensurability thesis, how can he assert that two different theories are logically incompatible. But, this problem has no bearing on the anti-realistic argument against realism.

There are simply three options available to the anti-realist:

(1) He can flatly refuse to consider the problem, because if the incommensurability thesis holds, then realism cannot hold,

(2) He can use the underdetermination thesis conditionally, asserting that even if theories were commensurable, realism would not hold as a result of underdetermination among the many reasons available, and

(3) He can assert that the argument derived from underdetermination holds when given theories are so close in referential and other relevant senses that it allows us to speak of their relative commensurability.

Accordingly, the attacks on the underdetermination thesis on account of incommensurability do not touch anti-realism.

Relativism and the Standard Objections to Anti-realism

I would like to show that the problems with which divergent relativism is confronted are considerably easier than the problems realism is confronted with. I mean that this kind of relativism easily resolves all the objections to relativism and anti-realism in general.

Actually, beside the already discussed no-miracle argument, there are two following objections:

> the principle of the common cause as well as the fact that there is no sharp distinction between theoretical and observational entities.

The principle of common cause asserts that many regularities which coincide in different fields of science would be totally unexplicable without a presupposition about their common cause.[30] Thus, illustrating this principle, Salmon argues that, without such a hypothesis, a situation in which quite different experiments lead to the same value of Avogadro's number would be something completely mysterious, something equal to a strange cosmic coincidence.[31]

In realists' opinion, the principle of common cause affirms the realistic hypothesis by showing that real causes of relevant phenomena must exist in the scope of the unobservable; this means that scientific explanation is impossible if a real event cannot be identified in the scope of the unobservable.

This argument evidently represents only a variant of the argument derived from the inference to the best explanation (i.e., the abductive argument), and all of my answers to the abductive argument hold for this argument too. It is very important here that anti-realism does not assert the non-existence of causes in the nature; anti-realism only asserts that the whole situation concerning causes and consequences can be described in a variety of ways, none of which has priority over the others.

In 1962, Maxwell attacked empiricist anti-realism by arguing that the theory/observation distinction cannot be drawn (or, in other words, that the distinction between theoretical and observational entities is not sharp).[32]

This argument has been assumed, for a long time, to have a great weight and to be able to refute anti-realism. But such a supposition was mistaken. In fact, Maxwell only challenged one specific instrumentalistic view (Carnap's); and it is even disputable there whether his criticism was successful.[33] When Maxwell's argument is used against the relativistic conception of anti-realism such as this, or against other constructivistic views, it has no weight at all. The content of this argument is in accordance with the basic tenets of divergent relativism: the language of science cannot be divided into the theoretical and the non-theoretical components; objects and events cannot be clearly classified into observable and unobservable (i.e., observation even in its most elementary form is theoretically determined). One of the main points of relativism is in fact that we always formulate and construct some specific kind of observation dependent on an accepted framework.

The Implausibility of the Realistic Assumptions

In this section I want to defend the thesis that the empirical, epistemological and metaphysical assumptions which lie behind realism are less acceptable than those behind divergent relativism.

I think it is clear from what was said above that the question which ought to be raised is not simply "How is the empirical success of science possible?" but - "How is the empirical success of science possible, even though :

(1) Methodology as a whole and observations are essentially theory-dependent (making a theory-independent description of reality very improbable); the more so because in the transitions from one paradigm to another no rationality is observable.

(2) Radically different theories seem to be incommensurable.

(3) The underdetermination thesis holds (at least for theories which are not radically different).

(4) Plausibility of the correspondence principle cannot be shown by any means; there is not a satisfactory definition of versimilitude at all.

(5) Some non-referential theories in the past were evidently empirically successful?.

Statements (1) - (5) are the result of very detailed study of the history of science and actual scientific practice.

Accordingly, realists can satisfactorily explain the empirical success of science only if their explanation takes into consideration statements (1) - (5). Also they must explain how the approximate truth may account for the success of science.

Of course, the majority of convergent realists try to do this. Thus, they introduce a whole variety of explicit or implicit assumptions. These assumptions, among other things, are aimed at annulling the consequences of statements (1) - (5) which are harmful to realism. Now, I will try to show that such realistic assumptions are awkward and implausible. To do this, I will briefly examine some of them without trying to be systematic and complete.

One of the first implausible assumptions is that the universe such as it really is, can be described without any framework. This complex assumption consists of one metaphysical and two epistemological sub-assumptions: (a) that the universe in itself exists in one and only one way, (b) that some infinite intelligence would have to know just this way, and (c) that the species homo sapiens can at least move toward such knowledge.

But even some elementary considerations in the philosophy of science show that scientific knowledge cannot be determined by the theory-observation relation; scientific knowledge is, in fact, formed within this inter-relationship: the interpretative theory – the explanatory theory – observations (for all that, these three factors depend one upon the other, and are theoretical by nature). An interpretative theory (or a framework) is essentially arbitrary and necessarily depends on a set of conventions. Every change in an interpretative theory automatically changes the explanatory theory as well as observations. The universe cannot be imagined with a priori acceptance of a system of reference; it even does not exist in only one way which could be described by some infinite intelligence, because a description always presupposes a framework. This is well illustrated in the numerous discussions about the real geometry of the world. Also, the relativistic revolution in physics dramatically points at the absurdity of the question of how the world really seems. Our picture of the world will be changed together with a change of the systems of reference. It is non-sensical to demand answers about absolute space, time or mass, etc.

Besides, it seems to me that realism assumes that homo sapiens is sense-wise equipped to grasp the totality of the universe. But this is very difficult to defend. More probably, a man gets only a small portion of the information about environment. Also, this information is dictated by biological needs, limited by them and mediated by a complex psycho-biological structure which determines the manner of interpretation of stimuli from the environment.

A number of realists accept the necessity of the existence of a framework, but introduce an assumption of a preferred framework (or some kind of a natural framework). This assumption in itself cannot save realism. It only replaces a framework-relativism with a framework-determinism.[34] However, together with some additional assumptions this conception can lead to realism.

I think that there is a whole range of good arguments against such a claim. Divergence, and not convergence, manifests already itself in the evolutionary plan. The non-existence of a preferred framework is evident even on the linguistical level.[35] The story about a real geometry of the world shows this.

In the light of the above, it is extremely difficult to justify the relativistic assumption that observation can be a neutral arbiter in deciding among rival theories.

By rule, realists, explicitly or implicitly, introduce the following three assumptions:

(1) The principle of correspondence holds although this cannot always be shown,

(2) Even the most radically different theories are commensurable, although it is not possible to show this now, and

(3) The massive underdetermination[36] in a non-trivial sense is not possible, although this cannot be proved at this moment.

The expression of these assumptions (given without stating reasons) has the structure of logical fallacy – the appeal to ignorance. Realists claim that the given assumptions hold although cannot be proved, and demand from anti-realists to prove that they do not hold. Obviously, the burden of proof rests only fairly on the realists.

Also, the realists' claim that the approximate truth can explain the empirical success of science is completely dubious. It seems that this claim is closely connected with the correspondence principle, and that it falls with such a principle. Namely, if the correspondence principle holds and if some portions of older theories can be preserved, then the claim that approximate truth explains empirical success would be relatively understandable. But without such a condition being fulfilled, it is mysterious. After all, it seems that the relativist's assumptions are more probable than the realists. The assumptions are in full accordance with the considered methodological facts, and assert much less about the world than the realistic ones.

Conclusion

The aim of this article was to show that at least one anti-realistic position (what I call divergent interactionistic relativism) represents a more plausible epistemological position, when evaluated by standard criteria for hypotheses, than the realistic position.

Besides, I also attempted to show two things.

First, that relativism explains the empirical success of science at least as well as realism.

Second, that this kind of relativism by no means claims that scientific theories are not in any contact with a mind-independent reality. Successful scientific theories are in a kind of contact, but they do not describe reality. Actually, they describe reality mediated by human cognitive instruments. Because of the nature of the interactions of these instruments and the environment, there can be a whole range of such descriptions which are different but equally successful. This explains why different theories speaking about the same facts can be empirically equally successful.

NOTES

1. See Richard N. Boyd, 'Realism, Underdetermination and a Causal Theory of Evidence', Nous (1973), 'Scientific Realism and Naturalistic Epistemology' in Asquith & Giere (eds.), PSA 1980, Vol. II, East Lansing: Philosophy of Science Association, 1981, William H. Newton-Smith, 'Trans-Theoretical Truth Without Transcendent Truth', manuscript (the part of In Defence of Truth, forthcoming), Abner Shimony, 'On Two Epistemological Thesis of Thomas Kuhn', in Cohen, Feyerabend & Wartofski, Essays in Memory of Imre Lakatos, Dordrecht/Boston: Reidel, 1976, Hilary Putnam, Meaning and the Moral Sciences, Londan and Henley: Routledge and Kegan Paul, 1987. (Later Putnam has left realism), and Ilka Niiniluoto, 'Scientific Progress', Synthese, 45 (1987). Similar positions can also be found in Wilfrid Sellars, Science, Perception and Reality, New York: Humanities Press, 1962, and J.J.C. Smart, Between Science and Philosophy, New York: Random House, 1968.

2. A similar position is given by Gonzalo Munevar in his Radical Knowledge, Indianapolis: Hackett Publishing Company, 1981.

3. In this case, realism and relativist anti-realism are simply treated as empirical hypotheses and evaluated on the basis of standard criteria for empirical hypotheses.

4. Alternative names for this relation are retention, convergence and cumulation. Every new theory (T_2) in the succession of theories referring to the same domain retains and preserves something of the earlier theory (T_1), that is the basis of the continual increase of versimilitude or truthlikeness. There is a diversity of guesses what this can be.

5. I.e., that the central theoretical terms genuinely refer.

6. This realistic, naturalistic and dialectical conception is developed by R.N. Boyd, especially in his 'Scientific Realism and Naturalistic Epistemology', PSA 80, Vol. II.

7. Geoffrey Hellman, 'Realist Principles', Philosophy of Science, 50 (1983), pp. 234-245.

8. Hilary Putnam, Meaning and the Moral Sciences, Boston, London and Henley: Routledge & Kegan Paul, 1978, p. 19.

9. Abner Shimony, 'On Two Epistemological Theses of Thomas Kuhn', in Cohen, Feyerabend and Wartofski (eds.), Essays in Memory of Imre Lakatos, Dordrecht/Boston: Reidel, 1976, pp.577-578.

10. See Ilka Niiniliuto, Is Science Progressive?, Dordrecht/Boston/Lancaster: Reidel Publishing Company, 1984, p.

11. Very often the following words are used to designate this structure: paradigm, conceptual scheme, research program, research tradition, background, tacit knowledge, world-view, theoretical tradition, etc.

12. In fact, this means that the segment of the framework (which is incorporated in our physiology) can be changed only by evolution - i.e., over an evolutionary significant period. The same holds for the psychological segment, etc.

13. Nevertheless, some patterns can be identified: these patterns which originate in the framework and show themselves of such an origin in the situation of replacing one framework by the other. So, it can be shown that different geometrical concepts, very general statements about time and space, ways of discribing reality that are dictated linguistically, etc., are framework-dependent.

14. The situation described in (5) and (6) may be illustrated with some B.L. Whorf's statements on language (which is a basic cognitive instrument): There is a variety of different (even untranslatable in the strict sense) linguistic systems which are equally adaptively successful. The origin of these systems is conditioned neither by environment, nor by a specific way of life. They are arbitrary, etc. (see Benjamin Lee Whorf, Language, Thought & Reality, Cambridge, Mass.: The M.I.T.Press, 1956). Besides, it is very important to note that point (5) does not assert that all frameworks are equally good. It asserts only that always the variety of equally good frameworks may exist.

15. See Konrad Lorentz, Studies in Animal Behaviour, Harvard: Harvard University Press, 1981, p. 290.

16. William Newton-Smith, 'Trans-Theoretical Truth without Transcendent Truth?', the manuscript, p. 6

17. See ibid, pp. 6-7.

18. A famous example is the expanding universe as a consequence of the general theory of relativity. Einstein and de Sitter (independently) discovered this implication of Einstein's theory in 1917. Einstein tried to modify his theory to escape the consequence. But this consequence was later dramatically confirmed by the so-called red shift. To-day, it is considered one of the principle proofs of the general theory of relativity.

19. See, for example, L. Laudan, 'A Confutation of Convergent Realism',p. 30.

20. By classical instrumentalism I primarily mean empiricistic anti-realism, whose basic thesis is that we can always organise available observational material in different ways preserving the same empiricl success (theory is simply an instrument for co-ordination of sense impressions).

21. See D. Miller, 'Popper's Qualitative Theory of Versimilitude', and P. Tichy 'On Popper's Definition of Versimilitude', the both articles in British Journal for the Philosophy of Science 25 (1974). An argument given by Miller and Tichy was anticipated by Carl Hempel. See p. 34. Joseph Agassi, 'Versimilitude: Popper, Miller and Hattiangadi', in Przelecki, Szaniawski & Vojcicki (eds.), Formal Methods in the Methodology of Science, Dordrecht/Boston: Reidel, 1976, p. 340.

22. See W.H. Newton-Smith, The Rationality of Science, p. 58.

23. Larry Laudan, 'A Confutation of Convergent Realism', p. 37.

24. The classical arguments against reductionism and commensurability are given in Thomas Kuhn, The Structure of Scientific Revolutions: Princeton: Princeton University Press, 1962, and Paul Feyerabend, 'Explanation, Reduction and Empiricism', in Feigl & Maxwell (eds.), Minnesota Studies in the Philosophy of Science, Vol. III, Minneapolis: University of Minnesota Press, 1962.

25. See L. Laudan, 'A Confutation of Convergent Realism', p. 33.

26. T.S. Kuhn, The Structure of Scientific Revolutions, Chicago: Chicago University Press, 1978, pp. 206-7.

27. This dilemma is rather unprecisely formulated. In fact, it is not a question of two geometries; indeed, the Euclidian one may be considered as one, but the non-Euclidian geometry represents the infinite set of possible alternative geometries.

28. See W.V.O. Quine, 'On Empirically Equivalent Systems of the World', Erkenntnis 9 (1975). In presenting this type of reasoning, Quine in no way wants to assert the validity of the underdetermination thesis, but only to show why it sounds convincing.

29. By evidentially equivalent theories Newton–Smith means observationally equivalent theories which fare equally well on any justifiable principles of evidence. See W.H. Newton–Smith, 'The Underdetermination of Theory by Data', Aristotelian Society Supplementary Volume LII, and his The Rationality of Science, pp. 39–43.

30. Hans Reichenbach formulated this principle in his The Direction of Time, Berkeley and Los Angeles: University of California Press, 1963.

31. See W.C. Salmon, Scientific Explanation and the Causal Structure of the World, Princeton: Princeton University Press, 1984, pp. 217–219.

32. G. Maxwell, 'The Ontological Status of Theoretical Entities', in Feigl & Maxwell (eds.), Minnesota Studies in the Philosophy of Science, Vol. III, Minneapolis: University of Minnesota Press, 1962.

33. Boyd points out that this argument is not too strong even used against the empiricistic anti–realism. See R. N. Boyd 'On the Current Status of the Issue of Scientific Realism', Erkenntnis 19 (1983) pp. 46–56.

34. A framework–determinism is in fact the Kantian position. Imre Lakatos explains the difference between what I call a framework–determinism and a framework–relativism in a lucid manner: 'The idea that we live and die in the prison of our 'conceptual framework' was developed primarily by Kant; pessimistic Kantians thought that the real world is for ever unknowable because of this prison, while optimistic Kantians thought that God created our conceptual framework to fit the world. But revolutionary activists believe that conceptual frameworks can be developed and also replaced by new, better ones; it is we who create our 'prisons' and we can also, critically, demolish them'. (I. Lakatos, 'Falsification and the Methodology of Scientific Researh Programmes', in Lakatos & Musgrave (eds.), Criticism and the Growth of Knowledge, Cambridge: Cambridge University Press, 1970, p. 104.

35. Here I allude again to the Whorfian relativistic thesis which is very relevant for the problem.

36. By massive underdetermination, Newton–Smith means the possible existence of a pair of incompatible evidentially equivalent total theories of nature. See The Rationality of Science, p. 41.

DELIBERATION, PRACTICAL WISDOM AND THE SELF IN THE NICHOMACHEAN ETHICS

HEDA ŠEGVIĆ

In the Nicomachean Ethics, Aristotle strongly emphasizes the importance of deliberation for virtuous conduct. I will start by mentioning a difficulty which arises in connection with his characterization of deliberation. In my opinion, it was this difficulty - among other things - that motivated a recent interpretation of Aristotle's views on virtue and, in particular, the role of intellect in virtuous conduct. Some of Aristotle's thoughts concerning the kata pathos way of living, the thought that virtuous actions are engaged in for their own sakes and his characterization of deliberation as a sort of zētēsis suggest, however, a different conception of the role of intellect in virtue. Aristotle's treatment of these topics, I will claim, points to a view of practical rationality in which a conception of the self or person plays an important role.

I. Deliberation and Perception: Preliminary Remarks

(1) A characteristic mark of the man of practical wisdom, according to Aristotle, is his excellence in deliberation (euboulia). It appears at first that deliberation (bouleusis, boulē) is understood as a process of reasoning that one actually goes through before acting [EN 1142b2-5]. But many actions Aristotle would count as chosen - that is, as resulting from deliberation [1] - turn out not to be preceded, at some time or other, by the relevant piece of practical reasoning. In particular, all virtuous actions of a practically wise person are in Aristotle's view chosen. But to think that they are all arrived at through reasoning, or even preceded by some relevant piece of reasoning, would be implausible as well as contrary to some remarks on virtuous actions made by Aristotle himself.[2] These considerations may lead us to think that it would be better to understand deliberation primarily as a procedure for explaining or justifying actions rather than a procedure the agent himself actually goes through in order to determine what to do. For the most part, "deliberative reasoning" will apply to the agent in an as if manner. When the agent has reasons for acting then, for the purposes of explanation or justification, his action can be represented as a result of some suitable piece of hypothetical reasoning. And this is so independently of whether he arrived at the decision to act through some reasoning or not.

A. Pavković (ed.), Contemporary Yugoslav Philosophy: The Analytic Approach, 283–309.

If we consider a further important point Aristotle wants to make about the man of practical wisdom, the thought that "deliberative reasoning" can sometimes be attributed to the agent only hypothetically appears almost inescapable. The phronimos is said to be able to deliberate well about "what sorts of thing conduce to the good life in general" [EN 1140a25–28]. He is characterized as the euboulos, the man who is good at deliberation, the euboulos being understood as "the man who is capable of aiming in accordance with calculation at the best for man of things attainable by action" [EN 1141b13–14]. Thus, not only will all actions characteristic of a practically wise person – which turn out to be, on Aristotle's view, virtuous actions – be chosen, but they will be chosen with a view to attaining the good life (eudaimonia). In this case, we would expect a fully explicit practical reasoning to involve a premise specifying the conception of eudaimonia involved. Indeed Aristotle at 1144a31–33 refers to inferences which contain the "starting point" of acts to be done suggesting that they run as follows: "Since the end, i.e. what is best, is of such and such a nature...".[3]

The picture then seems to be the following. If an action is truly virtuous, a piece of practical reasoning associated with it starts from a conception of eudaimonia and leads to a conclusion that the action, or an action of that type, ought to be performed. But this piece of reasoning should certainly not be thought of as a description of any sequence of mental states that takes place in phronimos's mind before he acts. For, it would be ludicrous to assume that a man could not act virtuously unless he goes through a long chain of inferences of the sort described. Aristotle would presumably agree with us that a man who is practically wise and virtuous need not even have a fully explicit conception of eudaimonia, or of particular virtues, let alone invoke it and reason from it each time he acts virtuously.[4]

(2) We ought to notice, however, that these considerations do not compel us to view bouleusis as merely a theoretical construct aimed at explaining actions. Deliberation can still be understood as the way in which decisions are actually reached.

It might be useful to distinguish between practical arguments and practical reasoning. We have already suggested that practical arguments have a role to play in explanation or justification of actions. The major premise of such an argument can be thought of as making explicit some practical commitment an agent has, e.g. to overthrow a tyrannical government. The minor premise would then make explicit his realization of an appropriate way – only occasionally, the best way – in which this commitment could be promoted, e.g. by acts of civil disobedience. Finally, the conclusion would express the decision which was acted upon.[5] We may then add, if we wish, that the agent acted as if he had reasoned exactly as our argument runs.

When talking about the agent's practical reasoning, we may only think of this hypothetical chain of mental states. But, the use of "as if" clauses and the talk about "hypothetical" reasoning should not mislead us into thinking that all we want to say is just that the action is compatible with the piece of practical reasoning in question. If the explanation of the action through a practical argument is a good one, the argument will single out those considerations which had actually motivated the agent to act as he did. On the other hand, when we invoke practical reasoning in explaining actions, we should do so with an understanding that nothing is added to the picture of the role of practical arguments just adumbrated.[6]

Aristotle himself consistently speaks of bouleusis as the manner in which decisions are made [see, in particular, EN III,3]. There is no reason not to take this at face value. The problem we faced above arises only on the assumption that bouleusis itself (as a way in which decisions are reached) is the actual practical reasoning. Or, in other words, the problem arises if we assume that practical arguments are meant to be literal transcripts of the agent's mental states some time prior to his action. But Aristotle need not be saddled with this view. If bouleusis is understood as the way in which the decision to act is reached, practical arguments and practical reasoning, can both be seen as representing some aspects of bouleusis for the purposes of explanation, although we should not think of them as being accurate transcripts of bouleusis.

If this line of thought is correct, we need not be disturbed much with Aristotle's practice of using bouleusis to refer to actual ways we come up with decisions as to which courses of action to pursue and which to avoid. This, of course, leaves it entirely open as to how exactly bouleusis is to be understood.

A proper account of an action typically involves some story about antecedent conditions that lead the agent to the action in question. But, not every such story will refer to deliberation. Some actions, according to Aristotle, do not result from deliberation. On the other hand, those actions that can be properly described as results of deliberation cannot always be traced back to some reasoning that the agent had actually engaged in before acting. We did not find it in the least puzzling that a practical argument may be applicable in the absence of such reasoning.[7] But we would like to know more about the conditions under which an action can be thought to result from deliberation. In particular, under what conditions is a practical argument with a major premise concerning eudaimonia applicable?[8]

(3) On Aristotle's view, deliberation is the work of practical intellect (dianoia). Hence, actions he would describe as results of deliberation - among others, all virtuous actions - are thought to involve some exercise of practical intellect. We are also told that the distinguishing feature of practical intellect, as opposed to theoretical, is that it "moves" [EN 1139a35-36; cf. also DA 433a17-18]. On the other hand practical arguments, as explicated thus far, serve to bring to light those concerns we think were responsible for the agent's acting in the way he did. Clearly, by producing a practical argument we do not display what the work of practical intellect

exactly consisted in. At most, we can conclude that being motivated by certain concerns appears to be constitutive of all excercises of practical intellect, whatever these exercises exactly consist in.

In what sense, however, can actions involve an exercise of practical intellect (dianoia, or in some of its uses, nous)? It is necessary to point out here the scope of Aristotle's dianoia. Not only reasoning, making inferences or calculating count as exercises of intellect, but also perceiving. [9] The importance of perception (aisthēsis) for virtuous conduct is particularly stressed in the Nicomachean Ethics.

Aristotle's remarks on perception, however, are somewhat cryptic. They are often coupled with the claim that practice is concerned with particulars (ta kat' hekasta) and with what is ultimate (to eschaton). And particulars, we are told, are the proper domain of perception.[10] But perception needed for practical wisdom is said to be different both from sense perception and from some exercises of perceptual capabilities in the theoretical sphere [see 1142a27–30].

These remarks may be related to the following general question. Aristotle assumes that a full practical argument giving an account of a virtuous action begins with a premise of the form: 'The good life is of the following sort...'. The underlying thought seems to be that a person acts virtuously only if he is guided in his action by a certain conception of the best way to live one's life. But what exactly is it to have a conception of this sort, and when is a person guided by such a conception? We have noticed already that a virtuous action need not be arrived at through any reasoning, or preceded by reasoning, which would provide a like between his actions and the practical ideal of the good life. Nor is the virtuous agent supposed to be able to make explicit the concerns which motivate him to act as he does. It suffices perhaps that the agent perceives the situation in which he finds himself as one which calls for an action of that type. Thus, to sufficiently characterize what motivated a person to go to a war, when this is a virtuous action, it may not be necessary to say more than that he saw a certain aspect of the situation as the most prominent one. He perceived for example, that the country is in danger and needs to be defended.[11]

When Aristotle is stressing the importance of aisthēsis for virtuous conduct, he may be emphasizing the connection between a person's conception of the life most worth living, a conception which certainly appears to be a rather general one, and particular concerns the person is pursuing in various circumstances of his life. But it would be quite misleading to say that recognition of particulars (ta kat' hekasta) is required for virtuous conduct in addition to recognition of universals (ta katholou).[12] On the one hand, the required "recognition of particulars" or perception, as understood above, consists in seeing particular situations as embodying some more general practical concerns. On the other hand, having the conception of virtues and of the life most worth living is not independent of acting on such concerns on particular occasions. It is these two interconnected claims that I wish to examine in more detail.

II Kata Pathos Zēn and Perception of Situations

(1) In the light of the considerations presented in I (3), understanding the connection between perception of situations, and the conception of the life most worth living, appears to be crucial for the correct grasp of Aristotle's notion of practical wisdom. But, some of the thoughts I found important to stress in I (3) seem to lead too easily to an interpretation which, although an attractive one, turns out upon closer examination to be singularly un-Aristotelian. The core of this view can be stated as follows. To have a conception of the good life is to act, situation by situation, on some practical concerns. A person's having practical concerns consists in his perceiving, occasion by occasion, certain aspects of situations as relevant (or prominent, or salient).[13]

The first question which is addressed here is this. How should we understand the claim that going to a war, as a virtuous act of a virtuous person, involves acting under some conception of the life most worth living? Obviously, a virtuous person's way of seeing particular situations need not involve the thought that acting as the relevant features of the situation require is a part of the pursuit of the good life. All that is needed, on the line of interpretation considered here, is that a virtuous person conceives of his acting thus and so as actiing well. In other words, the suggestion is that seeing the goodness of the action he is contemplating doing, or is about to do, is a part of the virtuous person's perception of the circumstances. If practical arguments were to be produced to account for his actions, we would probably detect considerable consistency in the pursuit of some concerns. We would characterize a virtuous person as somebody who can be relied upon to act in a certain way. But to characterize the "exercise of intellect" which his virtuous actions involve, according to this interpretation, it is sufficient to say that the virtuous person acts, occasion by occasion in a way which strikes him as good.

Let us recall now the way of living Aristotle wants to contrast with the life of a practically wise and virtuous person. Aristotle's formulas kata pathos zēn - usually translated as: to live in accordance with passion, or: as passion directs - and kata logon zēn - to live in accordance with reason, or as reason directs - repeatedly occur throughout the Nicomachean Ethics. These expressions convey a very general contrast between a way of life which involves practical wisdom to some considerable extent, and the way of life characterized by the absence of practical wisdom.[14]

Now, any human life involves both exercises of reason and the presence of passions, affection or emotions (pathē). A better way to characterize the way of living Aristotle intends by his formula kata pathos zēn may be as living at random.[15] To live at random, to a full extent, would be to act as it strikes one to be suitable on each particular occasion, with the proviso that insofar as the agent's own motivation is concerned, each of these occasions is independent from others. In other words, no concern which transcends a particular situation is responsible for his acting (or doing things) as he does. It may well be that a life of this sort could not justifiably be called human.[16] But this characterization may apply in a somewhat relaxed way: If the person's actions in a significant number of his life situations satisfy the above characterization of randomness, he may be said to be living kata pathos. Aristotle's own instances of the kata pathos way of living mostly[17] include young persons, and the many.[18] He considers cases of human lives which can be in some interesting way compared with the life embodying practical wisdom, and which are not anomalous enough to put in doubt the applicability of the concept of agency.[19]

To live kata pathos may involve living in this way: assailed by strong desires, occasion by occasion, a person acts to fulfil them without giving much thought to anything else. To give a full account of this sort of actions, one need not do more than cite the relevant desires.[20] But we may also invoke here the agent's way of perceiving the situations in question. On a number of occasions, certain features of situations strike the agent as salient, namely, those that he sees as likely to provide for the satisfaction of his desires. An agent who acts in this way is Aristotle's paradigmatic example of a person who lives "by passion" or "by perception",[21] that is to say, at random.

We may now formulate the problem which is faced by the interpretation I have previously called un-Aristotelian. A man stricken with paranoia, a person who acts compulsively, and a young person who, on Aristotle's account, does not yet have a developed conception of the good life, act not only in accordance with the way they perceive the circumstances, but may also exhibit a striking degree of coherence in their behaviour. Aristotle's favourite case is of the young people who act as if they were hedonists of some sort, but do not do so as a result of some reflective scheme of values they are committed to. In cases like this - and some of the "morbid" cases as well those, for instance, characterized by euphoria - the agent's mental state may well justify us in ascribing the thought to him: This way of acting is a good one, or: This is the thing to do! Neither the mere coherence we can detect in sombody's actions, nor the perceived goodness of particular actions on the part of the agent,[22] will serve to characterize what is distinctive about the lives of the practically wise.

The question we addressed was: when is a person guided in his actions by a conception of the life worth living? Or, what comes to the same thing: when is it justified to apply a practical argument which includes a premise of the form: 'The good life is such and such'? We will not succeed in singling out what is characteristic of a man who is effectively guided by such a conception, as a practically wise person is, if we insist that such a man acts in a way that strikes him occasion by occasion, as good. For, this will not help us at all to distinguish such a person from those who, on Aristotle's account, are lacking a conception of eudaimonia altogether.

(2) I have argued thus far that focusing just on these two features of virtuous conduct will not do: (a) the perceived goodness of action, and (b) some sort of coherence that we can detect among those aspects of situations which are perceived, occasion by occasion, as salient.

However, the line of interpretation I wish to reject assigns a more important role to perception of salient features of situations. Moral characters are thought to be ultimately distinguished by their irreducibly distinct ways of perceiving situations.[23] In other words, if there are morally relevant differences to be drawn between human characters, they stem from different ways in which situations are perceived. The starting point of this interpretation is the sound observation made several times above: a virtuous person need not be capable of making explicit the values that he can be justifiably seen as pursuing. But on this interpretation, a practical argument applies to a virtuous action of a virtuous person on account of his mental states prior to the action. Such mental states involve perceptual appreciation of the situation which, in turn, leads to action. A practical argument then supposedly spells out those practical concerns the awareness of which constitutes the mental states in question. It makes explicit those features of the situation which had struck the agent as relevant to, and sometimes as necessitating, his action. Although the mental states that help explain virtuous actions need not be characterized by explicit reasoning, they can nevertheless be understood as a kind of cognitive states.

The relevant perceptual capacity is understood here as a capacity to be impressed by certain features of situations. Further, differences in relevant mental states prior to or accompanying actions are thought not only to help explain the actions, but to constitute their sufficient explanation. In other words, if in the same circumstances two agents act differently, we ought to conclude that their perceptions of the situation had been different.

This view has some curious consequences. A person who is to a considerable extent virtuous may on an occasion be strongly tempted by some feature of a situation he is not usually tempted by and, consequently, he may act incontinently. On the view in question we are comitted to saying that on this occasion the person had perceived the circumstances in a substantially different way than he usually perceives circumstances of this sort. This will strike many as implausible. But the question I wish to raise is this: why should we think, in the first place, that there is such a perfect match between one's perception of circumstances and one's actions?

It seems to me that motivation for this view lies in the verificationist heritage of some recent moral philosophies. Invoking the agent's mental states to account for his actions does not appear to be a favourite verificationist strategy. But one ought to notice the use made here of the rather carefree talk about mental states. "Practical" mental states are imputed to a person only on the basis of an action he engaged in or abstained from. The question underlying this view seems to be this: what would count as establishing that a certain person in particular circumstances had some "practical" mental state? In other words, what would count as establishing that the person was motivated to engage in some course of action?[24] It may well be that unless the person acts in the way considered, other evidence may not help establish any facts about his motivation. However, the view in question implies not only that in the absence of an appropriate action it will not be possible to establish such facts, but also that in such circumstances there are no facts of the matter concerning the agent's motivation to engage in an action of that sort. How plausible this view seems to us depends on how plausible we find these underlying assumptions to be. But, more importantly, the picture this interpretation gives of a virtuous and practically wise person goes against exactly those features of virtue and practical wisdom which Aristotle was at pains to stress. It is to these that I will turn now.

III. An action "for its own sake" and the conception of oneself

Aristotle characterizes a virtuous action as one which is fine (kalon), and done for the sake of the fine [e.g. EN 1120a23–24]. He also says that virtuous actions are engaged in for their own sakes (di' auta) [EN 1105a32].

The thought that virtuous actions are engaged in for their own sakes is familiar. But, for no less familiar reasons, thinking that some such feature of one's motivation will single out those among actions which are virtuous will not do. As it stands, it will not even distinguish the actions which Aristotle would want to call chosen from the rest. Is not a young person who goes in for some pleasant thing or companionship engaging in his action for its own sake?[25] As we noticed above, acting with the thought that that very way of acting is good is not peculiar to virtuous actions. Nor is acting "for the goodness" of one's action. Yet, Aristotle's thought that a virtuous action is done for its own sake deserves further consideration.

In connection with this it is instructive to consider a passage in the Rhetoric. At 1366b20–22 Aristotle defines practical wisdom as follows: "Practical wisdom is that excellence of intellect that enables men to deliberate well about the relation to eudaimonia of the goods and evils that have been previously mentioned". When we look back at what "goods and evils" had been mentioned, it turns out that all sorts of things that can appear to agents in favourable light under some circumstances are counted among the "goods".[26] Moreover, the general discussion of goodness in the Rhetoric 6 and 7 lists indiscriminately all sorts of criteria on the basis of which something can be called good.[27] The purpose of the list, in the context of the Rhetoric, is to give us the grounds on the basis of which we can argue for the goodness or badness of something, or as Aristotle puts it, to give "the sources from which we must derive our means of persuasion (pisteis) about good and utility" [Rhet. 1363b3-4].

This passage gives us, so to speak, only the material for persuasion. Clearly, the criteria are not meant to apply all at once. The Rhetoric leaves it open under which conditions and with which qualifications each of the criteria is to be used. What is important is the role assigned to the practically wise and virtuous person with respect to the "goods and evils". No particular class of things or features, and no particular criterion is singled out as the one which has some independent worth, so that other goods or criteria of goodness should be measured against it. What the phronimos is seen as good at is judging each of the things or doings that may be called good, or the applicability of each of the criteria, with respect to eudaimonia. The multiplicity of goods - or if we want, goodness - is fully acknowledged.[28]

To judge the goodness of things "in relation to eudaimonia", I will suggest, is to judge them against a conception of oneself. This can give us a clue as to how engaging in an action "for its own sake" should be understood. The characterization of virtuous actions as those which are being engaged in "for their own sakes" faces at least two problems. The first is: Why should we not say that a young pleasure-seeker is engaging in his actions for their own sakes? Prima facie, such a person's actions are as good candidates for this class of actions as we can hope for. Is not a young pleasure-seeker exactly characterized as somebody who acts because he expects that his action will have a certain feature - that is, pleasantness - and not because he wants to achieve something through his action that goes beyond it, i.e. something else?[29] The second problem is the following: We have just observed that the phronimos engages in his virtuous actions as a result of his judgement of how things and actions lead to eudaimonia. If a virtuous action is engaged in with a view to attaining eudaimonia, what should we give to Aristotle's claim that it is done for its own sake?[30] We could argue that to engage in an action for its own sake does not preclude doing it also for the sake of something else. But, if "for its own sake" does not mean non-instrumentally, not with a view to attaining something else, it seems that the meaning of the phrase is eluding us.

However, Aristotle seems to have something specific in mind when he describes an action as being engaged in "for its own sake". The context against which this phrase is to be understood concerns the relation that holds in Aristotle's theory between human actions and human selves or persons. In the first place, the sort of actions one engages in is seen as partly determining the sort of person one is. Moreover, Aristotle's view is somewhat stronger. A person who did not engage in virtuous actions would not even stand a chance of being virtuous.[31] One cannot be virtuous without leading a certain sort of life, indeed a certain sort of an active life. On the other hand, actions are subject to something we may call the context principle. An action does not have a determinate moral value except through a contribution it makes to the moral worth of a person.

If we bear this context in mind, the relation between engaging in an action "for its own sake", and having the "final end" of eudaimonia in mind in one's action, may appear in a different light. Saying that one engaged in an action for the sake of that action itself implies that the agent conceived of some features of the action as in some way valuable or desirable. It also implies that he engaged in the action for the sake of the value attached to it, whatever this may be. But this is not all. To perform an action for its own sake means, for Aristotle, to perform it for the value attached to it when this evaluation is subject to the context principle. The desirable features of the action which motivate an agent to engage in it are seen by him against the background of the conception he has of himself and of the sort of life he want to lead.

This point can be put in a different way. We do not only act because we are attracted to or repelled from some features of situations we happen to be in. Sometimes, as it is said, we stand behind our actions as full-blown persons. The concept of standing behind one's action as a full-blown person would certainly require a further elucidation. One approach would be to say that in a full explanation of an action a recourse has to be made to the agent's conception of himself.[32] Given Aristotle's view on the relation between human actions and persons, it should not surprise us that engaging in an action "for its own sake" is understood in this way. A person is not engaging in his action for its own sake if we could explain his action appropriately by saying: He just felt like doing it. The idea behind all this is that when something is done di' auto this is not a matter of taste, but a matter of the conception that one has of oneself.[33]

On this interpretation, engaging in an action for its own sake is not only compatible with engaging in it "under some conception" of eudaimonia, but the former presupposes the latter. It will be noticed that engaging in an action for its own sake, as understood here, is not a distinctive feature of the virtuous actions. A wicked action may satisfy the description as well.[34] However, the actions of a young pleasure-seeker typically would not satisfy this description. This is not because the young, by focusing on the pleasantness of their actions, are missing some more "genuine" aspects of actions. Rather, the young in Aristotle's opinion do not have a developed conception of themselves and the sort of life they want to lead. They are not yet full-blown persons.[35]

IV. Deliberation as zētēsis

To act virtuously, one has to be a full-blown person. This involves having a conception of oneself and acting in such a way as to put this conception into effect in one's life. From my arguments above it follows that having such a practical conception cannot entirely consist in acting occasion by occasion on particular practical concerns, even if we could detect considerable consistency in one's pursuit of such concerns. For Aristotle is concerned most with the difference between acting on a deeply held and reflective scheme of values and acting on what we may call values in the thin sense. A young pleasure-seeker, although he does not have a developed scheme of values, is concerned with something in his actions, i.e. with the pleasantness of the actions, things or features of situations. It is concerns of this sort that we may call values in the thin sense.

Aristotle attempted to draw the line between thin and thick values by invoking the concept of deliberation and choice. If an action is chosen, it embodies one's values in the thick sense of the word. The problem we encountered above was this: a test for an action being chosen cannot consist in its being arrived at through reasoning. But, it does not seem that the lack of some such test for chosen actions renders the distinction useless and threatens to collapse all values into thin values.

However, we face a more serious problem in the attempt to give an account of thick values as chosen. The problem is that a certain obvious and attractive account we may be tempted to give here does not seem, upon reflection, to be appropriate. The answer runs along the following lines. Phronēsis can be understood as a sort of competence. Competence, we know, is usually understood as knowledge of some kind, possibly implicit knowledge, postulated to explain successful practice in some area. Now, in Aristotle's view, not only particular areas of human endeavour, but human conduct in general can be characterized as successful or unsuccessful: there are good and bad ways of living one's life. A person's success in this area is then thought to be related to practical wisdom in a way analogous to that in which a successful use of language, for example, is related to linguistic competence.

This view of the role of practical wisdom is not quite unappealing. The parallel draws its force from the following considerations. In the first place, if the agent's judgements as to what ought to be done in particular circumstances are reliably good, then it seems fair to say that, on Aristotle's view, it is practical wisdom that accounts for this. Secondly, practical wisdom is meant to be a complex set of dispositions ascribed to a person on the strength of some sort of success he continuously displays in his practice.[36] Finally, the parallel appears to make good sense of the claim that the man of practical wisdom is a standard of "practical truth" [see EN 1113a31-33, 1176a15-19]. Such a man is competent about those things which are most conducive to the achievement of the good life and, as such, he is a standard of correctness of choice as much as any competent person is a standard in the area of his competence.

On this view, an action can count as chosen if its explanation would appeal to some practical knowledge - understood here as knowledge of moral principles - which may well be implicit knowledge. However, this understanding of Aristotle's practical wisdom is in some important ways rather misleading. To see why, we ought to turn to Aristotle's own remarks on practical wisdom and deliberation.

Aristotle claims that practical wisdom should not be assimilated either to theoretical knowledge or to practical arts or skills.[37] Two of the reasons given for the first claim are the following. In the first place, practical wisdom is said to be of particulars, "of which there is no knowledge, but perception" [EN 1142a26-27]. Further, deliberation is described as some sort of search or inquiry (zētēsis), and denied the status of knowledge, since "men do not inquire about the things they know about" [EN 1142a34-42b1]. We may be tempted to dismiss Aristotle's characterization of excellence in deliberation as a kind of zētēsis because it seems to bring us back to the view that chosen actions are those which are arrived at through reasoning leading to a decision to act. But Aristotle's point here might be somewhat different.

Deliberation is understood by Aristotle as a kind of zētēsis,[38] contrasted with other kinds. He is concerned, in the first place, with a contrast between the sort of inquiry deliberation is and what he calls mathematical inquiries [EN 1112b21-23]. An example of a mathematical inquiry is an attempt to find a solution to a geometrical problem.[39] Perception is also needed to solve a geometrical problem, but this perception is said to be of a different sort than the one characteristic of practical wisdom [EN 1142a29-30].

The difference Aristotle has in mind may be of the following sort. Once a successful solution to a geometrical problem is found, it is possible to write down a proof, including the principles which justify each step of the proof. Let us think now of somebody who knows all the principles required for the proof when we confront him with a particular geometrical problem. Finding a solution may still require some ingenuity. For example, one has to "perceive" that the figure at hand can be analyzed into triangles and come up with a solution to the problem using some known principle concerning triangles. Now, there are areas of mathematical knowledge where a few principles suffice for the solutions to all the problems that one may encounter in that area. We may want to credit somebody capable of solving a large number of such problems with explicit or implicit knowledge of the theory, or a part of the theory, in this case, elementary geometry. This, I believe, is an example of the kind of knowledge Aristotle would call exact (akribes).

Things stand differently, in Aristotle's opinion, with respect to virtue and practical wisdom.[40] There is no area of human virtuous conduct that can be captured by some list of principles which would suffice to give guidance for practical conduct in that area. One cannot single out a particular virtue, say justice, and provide principles which would suffice - provided that the agent had knowledge of them and were capable of applying them - for just conduct.[41] It is here that the discussion of the so-called hard cases becomes important.

The hard cases are exactly those where a virtuous person's perception of the salient features of situations does not supply him immediately with an answer as to which course of action to take. It may also happen that the principle - or, in other words the values - he is committed to suggest an answer which strikes him as wrong. The problem here, according to Aristotle, need not concern application of the principles. The principles that may justifiably be imputed to the agent at the time may simply not suffice to give a satisfactory answer to the problem, or may actually suggest an answer that appears to him to be wrong. In a case of this sort, comparing the principles of a virtuous person with the law, Aristotle notes: "the error is not in the law nor in the legislator but in the nature of the thing, since the matter of practical affairs (hē tōn praktōn hylē) is of this kind from the start." [EN 1137b17-19].

The suggestion here is that the very existence of hard cases, and what is necessary for us to be able to deal satisfactorily with them, is not a matter of marginal significance, but is due to certain pervasive features of moral conduct. One of these features, we may suggest, is an incomplete or open-ended character of moral values. In some circumstances a virtuous person's ability to act - upon being impressed by some features of situations - in a way which appears satisfactory to him, gives out. In dealing with circumstances which are thus recognized as hard cases, a virtuous person may enrich or modify the values he had been following thus far. The search which dealing with hard cases involves need not be a search into the depths of one's soul to find something that was already there.

Admittedly, it is with respect to the hard cases that the parallel between deliberation and working towards solution of mathematical problems sounds most appropriate. It is also true that dealing with a hard case, as well as working on a mathematical problem, often involves making one's values, or mathematical principles, explicit enough to bring them to bear to the case at hand. But, it is not this explicitness of reasoning that gives a point to the description of either of the two processes as kinds of zētēsis. Coming to see what will constitute a satisfactory solution to the problem often requires, in both cases, imagination and experience. Mental processes involved in coming to a solution may not resemble much going through the lines of an argument presented as a justification of the solution.

However, it is the differences, and not the similarities of the two cases that Aristotle is at pains to stress. It is useful to consult here again the paragraph in EN V where Aristotle compares justice with equity. He points out here that "about some things it is impossible to lay down a law, so that a decree (psēphisma) is needed", for "when the thing is indefinite the rule also is indefinite" [EN 1137b27-30]. It may help us here to reflect on what this legal metaphor suggests about the nature of deliberation. To solve a mathematical problem, we feel, a decree is not needed. But, why is this so?

A decree gives guidelines as to how one should deal with a particular problem not covered by law, or not yet covered by law. In solving a mathematical problem, it seems, something similar may happen: a general principle may not be of much help in dealing with a particular case. It is true that, unless a more specific principle is brought to bear on the problem, we may not be able to find a solution. But in the mathematical case, a specific principle is in some sense dispensable. It is dispensable insofar as it follows from a more general one. In saying that decrees are indispensable in practical affairs, Aristotle seems to imply that the principles governing virtues do not form a set of deductively related principles.[42]

Decrees, rather than being deducible from the principles that make for the law, give further substance to these principles. It is this feature of the legal metaphor of psēphisma that is meant to be comparable to the case of virtuous conduct.[43] Virtuous conduct brings along progressive refinement and modification of values. In this respect the hard cases are not special cases. With regard to the hard cases, the open-ended character of values may just be more visible than otherwise.

Aristotle's claim that practical wisdom is not a skill or art (technē), may be related to another feature of virtuous conduct, which we have already discussed above. Aristotle says that practical wisdom cannot be a skill because "action and making are different kinds of thing" [EN 1140b3-4]. 1139b1-4 offers a gloss on this: "[...] everyone who makes makes for an end, and that which is made is not an end in the unqualified sense (but only relative to something, i.e. of something) - only that which is done is that: for acting well (eupraksia) is an end, and desire aims at this."

Our remarks from III may be of interest here. Engaging in an action for its own sake seems to be at least a part of what acting for the sake of eupraksia means.[44] Exercising one's practical wisdom in one's conduct, or making choices, involves acting from a certain conception of oneself.[45] Having in mind the close connection between having such a conception and acting on particular concerns, and the conclusions we have drawn from characterization of deliberation as a kind of zētēsis, we may conclude that in the course of virtuous conduct the conception which the virtuous person has of himself becomes modified and enriched as well.[46]

V. Persuasion

The discussion thus far suggests that, for Aristotle, it is only against the background of some conception of the self or person and of the sort of life that is worth living that the question of what is right or wrong in human conduct, or what ought or ought not to be done, can be significantly posed and answered. A question arises how effective, on this view, moral arguments can be.[47]

The conception a person has of himself is seen above as including values he is committed to and which he pursues in his actions. But if something being right or wrong is systematically dependent on such values, it does not seem that those who do not share a particular moral outlook will be responsive to arguments based on it. In particular, on the line of interpretation discussed in II, having a practical conception of oneself, being guided by it and being a certain sort of person coincide. All are analyzed in terms of a distinctive way in which situations related to one's actions are preceived. If, however, different moral characters have irreducibly distinct perceptions of circumstances, the effectiveness of moral arguments seems to be quite seriously threatened.[48]

Indeed, Aristotle points out repeatedly that in some cases moral arguments have little impact on those to whom they are addressed. In the first place, argument is said not to be very powerful with young people.[49] It is also not very powerful with those who are thoroughly wicked [cf. EN 1151a14].

Here, the context of the Rhetoric becomes important once again. The question as to how effective moral arguments are is the question as to what it takes to persuade somebody into something - or more appropriately to persuade him into something else (metapeisai) - e.g. that something is good, or ought to be done, which he did not previously consider to be such. Thinking now about the grounds for persuasion given in the Rhetoric, it seems puzzling why those who live kata pathos should be less susceptible to arguments than others. Living kata pathos, as explicated above, involves a lack of a full-fledged conception of oneself. Should it not be easier to persuade somebody who is in this condition to accept that some course of action is good since it is, say, desired by the majority, or is contrary to that which his enemies desire, or on some other from among the "grounds of persuasion"? Presumably, having a full-fledged conception of oneself imposes some restrictions on what can be held good in which circumstances. Those who do not have such a conception, it seems, should be more liable to persuasion, including persuasion by sophistic arguments.

It is noteworthy that Aristotle sometimes refers to choice (proairesis) as conviction, or a state of one's being persuaded into something (to pepeisthai). In the exposition of a sophistic argument at EN 1146a31-46b2,[50] and the answer to it at 1151a10ff., the morally weak person is said to pursue bodily pleasures not on conviction (mē dia to pepeisthai), while the self-indulgent man "is convinced (pepeistai) because he is the sort of man to pursue them [sc. such pleasures]". We know that, on Aristotle's view, it is his own choice that a morally weak person acts against. Clearly, to successfully persuade somebody into something else (metapeisai), is construed here rather strongly. It seems to require that the persuaded party considers adopting the course of action he is being persuaded into as a part of his own choice. Moreover, something even stronger is implied: a morally weak person already has a choice that would require him to act in the way he is now being persuaded to act. Hence, to successfully persuade him into something else (metapeisai) is to succeed in making him act on the basis of his choice.

To appreciate properly the force of Aristotle's warnings about the ineffectiveness of "argument and teaching" [e.g. at 1179b23-24], one has to consider what in each case counts as success in persuasion. When he says that those who live kata pathos will neither hear nor understand the arguments which are meant to "turn them away" from something, and will not be easily persuaded into something else (metapeisai - 1179b26-28), we ought to understand the force of metapeisai to be as strong as I have outlined above (i.e. as strong as in 1151a10ff.). Those who live at random, we shall recall, are those who have not organized their lives around some scheme of values. To persuade them to turn away from something in the relevant sense of metapeisai, is to persuade them into avoiding to do things they do, and which they may have a strong inclination to do, on the basis of their own choice. But again, having a choice presupposes having a stronger conception of oneself than the one who lives kata pathos presumably has. It ought not to be surprising that Aristotle suggests, in the passage discussed, that persuasion of this sort will be quite difficult to achieve.[51]

Yet, if arguments are ineffective on this strong conception of success in persuasion, this does not imply that there may not be successful persuasion of a less ambitious sort. Aristotle's recognition that moral education may be effective can be understood as recognition of success of these less ambitious varieties of persuasion. In particular, we know that persuading youth to do what is "right" on the strength of authority, or by imparting a sense of shame at what is done, are thought by Aristotle to be effective tools of moral education.

It is important to recognize that a weaker conception of what constitutes success in persuasion - or effectiveness of moral argument, if it is through argument that persuasion takes place - is at work here. Yet it ought to be noticed that if our aim is to persuade somebody into virtuous conduct, then our aim is to persuade him, in the long run, in the strong sense of the word. Hence persuasion which succeeds in a limited sense must somehow be related to persuasion which succeeds in the strong sense. An indication as to what is required for this connection to be established can be found at EN 1095b6-13. There it is suggested that following a piece of advice that something ought to be done (to hoti) can lead, in due time, to virtuous conduct (i.e. conduct characterized by appreciation of the proper grounds for acting so, to dioti), only if the advice is, in Hesiod's words, "laid to heart" by a person who takes it.[52] The passage can be understood to imply two different ways in which one can act by following the advice of an authority.[53] If acting on such advice is to have a role in moral education, a person must be brought to care about those grounds which the authority employs in persuading him to take the course of action in question. Perhaps he may not ultimately act on the grounds offered.[54] But what is crucial here is the ability to care about grounds presented to one, whether it is for this reason that one ultimately acts or not.

The possibility of taking advice in this way may be an important element in understanding how, on Aristotle's view, acting on advice of an authority may, in due course, lead to autonomous moral choice.[55] The question I am primarily interested in here is not about moral education, but about the effectiveness of moral arguments in general. But Aristotle's recognition of the possibility of moral education through persuasion which succeeds only in a limited sense suggests that he would not take acceptance of a moral outlook to be a precondition for the effectiveness of moral arguments. One does not have to be committed to the values underlying an argument to be successfully persuaded to follow a course of action. However, just being willing to lend one's ear to the other side is not enough. If an attempt to persuade somebody to take a virtuous course is to succeed even in a limited sense, the person in question must be able to take a vivid interest in the way in which certain considerations, which perhaps typically do not motivate him, motivate other people.

*

* *

Aristotle's views on practical wisdom - or, as we may call it, practical rationality - were seen above to be closely linked to his views on human selves or persons. I shall conclude with several remarks on Aristotle's conception of practical rationality which are strongly suggested by, but may well go beyond, my discussion above.

(1) For Aristotle, being practically rational implies having an ability to make and revise a conception of the sort of person one wants to be. With regard to the kind of conception a practically rational person is supposed to have of himself, and the kinds of revisions he is expected to make in his conception, this requirement may strike us as being a fairly strong one.

In the first place, Aristotle does not seem to envisage the possibility of the practically wise person making radical revisions in his conception of himself. A practically rational, full-blown person is taken to be steadily committed, over his life-time or a considerable part of his life-time, to a certain scheme of values. Practical rationality calls for progressive refinement of these values, but does not seem to allow for radical changes of values. Secondly, one would think that a person's wish to be impulsive, to be immersed in the present, to do things as he likes on a given occasion, may be a part of his reflective conception of himself. Aristotle does not discuss this sort of case, but it appears that he is committed to saying that a person with this kind of reflective attitude has failed to make himself into a full-blown person, and hence cannot be considered practically rational.[56] In these two respects at lease, Aristotle's conception of the kind of reflective attitude towards oneself which a person has to take to be practically rational seems to be quite strong.

(2) One's conception of the sort of person one wants to be is a practical conception. This seems to mean for Aristotle, among other things, that it is concerned with the sort of life one wants to lead, the kinds of activities one finds worthy to engage in. A person's practical conception of himself is partly constituted by those concerns he is responsive to in his actions. But, Aristotle does not seem to imply that only those concerns that are in fact acted upon form a part of the practical conception of oneself. There is room for genuinely practical concerns which, for a variety of reasons, do not get acted upon.[57]

If this is so, we should revise our initial characterization of practical arguments. They do have a role to play in explanation of actions, but they should not be construed merely as devices for explanation or justification of actions. They may give an account of deliberation which does not, for whatever reason, lead to action.

(3) Related to this is a further important feature of Aristotle's view of practical rationality. It can be derived in an interesting way from his views on moral strength and continence. The continent man, we shall recall, is the one who consistently pursues the course of action he considers the best in spite of his strong desires to act otherwise. The virtuous person is free from this sort of internal conflict. It is often noticed that this conception of virtue may be unusual from our point of view. Indeed, we are not inclined to think that one among possible dangers that the plea et ne nos inducas in tentationem is trying to avert is the loss of full virtue that could result from a mere presence of temptation. But, the implications for Aristotle's views of practical rationality are much more interesting. We know that, according to Aristotle, a person who is merely continent is not fully practically rational. This may seem quite puzzling. We might think that a judgement as to whether a person is practically rational or not is based on reasons he has for his actions and the structure of these reasons. But, if this is so, the continent man's plight should not affect our judgement of this practical rationality: however strong his temptations are, they do not enter into explanation of his actions.

The view that the continent person is not fully practically rational may seem less puzzling if we notice two requirements that enter into Aristotle's conception of practical rationality. Being practically rational requires, in the first place, an ability to make and revise a conception of the sort of person one wants to be. But it also requires an ability to be the sort of person one wants to be.[58] The two may fall apart in a variety of ways of which the so-called moral weakness is but one kind. Less obviously, Aristotle also thought that being strong enough to act in accordance with one's considered judgement may not in all circumstances amount to being the sort of person one wants to be.

NOTES

1 The object of choice (*proairesis*) being "what has been decided upon as a result of deliberation" [EN 1113a4-5].

2 For a formulation and discussion of this problem, see Cooper: 5-10.

3 EN 1144a 31-3: οἱ γὰρ συλλογισμοὶ τῶν πρακτῶν ἀρχὴν ἔχοντεσ εἰσιν, ἐπειδὴ τοιόνδε τὸ τέλοσ καὶ τὸ ἄριστον[...]. I prefer the reading according to which τῶν πρακτῶν goes with αρχήν rather than with οι συλλογισμοί. Ross's translation, however, reads: "for inferences which deal with acts to be done are things which involve a starting point, viz. 'since the end, i.e. what is best, is of such and such a nature'" [CWA, p. 1807]. The sentence is difficult and this translation may be problematic in many respects. τὸ τέλοσ καὶ τὸ ἄριστον is traditionally understood as referring to the final end of *eudaimonia*. Hardie, however, thinks that the end here may be of a less general sort: "The end in view is not the ultimate human good but the kind of end a soldier has in view when he sets himself to behave well in a battle..." [Hardie, p. 252]. I do not find enough evidence for this view and, hence, I follow the traditional understanding. Since a *full* account of a virtuous action involves a conception of *eudaimonia*, the complications I discuss here have to be faced, in some way, even on Hardie's view.

4 Similarly, it does not seem any more plausible to impute to a *phronimos* implicit reasoning of this sort.

5 I propose this as a sketch of a paradigmatic instance of a practical argument. The variety of forms of practical arguments Aristotle offers at various places is notorious.

6 In other words, a piece of practical reasoning gives an adequate account of the action just in case the corresponding practical argument does. This leaves it open as to what the state of the *phronimos*'s mind was at any time before his action. The relevant processes may involve reasoning, whether explicit or implicit, direct appreciation of the circumstances, or whatever.

7 It would indeed be surprising if it were otherwise. If a practical argument were not applicable in such circumstances, this would imply that no action about which the agent had not reasoned before engaging in it could be explained in terms of his beliefs, desires or values.

8 The query also remains as to why Aristotle calls some of the antecedent conditions that lead to actions bouleusis and classifies this as a sort of search or inquiry (zētēsis)? The terms bouleusis and bouleuesthai (as well as logizesthai with which bouleuesthai is explicitly identified at 1139a13–14) seem to suggest that some sort of reasoning, or something like reasoning, is involved. The use of zētēsis carries an implication again that deliberation is a process which takes time. But this query could only be addressed after some attempts are made towards answering the general question raised here.

9 I.e. exercises of aisthēsis. All of these are exercises of faculties which are kritika that is, concerned with making distinctions [see DA 427a19–21, MA 700b19–22]. One should probably add: and exercises of phantasia, in particular bouletikē phantasia [see DA 434a5–10], but I will not discuss this topic here.

10 See esp. EN 1141b14–16; 1142a25–27; 1147a25–26.

11 This line of thought, in some form or other, is much emphasized in recent interpretations of Aristotle. See particularly, Wiggins 1980, McDowell 1979.

12 As Aristotle seems to be doing at EN 1141b14–16.e

13 I will consider in what follows a certain general line of interpretation without paying much attention to the details of any particular view. Some elements of this view can be found in Wiggins. McDowell comes closest to the view I will be concerned with.

14 The contrast is best understood, I believe, as being between two ways of living broadly understood, than as being between two well-defined kinds of lives. In a sense, only an ideal phronimos may be said to live in accordance with the "right" reason [kata ton orthon logon, see EN 1144b21–24], or rather to live in a way which involves the presence of the "right" reason [meta tou orthou logou 1144b26–28]. But broadly understood, anybody who can be credited with practical wisdom to a high degree and in relevant respects (so perhaps even an enkratic person), can be said to live kata logon. The other way of living, kata pathos may also be understood broadly, as will transpire below. But, I take Aristotle's concept of practical wisdom to be such as to justify us in saying that a life poorly organized may be said to lack practical wisdom. (Hence the asymmetry one can notice in the sentence above.)

15 This phrase occurs, although in a different context, at 1124b5-6: "For the proud man despises justly (for he has true opinions), but the many do so at random (tuchontos)".

16 It probably would not satisfy Aristotle's requirements for human agency.

17 See however fn. 21 below for special cases.

18 See EN 1095a3-8, 1128b16-18, 1156a31-33, 1179b10-16, 1179b23-29.

19 Recall that very young children and animals can be said, according to Aristotle, to do things voluntarily, but their doings do not qualify as chosen actions. Lives of humans Aristotle would describe as brutish or morbid way provide other instances of kata pathos zēn. Clearly, if Aristotle's aim is to give an informative account of the life involving practical wisdom, the examples which involve normal human agency in an umproblematic way will be more relevant. (Aristotle's remark at 1141a26-28 is noteworthy: even some of the lower animals are sometimes called phronima, insofar as they have dunamin pronoētikēv concerning their own life. Aristotle clearly does not think that a human life which involves practical wisdom can be profitably discussed in comparison with animal behaviour. But the remark itself indicates an important aspect of the concept of practical wisdom.)

20 It is important to notice that this way of acting need not be described as acting without reasons, but is not a result of deliberation in Aristotle's sense. However, if an agent has reasons for acting, we may expect that a practical argument giving an account of his action would be applicable. Thus, if we think that on Aristotle's view a practical argument applies just in case an action is a result of deliberation, we should be careful to narrow down the concept of reasons for action. On this view, "Fulfilling this desire here and now is good" could not serve as a premise of a practical argument, nor does an expected fulfillment of a desire alone constitute a reason for acting. If, however, one allows for practical arguments of this sort and corresponding reasons, the presence of deliberation will not overlap with applicability of practical arguments. It does not seem to me that the first line - narrowing down the concept of reasons for action - is particularly promising. But in either case, drawing a line between actions which result from deliberation and those which do not is crucial.

21 It is instructive to look at EN 1149a4-12. If my interpretation of the ways of living which can be called kata pathos is correct, living by perception, tē aisthēsei zēn, at 1149a10 is a phrase equivalent to "living in accordance with" or "by passion" (pathei zēn occurs, for instance, at EN 1128b17-18). It is not essential to these ways of living that one pursues pleasure, or satisfies appetites. It appears that Aristotle thought most young people do exactly that. But living kata

pathos may be of a quite different sort. 1149a9–12 discusses those among the foolish people who are by nature thoughtless (or lack reason, alogistoi) and live by perception alone – these are said to be either brutish or morbid. An example of the recurrent episodes of this sort is of "the man who feared a weasel" and did so "as a result of disease" (a8–9). A life characterized by compulsive behaviour or phobias can hardly be understood as a life devoted to the pursuit of pleasure. It is important to stress that Aristotle is not assimilating various other ways of living or life episodes that lack practical wisdom to those that can be called "morbid" or "brutish". Rather he is at pains to demarcate states he calls human, e.g. human wickedness, self-indulgence or incontinence (see 1149a16–20), that is, those states that correspond to normal human desires and needs. My point here is that the cases of morbid or brutish ways of acting also belong to the broad category of kata pathos or tē aisthēsei zēn.

22 Cf. EN 1176b21–23: "for boys, too, think that the things that are valued among themselves are the best".

23 Moral characters are understood broadly: the akratic and enkratic person, as well as the virtuous and wicked person are among the varieties of moral characters.

24 Perceiving some features of situations as salient, where this is a "practical" mental state, was understood above as equivalent to recognizing some practical concerns and being motivated to act on them.

25 Or, at least, unless some further content is given to the phrase "for its own sake", it is unclear why this should not be so.

26 I am assuming here that the most plausible antecedent of ἀγαθῶν καὶ κακῶν τῶν εἰρημένων are the goods and relative goods (and their opposites) discussed in the Rhetoric, 6 and 7. It could be that the reference extends back to I, 5 where eudaimonia is discussed. But, the extensive list of things which are good, or better than others, and the criteria of goodness from I,6 and 7 come more readily to mind. If, however, reference is to I,5, the goods referred to belong again to the same comprehensive class: τά τ'ἐν αὑτῷ καὶ τὰ ἐκτὸσ ἀγαθά οὐ γὰρ ἔστιν ἄλλα παρὰ ταῦτα [1360b25–26].

27 For instance, that which is contrary to what our enemies desire, that which is sought after by everybody, that which is man's very own, that which is chosen by those we admire, that which we in fact desire....

28 This can be related to the distinction between the apparent goods and the goods in themselves. The point of the distinction is not that there is something that can be considered good on independent grounds – be it pleasure, or the class of the so-called external goods – while other

things are merely apparently good, or good to the extent to which they are instrumental in bringing about the goods-in-themselves. Everything that is good is in a sense an "apparent" good: what the virtuous find good, or what appears good to them, may be called good-in-itself. We may note here that since there is nothing that could be accorded a value on some independent grounds, there is nothing for a virtuous agent to maximize. It does not seem that maximization schemes could play any role in Aristotle's understanding of virtue and the human good.

29 One may want to insist that an action of this sort is not done for its own sake but instrumental in bringing about some mental state (i.e. pleasure). But this is streching the sense of "instrumental" a bit too far. It amount to abandoning the already tenous hope that the distinction for its own sake/for the sake of something else, will be of any use. In any case, acting for the sake of το ηδυ is best understood as acting for the sake of the pleasantness, either of actions or of objects of one's actions, rather than acting for the sake of pleasure.

30 We may also notice that at 1151b19-22 a virtuous action is described as being done di' hedonēn, and that this description - acting for the sake of pleasure, or pleasantness of something, but a noble pleasure - is implicitly extended to other virtuous actions. Hence, we need an account of engaging in acts di' auta which is compatible with these remarks.

31 Cf. 1105b9-12. Aristotle says here that "without doing these [just and temperate acts] no one would have even a prospect of becoming good". That the thesis holds for Aristotle in the form I have presented it can be seen only after we come to realize how closely, in his view, one's being good is related to one's becoming good. As I will point out below, being good involves making oneself continuously into a good person.

32 Recall the practical arguments starting from the premise concerning the nature of eudaimonia.

33 A judgement as to what sort of person one is will depend to a great extent on the actions we assume are performed for their own sakes. Compare here EE 1227b11-12: "For the man who is fond of sweet things, or the one fond of bitter ones, is not said to be a certain sort of person in regard to character". But, this does not exclude the possibility that an agent may engage in his acts di' auta although we would think that acting on the grounds he has should just be a matter of taste. I think that snobbery gives an example of this.

34 Engaging in an action for its own sake is only a part of Aristotle's characterization of doing something fine for the sake of the fine. See the Rhetoric definition of the fine or the noble: "The noble is that which being desirable for its own sake is also worthy of praise"

[1366a33-34]. The concept of the fine is clearly richer. It corresponds to the objects of a good choice.

35 It is unclear to me whether Aristotle's young pleasure-seeker is a fair representative of youth. But the above remark about youth seems sound, no matter what exactly it is that youth pursue. As for the many, their situation ought to be somewhat different. But the following remark gives evidence that Aristotle thought they often do not have a firm conception of themselves and the life worth living: "[The many] differ from one another - and often the same man [identifies eudaimonia] with different things, with health when he is ill, with wealth when he is poor" [EN 1095a23-25].

36 Moreover, Aristotle seems to be impressed here by the possibility of making the right choices about what to do or not to do in ever varying circumstances of one's life. Here, the parallel with the linguistic an case may appear to be quite striking.

37 See 1140b1-2: [...] οὐκ ἂν εἴη ἡ φρόνησισ ἐπιστήμη οὐδὲ τέχνη [...]. In what follows, I shall try to explicate only some of the thoughts that Aristotle might have had in mind when making these two claims.

38 I.e. zētēsis is the genus of deliberation. See 1142a31-32.

39 It seems unlikely that Greenwood's and Ross's understanding of 1142a26-27 is correct. The text reads [...] αἴσθησισ, οὐχ ἡ τῶν ἰδίων, ἀλλ' οἵᾳ αἰσθανόμεθα ὅτι τὸ [ἐν τοῖσ μαθηματικοῖσ] ἔσχατον τρίγωνον. When read in conjunction with 1112b21-23, there is no doubt that the talk is here about perceiving that the "ultimate" figure required for a solution to a geometrical problem is a triangle, rathar than about the mere recognition that a particular shape in front of us represents a triangle. EN 1112b21-23 talks about mathematical inquiries as a kind of zētēsis contrasted with deliberation. It is significant that the passage immediately following the one from which we quoted here, takes up again the topic of deliberation as zētēsis. (It seems very likely that these two passages should be read together - as noted by Stewart, P. 76-77. Moreover, it seems to me that the passage preceding both of them, 1142a11-20, addresses essentially the same problem: the difference between mathematical knowledge and the ability to solve mathematical problems, and practical wisdom.) The reading preferred here is supported by Burnet, p. 274.

40 Throughout the Nicomachean Ethics one finds warnings that the domain of practical wisdom does not admit of "exact" analysis, that things happen here "for the most part" and are "indeterminate" [e.g. 1112b8-9, 1094b12-27].

41 One aspect of what I am describing here has to do with the so-called unity of virtue. But I will not specifically discuss this topic here.

42 This points perhaps to even deeper differences between moral deliberation and solving mathematical problems. The concept of a "solution", in fact, may be quite different in the two cases.

43 There are all sorts of differences, of course, between the two areas. For example, not only are legal principles made explicit, but acting in accordance with them or against them depends to a great extent on their being made explicit.

44 Acting for the sake of eupraxia may well involve more than this.

45 This feature will hence be crucial in distinguishing deliberation in thc poetical sphere (i.e. the sphere of arts or skills) from strictly practical deliberation (deliberation required for, e.g., virtuous conduct). (That there is room for deliberation in arts, see EN 1112b2-8.)

46 The claim above that being good involves making oneself continuously into a good person (fn. 31) was meant to capture, in the first place, this particular feature of virtuous conduct. The second component will be mentioned below.

47 I am interested in this question here only to the extent to which it relates to the issues already discussed.

48 It would be necessary for somebody who wanted to argue in favour of the ineffectiveness of moral argument to spell out more clearly how exactly "irreducibly distinct" perceptions and "systematic dependence" on moral outlook are to be taken. I will satisfy myself here with outlining Aristotle's view in these matters relying on a sufficient understanding of these vague expressions.

49 See EN 1179b26-28: "For one who lives in accordance with passion (ho kata pathos zon) will not hear argument that attempts to turn him away from something (logou apotrepontos), nor understand it if he does; and how can we persuade one in such a state to change his ways (metapeisai)?". As for youth cf. also EN 1142a11-20 and 1095a2-6.

50 The argument is to the effect that it is easier to persuade the self-indulgent man, i.e the one who pursues such pleasures as a matter of choice, to change his ways, than to persuade a person who pursues them through moral weakness.

51 Or, similarly, in EN 1095a5 that listening to lectures concerning practical affairs will for a young person be "vain and unprofitable".

52 A good discussion of this passage, on which I rely for the most part, can be found in Burnyeat, "Aristotle on Learning to be Good", pp.71–73.

53 The context of the passage allows us to understand that those who do not "lay to heart what another says" are either not following advice at all (i.e. neither to hoti, nor to dioti), or are acting in accordance with advice but not laying to heart the grounds on which it is offered. In either case, the important contrast is with those who "lay to heart what another says", when this includes both features of advice. I have concentrated on the two ways in which advice may actually be followed in action, in order to bring out the most important aspect of the contrast more clearly.

54 Even in the best possible case, a young person will not have a fully-blown conception of himself which is needed, in Aristotle's opinion, for the action to be properly grounded.

55 I.e. how one's being persuaded by somebody (with limited success) can lead to a state of one's being persuaded into something (to pepeisthai), i.e. to one's own conviction or choice.

56 One way to defend this would be to say that insofar as the person's reflective attitude towards his desires, aspirations or whims, consists in wishing that they would have their way, his conception of himself is not doing any real work with respect to his life. Yet, one could reasonably object that this sort of life ought not to be assimilated to the kata pathos way of living.

57 Here I propose to take seriously Aristotle's repeated observation that the morally weak person acts against his choice, that his choice may be a good one, i.e. the same as the choice the virtuous person would make in the same circumstances. Aristotle recognizes, against Socrates, the possibility of genuine moral weakness. He also acknowledges that there may be persons who are consistently weak in some respects. The view I have been considering above would force us to say – against Aristotle's considered judgement – that there is no specific choice (e.g. not to commit adultery) against which a person who is consistently weak acts. (Among other places, DA 433a1–3 strongly suggests that a failure to act may be accompanied by genuinely practical mental states.) It seems that an adequate account of Aristotle's views on moral weaknes, whatever exactly it will look like, has to accommodate this central claim.

58 Discussion of the second requirement falls far beyond the scope of this paper. But I may point to a second component which was meant to be included in my claim above that being virtuous implies making oneself continuously into a good person. Without making any attempt to analyze it, the second component may be stated thus: being virtuous involves

living in such a way that both one's praxeis and one's pathē come to reflect the progressively refined conception of the sort of person one wants to be.

REFERENCES

J. Burnet, The Ethics of Aristotle, Methuen and Co., London 1900.

M.F. Burnyeat, "Aristotle on Learning to be Good", in: Essays on Aristotle's Ethics, ed. by A.O. Rorty, U. of California Press, Berkeley, etc. 1980.

J.M. Cooper, Reason and Human Good in Aristotle, Harvard U.P., Cambridge, etc. 1975.

L.H.G. Greenwood, Aristotle's Nicomachean Ethics, Book Six, Cambridge U.P., Cambridge 1909.

W.F.R. Hardie, Aristotle's Ethical Theory, Clarendon Press, Oxford 1968.

J. McDowell, "Virtue and Reason", Monist, Vol 62, No. 3, July 1979

J.A. Stewart, Notes on the Nicomachean Ethics of Aristotle, v. II, Clarendon Press, Oxford 1892.

D. Wiggins, "Deliberation and Practical Reason", in: Essays on Aristotle's Ethics.

CWA: The Complete Works of Aristotle, vol. II, ed. by J. Barnes, Princeton U.P., Princeton 1984.

EN: References to the Greek text of EN are to Bywater's edition in Oxford Classical Texts.

A SELECT BIBLIOGRAPHY OF YUGOSLAV ANALYTIC PHILOSOPHY

Arsenijević, Miloš. "Kantova teorijska filozofija" (Kant's Theoretical Philosophy), Filozofske studije, vol.6, Beograd, 1976.

Arsenijević, Miloš. "Dodirivanje" (On Touching). Filozofske studije, vol.10, Beograd, 1978.

Arsenijević, Miloš. "Jedna aporija kosmičke beskonačnosti" (An Aporia concerning Cosmic Infinity), Gledišta, no.6, Beograd, 1980.

Arsenijević, Miloš and Laban, M. "Kockarski paradoks", (Gambler's Paradox), Dometi, no.5-6, Rijeka, 1983.

Arsenijević, Miloš. Prostor, vreme Zenon (Space, Time, Zenon), Beograd and Zagreb, 1986.

Arsenijević, Miloš. "Za i protiv smrtne kazne" (For and Against the Capital Punishment), Sociološki pregled, no.1-2, Beograd, 1987.

Arsenijević, Miloš. "Einsteinov 'paradoks dvojčikov'" (Einstein's Paradox of Twins). Dialogi, Vol.23, Maribor, 1987.

Babić, Jovan. "Jedan pokušaj izvodjenja normativnih iskaza iz indikativnih" (An Attempt to Infer Normative Propositions from the Indicative Ones), Gledišta, no. 9, Beograd, 1976.

Babić, Jovan. "Jedan pokušaj svodjenja normativnih iskaza na indikativne" (An Attempt to Reduce the Normative Propositions to the Indicative Ones), Theoria, no.3-4, Beograd, 1976.

Babić, Jovan. "Searlova teorija govornih akata i 'jeste-treba' pitanje", (Searle's Speech Act Theory and the Ought-Is Question) Filozofske studije , vol.9, Beograd, 1977.

Babić, Jovan "Marksistička etika i relacija sredstvo-cilj" (Marxist Ethics and the Relation of Means to an End), Ideje, no.3-4, Beograd, 1979.

Babić, Jovan. "Istina i laž" (Truth and Lying), Gledišta, no.7-10, Beograd, 1984.

Babić, Jovan. "Obećanje" (Promise). Gledišta, no. 3-4, Beograd, 1985

Babić, Jovan. "Utilitarizam i sloboda: Ogled o J.S. Millu" (Utilitarianism and Liberty: An Essay on J.S. Mill) . Dometi, no.10, Rijeka, 1985.

Babić, Jovan. "Etički problem rata" (The Ethical Problems of War), Gledišta, no.9–10, Beograd, 1985.

Babić, Jovan. Kant i Scheler: Fenomenološka kritika Kantovog formalizma u etici (Kant and Scheler: A Phenomenological Critique of Kant's Formalism in Ethics, Beograd, 1986.

Bojadžijev, D. "Pomen kovencije T" (The Meaning of the Convention T), Anthropos, no.5–6, Ljubljana, 1983.

Borstner, Boris. "Ali Bogu lahko pripišemo intencionalnost?" (May the Intentionality be ascribed to God ?) Anthropos, no.3–4, Ljubljana, 1986.

Borstner, Boris. "Primjerava znanstvenih teorij in pojem napredka v znanosti" (The Comparison of Scientific Theories and the Concept of Progress in Science), Anthropos, no.3–4, Ljubljana, 1985.

Božičković, Vojislav. "Kripke o vlastitim imenima" (Kripke on Proper Names), Theoria no.1–2, Beograd, 1986.

Ćulum, Jovan. Filozofske beleške (Philosophical Notes), Beograd, 1967. (A collection of philosophical essays originally published from 1957 to 1966).

Djurić, Mihailo. "Vitgenštajnova logička analiza jezika", (Wittgenstein's Logical Analysis of Language), Delo, no.2, Beograd, 1958.

Gligorijević, Miroslav. "Shvatanje supstancije u XX veku" (The Conception of Substance in 20th Century), Dijalektika, Beograd, Part I in no.1–4, 1984 and Part II in no.1–4, 1985.

Grahek, Nikola. "Nepogrešivost i čulno opažanje" (Incorrigibility and Sense-perception"), Filozofske studije, vol 11, 1981.

Grahek, Nikola. "Lok, Hjum i problem ličnog identiteta" (Locke, Hume and the Problem of Personal Identity), Theoria, no.3–4, Beograd, 1982.

Grahek, Nikola. "Granice funkcionalističkog i fizikalističkog shvatanja mentalnih stanja" (The Limits of Functionalist and Physicalist Knowledge of Mental States). Filozofska Istraživanja, no. 18, Zagreb, 1986.

Hajdin, Mane and Damjanović, Zlatan. "Račun volje i pojam racionalnog subjekta" (The Volitional Calculus and the Notion of a Rational Subject), Dometi, no.10, Rijeka, 1982.

Hajdin, Mane. "Criminals as Gamblers: A Modified Theory of Pure Restitution", Dialogue (Canada), vol.26, 1987.

Hristić, Jovan. "Kratak spis o jeziku" (A Short Treatise on Language), Delo, Beograd, 1960.

Kron, Aleksandar. "Teorija verovatnoće i indukcije Dž.M.Kejnza" (Keynes' Theory of Probability and Induction) in Nauka i filozofija, Beograd, 1963.

Kron, Aleksandar. Odnos polivalentnih logika i teorije verovatnoće (The Connection Between Many-valued Logics and Probability Theory), Beograd, 1967.

Kron, Aleksandar. "Sistemi sa aspekta filozofije" (The Philosophical Aspects of the Theory of Systems) in Sistemi u prirodi i društvu, Beograd, 1970.

Kron, Aleksandar. "Filozofija matematike" (The Philosophy of Mathematics), Matematika, vol.4 (1972) and vol.5 (1973), Beograd.

Kron, Aleksandar. "Deduction Theorems for Relevant Logics", Zeitschrift für Mathematische Logic und Grundlagen der Mathematik, vol. 19, no. 1, 1973.

Kron, Aleksandar. "An Analysis of Causality" in Essays in Explanation and Understanding, eds. Manninen and Tuomela, Dordrecht, 1975.

Kron, Aleksandar. "Causal Ordering and Mill's Canons", Philosophia, no.3, 1987.

Jereman, Frane. Med logiko in filozofijo,(Between Logic and Philosophy), Ljubljana, 1971.

Jereman, Frane. "Meje formalizacije" (The Limits of Formalization). Anthropos, no.3-6, Ljubljana, 1984.

Jokić, Aleksandar. "Da li je bol uvek strašan?" (Is Pain always Terrible?), Dometi, no. 3-4, Rijeka, 1984.

Jokić, Aleksandar. "Analiza pojma otkrića" (The Analysis of the Concept of Discovery), Istraživač, no. 1, Beograd, 1986.

Jokić, Aleksandar. "Buduće generacije i njihova sadašnja prava", (Future Generations and Their Present Rights),Gledišta, no.7-8, Beograd, 1987.

Južnić, S. "Fizikalizem kot splošna znanstvena metoda" (Physicalism as Scientific Method), Anthropos, no.3–4, Ljubljana, 1985.

Knjazev, Svetlana. "Tipovi filozofskih iskaza" (Types of Philosophical Propositions), Filozofija, no. 1–2, Beograd, 1963.

Knjazev, Svetlana. Filozofija Lavovsko-varšavske škole (The Philosophy of Lvov-Warsaw School), Beograd, 1964.

Knjazev, Svetlana. "Istina kao korespondencija" (Truth As Correspondence), Praxis, no. 2, Zagreb, 1965.

Knjazev, Svetlana. Filozofija Bertranda Rasela, (The Philosophy of Bertrand Russell), Zagreb, 1966.

Knjazev, Svetlana. "Predmet i problemi teorije saznanja" (The Subject-matter and Problems of the Theory of Knowledge) in the Problemi filozofije marksizma, Beograd, 1967.

Knjazev, Svetlana. "Domet jednog argumenta", Filozofija, no.4, Beograd,1972.

Knjazev, Svetlana. "Etičke teorije i pojam 'žrtvovanja' (Ethical Theories and the Concept of Sacrifice), Filozofske studije, vol. 5, Beograd, 1974.

Knjazev, Svetlana. "Značenje termina 'istina'", Filozofske studije, vol.13, Beograd, 1981.

Knjazev, Svetlana. "Radikalni konvencionalizam in teorija resnice" (Radical Conventionalism and the Theory of Truth), Anthropos, no. 1–2, Ljubljana, 1985.

Lazović, Živan. "Da li je egzistencija predikat ?" (Is Existence a Predicate?) (Is Existence a Predicate?) Dometi, no.3, Rijeka, 1983.

Lazović, Živan. "Objašnjenje delanja" (The Explanation of Action), Dometi, no.7, Rijeka, 1984.

Lazović, Živan. "Sloboda, determinizam i odgovornost" (Freedom, Determinism and Responsibility), Gledišta, no.7–8, Beograd, 1985.

Lazović, Živan. "Practical Reasoning", Acta Analytica, no.3, Ljubljana, 1987.

Marković, Aleksandra. "Kant i postupci preko granica dužnosti" (Kant and Acts Beyond the Bounds of Duty), Gledišta, no.3-4, Beograd, 1986.

Miščević, Nenad. Filozofija jezika (The Philosophy of Language). Zagreb, 1981.

Miščević, Nenad and Smokrović, N. (eds.) Namera i čin (Intention and Action). Introduced by Nenad Miščević. Rijeka, 1983.

Miščević, Nenad and Potrč, Matjaž (eds). Ime i kontekst (Name and Context), Rijeka, 1986. (A collection of translations of recent articles on semantics).

Miščević, Nenad. Od misli do jezika (From Thought to Language), Rijeka, 1987.

Nikolić, Nenad. "Doktrina pokazivanja i teorija slike u Wittgensteinovom Tractatusu" (The Doctrine of Showing and the Theory of Picturing in Wittgenstein's Tractatus) Filozofske studije, vol. 5, Beograd, 1974.

Novaković, Staniša. Problem demarkacije nauke i metafizike u savremenoj empirističkoj filozofiji (The Problem of Demarcation of Science from Metaphysics in Contemporary Empiricist Philosophy), Beograd, 1965.

Novaković, Staniša. Problem metafizike u savremenoj analitičkoj filozofiji (The Problem of Metaphysics in Contemporary Analytical Philosophy), Beograd, 1967.

Novaković, Staniša. "Poperova kritika Marksa" (Popper's Criticism of Marx), Savremene Filozofske Teme, no. 1, Beograd, 1969.

Novaković, Staniša. "Two Concepts of Science and Humanism", Praxis, International edition, no.1-2, Zagreb, 1969.

Novaković, Staniša. "Metodološka i filozofska shvatanja K.R. Poppera" (Methodological and Philosophical Views of K.R. Popper), as the introduction to the translation of Popper's Logic of Scientific Discovery, Beograd, 1973.

Novaković, Staniša. "Is the Transition from the Old to the New Scientific Theory of a Sudden and Unexpected Character?", Boston Studies in the Philosophy of Science, vol.14, Dordrecht, 1974.

Novaković, Staniša. Hipoteze i saznanje (Hypotheses and Knowledge), Beograd, 1984.

Novaković, Staniša, "Savremena metodologija i istorija metodološke misli (Contemporary Methodology and the History of Methodological Thought), Filozofska istraživanja, no.3, Zagreb, 1986.

Pavković, Aleksandar. "Čulni podaci, iluzije i zli demon" (Sense-data, Illusions and the Evil Daemon), Theoria, no.1, Beograd,1976.

Pavković, Aleksandar. "Iluzije i demon - još jednom" (Illusions and the Daemon - Once Again), Theoria, no. 2, Beograd, 1977. (The above two are rejoinders to Sesardić (1975) and (1976))

Pavković, Aleksandar. "Opažanje i oseti" (Perception and Sensation), Filozofske studije, vol.9, Beograd, 1977.

Pavković, Aleksandar (ed.) Svest i saznanje (Mind and Knowledge), Beograd, 1980. (Introduction by the editor).

Pavković, Aleksandar. "Hume's Arguments for the Relativity of Sense-perception", International Philosophical Quarterly, vol.25, New York, 1985.

Pavković, Aleksandar. "What are Private Languages for?", Synthesis Philosophica, no.1, Zagreb, 1986.

Pavković, Aleksandar. Razlozi za sumnju: ogledi iz skepticizma (Reasons for Doubt: Essays in Scepticism), Beograd, 1988 (forthcoming).

Potrč, Matjaž. "Austinova teorija govornih dejanj" (Austin's Theory of Speech Acts), Anthropos, no.5-6, Ljubljana, 1977.

Potrč, Matjaž. "Russell in Strawson on položaju določnih opisov" (Russell and Strawson on Definite Descriptions), Anthropos, no.4-6, Ljubljana, 1981.

Potrč, Matjaž. "Filozofija običnog jezika". (The Philosophy of Ordinary Language,) Delo, no.9, Beograd, 1981.

Potrč, Matjaž. "Ime i kontekst", (Name and Context). Filozofska Istraživanja, no. 2, Zagreb, 1984.

Potrč, Matjaž and Miščević, Nenad. "Davidsonovo učenje o značenju (Davidson's Doctrine of Meaning), Treći program, no.61, Beograd, 1984.

Potrč, Matjaž. "Nekaj tez o fizikalizmu" (A Few Theses on Physicalism), Anthropos, no.3-4, Ljubljana, 1985.

Potrč, Matjaž. "Some Considerations on Belief and Language", Acta Analytica, no.1, Ljubljana, 1986.

Pražić, Aleksandar. "O metodi pobijanja teizma" (On the Method of Rebuttal of Theism), Theoria, no. 2-3, Beograd, 1981.

Pražić, Aleksandar. Reč o slobodi (A Discourse on Freedom), Beograd, 1982.

Pražić, Aleksandar. "Smisao iskaza 'Zločin ostaje zločin i kad je učinjen u ime revolucije'", (The Sense of the Proposition "The Crime Remains a Crime Even When Committed in the Name of the Revolution") Glediš ta, no.5-6, Beograd, 1984.

Pražić, Aleksandar. Priroda i teleologija (Nature and Teleology), Beograd, 1986.

Primorac, Igor. "Deontološ ka etika, sveci i heroji" (Deontological Ethics, Saints and Heroes), Filozofija, no.2, Beograd, 1972.

Primorac, Igor. "Utilitaristička greš ka" (The Utilitarian Mistake), Gledista, no.7-8, Beograd, 1974.

Primorac, Igor. "Etika i pozivanje na običan jezik" (Ethics and the Appeal to Ordinary Language), Theoria, no. 1-2, Beograd, 1976.

Primorac, Igor. "Utilitarianism and Self-Sacrifice of the Innocent", Analysis, vol.38, Oxford, 1978.

Primorac, Igor. Kazna i prestup (Punishment and the Offense), Beograd, 1978.

Sesardić, Neven. "O izvjesnosti" (On Certainty). Praxis, no.3-5, Zagreb, 1974.

Sesardić, Neven. "Problem vanjskog svijeta" (The Problem of External World), Theoria, special edition, Beograd, 1975.

Sesardić Neven. "Iluzije" (Illusions), Theoria, Beograd, 1976.

Sesardić, Neven. "O analitičkoj filozofiji" (On Analytic Philosophy), Pitanja, no. 7-8, Zagreb, 1978.

Sesardić, Neven. " The Heritage of the Vienna Circle", Grazer Philosophische Studien, no.9, Graz, 1979.

Sesardić, Neven. Fizikalizam (Physicalism), Zagreb, 1984.

Sesardić, Neven. Marxian Utopia?, London, 1985.

Sesardić, Neven. "Psychology Without Principle of Charity", Dialectica, no.4, Lausanne, 1986.

Sesardić, Neven. "U odbranu znanstvenog realizma" (In Defence of Scientific Realism), Filozofske studije, vol.18, Beograd, 1986.

Sesardić, Neven. "Reconciling Philosophy of Science with Sociology of Science", Synthesis Philosophica, no.3, Zagreb, 1987.

Stojanović, Vojislav. "Jedna analiza metafizičkih iskaza" (An Analysis of Metaphysical Propositions), Ideje, no. 6, Beograd, 1971.

Stojanović, Vojislav. "Istina i dobro" (Truth and the Good), Theoria no.1-2, Beograd, 1976.

Sindjelić, Svetozar. "Sapir Vorfova ligvističko relativistička hipoteza" (The Sapir Whorf Linguistic Relativist Hypothesis), Delo, no.8-9, Beograd, 1973.

Sindjelić, Svetozar. "Grunbaumova kritika Duhevomovog konvencionalizma i Einsteinove filozofije geometrije" (Grünbaum's Criticism of Duhem's Conventionalism and Einstein's Philosophy of Geometry), Filozofija, no.3-4, Beograd, 1973.

Sindjelić, Svetozar. "Istina kao negativna vrednost" (Truth as a Negative Value),Theoria, no.1-2, Beograd, 1976.

Sindjelić, Svetozar."Popperova koncepcija opovrgavanja naučnih teorija" (Popper's Concept of the Falsification of Scientific Theories), Filozofske studije, vol.7, Beograd, 1976.

Sindjelić, Svetozar. "Grinbaumova kritika konvencionalizma" (Grünbaum's Criticims of Conventionalism), Filozofske studije, Vol.XI, Beograd, 1979.

Sindjelić, Svetozar. "Metodologija istraživačkih programa" (The Methodology of Research Programs), Filozofske studije, Vol. XVIII, Beograd, 1986.

Ule, Andrej. Od filozofije k znanosti in nazaj (From Philosophy to Science and Back), Ljubljana, 1986.